Applied
Mathematical Models
in Human Physiology

SIAM Monographs on Mathematical Modeling and Computation

About the Series

In 1997, SIAM began a new series on mathematical modeling and computation. Books in the series develop a focused topic from its genesis to the current state of the art; these books

- present modern mathematical developments with direct applications in science and engineering;

- describe mathematical issues arising in modern applications;

- develop mathematical models of topical physical, chemical, or biological systems;

- present new and efficient computational tools and techniques that have direct applications in science and engineering; and

- illustrate the continuing, integrated roles of mathematical, scientific, and computational investigation.

Although sophisticated ideas are presented, the writing style is popular rather than formal. Texts are intended to be read by audiences with little more than a bachelor's degree in mathematics or engineering. Thus, they are suitable for use in graduate mathematics, science, and engineering courses.

By design, the material is multidisciplinary. As such, we hope to foster cooperation and collaboration between mathematicians, computer scientists, engineers, and scientists. This is a difficult task because different terminology is used for the same concept in different disciplines. Nevertheless, we believe we have been successful and hope that you enjoy the texts in the series.

Mark Holmes

Applied Mathematical Models in Human Physiology

Johnny T. Ottesen
Roskilde University
Roskilde, Denmark

Mette S. Olufsen
North Carolina State University
Raleigh, North Carolina

Jesper K. Larsen
Math-Tech Aps
Charlottenlund, Denmark

siam

Society for Industrial and Applied Mathematics
Philadelphia

10 9 8 7 6 5 4 3 2 1

Library of Congress Cataloging-in-Publication Data

Applied mathematical models in human physiology / [edited by] Johnny T. Ottesen, Mette S. Olufsen, Jesper K. Larsen.
     p. cm. — (SIAM monographs on mathematical modeling and computation)
  Includes bibliographical references and index.
  ISBN 0-89871-539-3 (pbk.)
1. Human physiology—Mathematical models. I. Ottesen, Johnny T., 1959-
II. Olufsen, Mette S. III. Larsen, Jesper K. IV. Series.

QP33.6.M36A67 2004
612'.001'5118—dc22

                                2003067274

siam is a registered trademark.

Contributors

Pernille Thorup Adeler
Rambøll
Bredevej 2, DK-2830 Virum, Denmark
pta@ramboll.dk

Viggo Andreassen
Department of Mathematics (IMFUFA)
Roskilde University, P.O. Box 260, DK-4000 Roskilde, Denmark
viggo@mmf.ruc.dk

Tine Guldager Christiansen
Sigerstedgade 8 4th, DK-1729 Copenhagen V, Denmark
tguldager@yahoo.com

Michael Danielsen
Langesund 13, DK-2100 Copenhagen Ø, Denmark
michaeldanielsen@hotmail.com

Claus Dræby
Mørkager 4, DK-2620 Albertslund, Denmark
claus@dr1by.dk

Jacob M. Jacobsen
Department of Mathematics (IMFUFA)
Roskilde University, PO box 260, DK-4000 Roskilde, Denmark
jmj@mmf.ruc.dk

Jesper K. Larsen
Math-Tech
Ordruphøjvej 54, DK-2920 Charlottenlund, Denmark
jesper@math-tech.dk

Mette S. Olufsen
Department of Mathematics
North Carolina State University
Box 8205, Raleigh, NC 27695
msolufse@math.ncsu.edu

Johnny T. Ottesen
Department of Mathematics (IMFUFA)
Roskilde University, P.O. Box 260, DK-4000 Roskilde, Denmark
johnny@mmf.ruc.dk

Contents

Preface

The purpose of this book is to study mathematical models of human physiology. The book is a result of work by Math-Tech (in Copenhagen, Denmark) and the BioMath group at the Department of Mathematics and Physics at Roskilde University (in Roskilde, Denmark) on mathematical models related to anesthesia simulation. The work presented in this book has been carried out as part of a larger project, SIMA (SIMulation in Anesthesia),[1] which has resulted in the production of a commercially available anesthesia simulator and several scientific research publications contributing to the understanding of human physiology. This book contains the scientific contributions and does not discuss the details of the models implemented in the SIMA project.

In order to develop an anesthesia simulator, it is necessary to model many aspects of the physiology in the human body. This book is devoted to presenting models reflecting current research relevant to cardiovascular and pulmonary physiology. In particular this book presents models describing blood flow in the heart and the cardiovascular system, as well as transport of oxygen and carbon dioxide through the respiratory system. The models presented describe several aspects of the physiology, and it is our hope that this book may provide inspiration for researchers entering this area of study and for advanced undergraduate and graduate students in applied mathematics, biophysics, physiology, and bioengineering. Each of the chapters presents a unique model that can be read independently of the other chapters.

Chapters 5 and 6 have been used in graduate level courses in applied mathematics at Roskilde University, Boston University, and North Carolina State University. Moreover, most of Chapters 2 to 8 have been used in project-organized and problem-based student activities at graduate and advanced undergraduate levels at Roskilde University. When using the book in a traditional organized course, we suggest that the students use the models as a collection of examples that can serve as inspiration in their own modeling. Since we find it important that, in a modeling situation, students be involved in both formulating and solving problems, we have not included traditional exercises at the end of each chapter.

The mathematical background used to derive our models is not presented in detail in this book but each chapter includes references to pertinent background material. It is expected that the reader has some knowledge of ordinary and partial differential equations, which are used in the models and are solved using numerical methods. The equations are not subject to analytical mathematical scrutiny, so a limited understanding of the methods under-

---

[1] SIMA comprises the Danish contribution to the Eureka project HPPC/SEA EU 1063 that ran from 1994 to 1998.

lying the use of these equations is sufficient. In addition, it is expected that the reader has a basic knowledge of physiology. We provide a chapter summarizing the main physiological results necessary for understanding the modeling assumptions and methodologies.

The introduction discusses the different levels of models presented in this book. Some models are simple *real-time* models that can be directly used in larger systems, while others are more detailed *reference* models that the reader can use to obtain a better understanding of the underlying physiological mechanisms and to provide parameters for and validation of simpler models. The second chapter presents an overview of aspects of cardiovascular and pulmonary physiology necessary for understanding the model assumptions and limitations used throughout the remaining chapters. Most of the information in this chapter is taken from standard textbooks in physiology, supplemented with more advanced concepts needed for modeling.

Chapters 3 to 8 describe six different models of the cardiovascular and pulmonary systems. These six models can be studied individually, so the chapters may be read in any order. Each chapter may be used as a separate case study of the relevant subject. We have chosen to present the chapters in the order that we find most natural with respect to human physiology. Chapter 3 presents a two-dimensional model of the pumping heart that is based on the Navier–Stokes equations. It describes contraction and blood flow through the left ventricle of the heart. The results obtained from the model are compared with data obtained from magnetic resonance imaging. Chapter 4 continues the discussion of the heart and presents a model describing the heart as a pressure source depending on time, volume, and flow. The model offers a separation between isovolumic (isolated) and ejecting heart properties, an approach that contrasts with traditional time-varying elastance models. After discussing the heart, we describe cardiovascular circulation using three models: a one-dimensional model that is able to predict blood flow and pressure at any location along the large systemic arteries (in Chapter 5) and two zero-dimensional models (in Chapters 6 and 7) that provide flow and pressure at a number of discrete locations in the entire cardiovascular system, including the heart, the systemic and pulmonary arteries, and the veins. The one-dimensional model is based on the Navier–Stokes equations with a constitutive equation relating pressure to the cross-sectional area of the vessels. The second model in Chapter 7, which describes baroreceptor control of blood pressure, is based on the circulatory model described in Chapter 4. After discussing the pumping heart and the circulatory system, the last chapter of the book (Chapter 8) presents a model of the respiratory system, which describes the exchange of the respiratory gases O_2 and CO_2 using a mechanical lung model and a blood transport model. In addition, an elaborate model of blood pH value is also presented.

The editors wish to thank Professor Stig Andur Pedersen at the Department of Philosophy and Science Studies at Roskilde University for starting the SIMA simulation project and the BioMath group at the Department of Mathematics and Physics at Roskilde University. We also thank Professors James Keener, James Sneyd, and Clyde Martin for reading this book and providing essential comments and suggestions. In addition we wish to thank Heine Larsen at Systematic and Denis Thompson at the BioMath Program at North Carolina State University. We thank Mr. Larsen for technical support with putting this book together. Without Mr. Larsen we would not have been able to keep track of the newest versions of texts, figures, and tables. We thank Mr. Thompson for revising this manuscript and improving its readability. This book is a result of collaborative work between the authors and editors, and as editors we have shared the work in getting this book finished. It is our hope that through these case studies, this book can serve as a background for discussions and provide new ideas for anybody interested in physiological modeling.

Acknowledgments

The work presented in this book was done with support from the High Performance Parallel Computing, Software Engineering Applications Eureka Project 1063 (SIMA—SIMulation in Anesthesia), Department of Mathematics and Physics, Roskilde University, Denmark, and Moth-Lunds fund, Denmark.

Chapter 1

Introduction

J.T. Ottesen, M.S. Olufsen, and J.K. Larsen

This book is about mathematical modeling of human cardiovascular and respiratory physiology at the systems level. In the introduction we give some background on the advances in the field through examples mainly drawn from our own experience and through a discussion of the modeling process. The field of mathematical cardiovascular and respiratory physiology is so vast that it is not possible to give a thorough description of all aspects in one book. For a general introduction to the subject we recommend the book by Hoppenstaedt and Peskin (1992), which includes a thorough introduction to processes involved in setting up realistic models that obey physical laws and describe the underlying physiology. A more comprehensive and advanced treatment of mathematical physiology can be found in the book by Keener and Sneyd (1998).

1.1 Background

The interdisciplinary field of applied mathematical modeling in human physiology has developed tremendously during the last decade and continues to develop. One of the reasons for this development is researchers' improved ability to gather data. The amount of physiological data obtained from various experiments is growing exponentially due to faster sampling methods and better methods for obtaining both invasive and noninvasive data. In addition, data have a much better resolution in time and space than just a few years ago. For example, some of the noninvasive measurements using magnetic resonance imaging (MRI) can provide information on blood velocity as a function of time and three spatial coordinates both in the heart and in arteries with a diameter of only a few millimeters. Another recent accomplishment is the ability to image neural activity in the brain by studying the changes in the oxygen level in capillaries. These studies depict not single vessels but small regions.

This large amount of data obtained from advanced measurement techniques constitutes a giant collection of potential insight. Statistical analysis may discover correlations but may fail to provide insight into the mechanisms responsible for these correlations. However, when it is combined with mathematical modeling of the dynamics, new insights into physiological mechanisms may be revealed. The large amount of data can make the models give not only qualitative but also quantitative information on the function they predict and may also be used to suggest new experiments. We think that such models are necessary for improving the understanding of the function of the underlying physiology, and in the long term mathematical models may help generate new mathematical and physiological theories. Some examples are as follows: Modeling the time delay related to the baroreceptor mechanism may lead to suggestions for what could be responsible for Mayer waves (certain oscillations in the mean arterial pressure) (Ottesen, 1997a). Modeling the propagation of the pulse wave along the aorta may explain the presence of the dicrotic notch in the pulse profile and how the dicrotic notch changes along the aorta (Olufsen and Ottesen, 1995b; Olufsen and Ottesen, 1995a; Olufsen, 1999; Olufsen et al., 2000). Modeling the dynamics of cerebral blood flow response to sudden hypotension during posture change from sitting to standing may provide a better insight into cerebral autoregulation (Olufsen, Nadim, and Lipsitz, 2002). Other examples can be found in the recent book by Keener and Sneyd (1998).

In addition, models can help to avoid confusion, misunderstandings, and wasted effort. Most if not all concepts can be clearly defined only by the use of mathematics. Without mathematical descriptions, obscurity and ambiguity will arise sooner or later. Such ambiguity was seen, for example, when a few researchers separately proposed different single indices characterizing the contractile state of the ventricle. Some of these indices depend on the vascular system in an essential way; thus instead of characterizing the contractile state of the ventricle, they characterize the interaction between the ventricle and the vascular system (Danielsen and Ottesen, 1997; Danielsen, 1998; Ottesen et al., 1999; Danielsen and Ottesen, 2001).

Likewise the use of mathematics often provides a tool for structuring the thoughts of the researchers who create the model and those who use it. For example, when a vein is occluded during surgery, the resistance to the blood flow is increased, and as a result a fall in cardiac output is usually observed. However, even though cardiac output usually falls, there are cases that show the opposite response: an increase in cardiac output. This apparently inconsistent response to the occlusion of veins is not understood by most physicians. The inconsistent responses make sense in light of the topology of the cardiovascular system model: While occluding the supply to a highly compliant organ causes an increase in cardiac output, occluding the supply to regions with low compliance causes a decrease in cardiac output (Ottesen, 2000).

Another point is that mathematical models frequently generate new and very important questions that could not be asked without the use of mathematical models. Examples are as follows: How does the topology of the vascular system influence the function of the system? Are vortices, which are created on the downstream side of the aortic valves when blood is ejected from the ventricle, responsible for the subsequent closing of the valves? If so, can these valves close without any blood flowing back into the ventricle? What is really meant by the phrase "contractility of the ventricle," and do people use this phrase consistently? Can hysteresis and other nonlinear phenomena of the baroreceptor nerves be caused by a single mechanism? Under what conditions is the cardiovascular system stable? Will minor perturbations in function cause only minor changes in the state of the cardiovascular system? How much can the function of the baroreceptor feedback mechanism, which controls how heart rate responds to arterial pressure, vary without vital failure? What factors are responsible for the creation of the dicrotic notch in the pulse profile? Why does the dicrotic notch change along the aorta? These and many other questions are easy to describe and make perfect sense to people who are not mathematically oriented. However, upon closer scrutiny, the questions above reveal that the exact meanings of single words or of the systems involved are based on particular mathematical models. In any case all of the above questions are direct results of mathematical models even though the underlying models are not mentioned explicitly.

Finally, the ongoing development of mathematical modeling in physiology has increased the use of models (and the insights they offer) in the medical industry. This is an area that has been used in engineering for decades. For example, who would imagine a pilot of a freight ship or an airplane who has not had extensive training in a simulator? However, the surgical team in an operating theater has usually not had training in a simulator—neither has the surgeon used a simulator for training or planning the surgery, nor have the anesthesiologists trained to respond to the rare reactions to anesthesia that may be fatal if not treated within minutes. Finally, the doctors, nurses, and technical personnel have not usually had any simulated training in how to operate the very advanced technical monitoring equipment as a team. The wish to establish a solid foundation for the development of an anesthesia simulator has been the main inspiration for the work described in this book. Appendix A has a short description of the anesthesia simulator SIMA, which is based on the models reported in this book.

1.2 Mathematical Modeling

This book will discuss various simple as well as more detailed models of human physiology that can be used to obtain a better understanding of parts of the underlying systems. The models described in this book are based on fundamental physical laws. A goal has been to achieve models that reflect correct qualitative and quantitative behavior. All of the models discussed in this book have been derived using advanced mathematics, and the quantitative results are obtained from implementing the models using a numerical approach where parameters have been estimated based on experimental measurements. As a result, the models require a computer system to run.

As described above, the inspiration for this book has been the development of an anesthesia simulator. This has inspired the development of two levels of models: comprehensive models that adequately describe the physiology in a detailed way but cannot run in real time and simple models that utilize the understanding obtained from the comprehensive models but are modified such that they can run in real time. When developing an anesthesia simulator, it is important to have models that can run in real time; hence we distinguish between the two levels of models.

The comprehensive models will be referred to as *reference models* and the simpler models as *real-time models*. The aim of this book is to describe these models, to show the importance of the two levels of models, and to give an idea of how to implement real-time models. Even though the reference models presented here are only a small subset of possible physiological models, they still represent the framework for building and validating simple models that have to run in real time.

The development of models with different layers of complexity reflects the difficulties that arise during scientific analysis of objects and phenomena as they appear to us in daily life. Such objects are often too complex and vaguely defined to be accessible to scientific scrutiny. As a result, the objects must be delimited and prepared in such a way that scientific concepts are applicable. This is a relatively complex process in which a concrete object is transformed into a generalized generic object that we may call the model object. Definition and interpretation of such objects are performed with some purpose in mind that will give rise to a certain level of detail in a derived mathematical formulation. For example, consider the transport model in Chapter 8. This model describes the transportation of various substances such as oxygen or anesthetic agents in the human body. In order to develop such a model, one must think of a generic human body in two distinct ways: as an average human being and as a simplified and idealized object. One simplification is that blood is a homogeneous fluid. Another is that various body tissues can be grouped together into a few compartments; e.g., in the respiratory model described in Chapter 8 the complex branching of the airways is lumped into four compartments depending on the diameter of the airways. In this way we think of the human body not as it is in reality, but as an idealized and simplified object. All measurements and mathematical models use such idealized versions of real objects.

It is important to notice that physical measurements must always be interpreted based on a set of assumptions that makes sense relative to the model object. For example, when an invasive blood pressure is measured in a large artery, it is assumed that the catheter is placed in a smooth laminar blood stream and that the blood is a well-behaving incompressible fluid. Impurities and interfering features are eliminated or included in a negligible amount of noise. Furthermore, fundamental laws of nature hold only for generic model objects and not for objects as they are in themselves. Consider, for instance, the incompressible Navier–Stokes equations. They can be derived from the Boltzmann equation of nonequilibrium statistical mechanics (Bardos, Golse, and Levermore, 1991) under some simplifying assumptions: that the flow field varies only in length and time scales much greater than the microscopic scales associated with the mean free path and mean free time of particles, and that velocities are much less than the speed of sound. Therefore, when we apply the Navier–Stokes equations, we tacitly assume that our fluid is a system of particles complying with such general assumptions. These equations are studied in detail in Chapters 3 and 5.

Consequently, when we embark on the task of building a mathematical model of a physiological process in accordance with fundamental laws, we have already made far-reaching generalizations. We are considering a generic human being as a model object. However, in order to construct a mathematical model that is computationally accessible, we must make even more radical idealizations and simplifications. As an example, consider the arterial tree. When a clinician is measuring the blood pressure somewhere in a large artery, it is assumed that there is a definite well-defined pressure at that place, e.g., that the blood flow is regular and smooth enough to allow a well-defined pressure. But when we want to construct a mathematical model of the arterial tree that can predict pressure and flow profiles, we must make a series of simplifications. First, we must make the assumptions underlying the Navier–Stokes equations. Second, we must simplify the arterial tree such that we have a manageable number of equations. Finally, the equations must be simplified so they are accessible to numerical solution. An example of such a model is discussed in Chapter 5.

When so many simplifications and idealizations are made during the creation of a manageable model, how can one have confidence in its validity? How can we be sure that the model still complies with basic laws? And how is it possible to compare the model, which describes a general model object, with physiological data? We shall not go into a full discussion of these problems here. But we emphasize that in our reference models all simplifications are made in such a way that basic laws, e.g., mass and momentum conservation, are not violated. Furthermore, we do not claim that our models are able to give detailed numerical predictions, but only that they account for the principal qualitative behavior of the system being modeled. And it is that principle qualitative behavior of the model that assists the underlying understanding. The understanding obtained from these detailed models can be carried over to the simpler models that can be directly used in a larger system. Our goal is that even these real-time models obey fundamental physical laws. However, achieving that goal is not always possible when creating a large realistic system. So in cases where it is not possible to develop physical models it becomes necessary to use shortcuts based on empirical, statistical, or even simple profile models.

The strategy outlined for building models of large physiological systems has been applied in the development of the anesthesia simulator SIMA. This development has been an interdisciplinary process involving mathematical modeling, physiological experimentation, numerical analysis, and design of hardware and software interfaces. Although it has been necessary to make compromises at vital points, the process has shown that it is possible to develop a full-scale anesthesia simulator based, at essential points, on realistic mathematical models of physiological processes. Our use of a combination of real-time models and reference models has been instrumental to our successes. We believe that many researchers in the mathematical, biological, physical, and biomedical communities will find this approach useful.

Biomedical modeling is an important discipline that has gained new impetus due to new developments in computer technology and mathematical modeling. By building better reference models and improving the existing ones, it is possible to develop advanced simulation environments that are essential for education and research, leading to a deeper understanding of human physiology.

Acknowledgments

The authors were supported by High Performance Parellel Computing, Software Engineering Applications, Eureka Project 1063 (SIMA—SIMulation in Anesthesia).

Chapter 2

Cardiovascular and Pulmonary Physiology and Anatomy

M.S. Olufsen, M. Danielsen, P.T. Adeler, and J. Larsen

2.1 Introduction

In order to develop physiologically correct models of aspects of the cardiovascular or pulmonary system, one must understand relevant parts of the underlying physiology. This book describes a number of models that can be used to study aspects of the heart and the cardiovascular and pulmonary systems. The background physiological knowledge upon which these different models are based overlaps, so it is all covered in this chapter. Most of the material in this chapter can be found in basic physiological textbooks such as *Textbook of Medical Physiology* by A.C. Guyton (1991), *Principles of Human Anatomy* by G.J. Tortora (1999), *Human Anatomy and Physiology* by E.P. Solomon, R. Smidt, and P. Adragna (1990), *Review of Medical Physiology* by W.F. Ganong (1975), and *The Mechanics of the Circulation* by C. Caro et al. (1978). In addition, however, this chapter includes more subtle details necessary for understanding all aspects of the model. Readers with sufficient physiological knowledge can skip this chapter and go directly to the modeling chapters. However, others will benefit from first reading this chapter.

Section 2.2 gives an introduction to the organization of the cardiovascular system. Section 2.3 offers an overview of the basic physiology and anatomy of the heart. The physiology of the systemic arteries is described in section 2.4, and cardiovascular regulation is described in section 2.5. Finally, section 2.6 gives an introduction to pulmonary physiology.

2.2 Cardiovascular Physiology

The human cardiovascular system is primarily a transport system in which oxygen, carbon dioxide, and nutrients are carried by the blood to and from the various muscles and organs.

The cardiovascular system consists of two separate parts: the systemic circulation and the pulmonary circulation.

These two parts are connected via the heart. From the left ventricle blood is pumped into the systemic circulation through the aorta (the largest artery in the body). The systemic arteries transport oxygen and nutrients to the various muscles and organs. At the capillary level oxygen and nutrients diffuse from the vessels into the muscles and organs. In the muscles and organs oxygen is partially exchanged with carbon dioxide, and as a result the blood becomes partly deoxygenated. From the capillaries blood is discharged into the venules and then into the veins. Finally, through the vena cava (the largest vein in the body), blood is transported to the right atrium and from there to the right ventricle. The right ventricle pumps blood into the pulmonary circulation, in which the partly deoxygenated blood is carried to the lung tissues, where carbon dioxide is exchanged for oxygen in the alveoli. Subsequently, the nearly reoxygenated blood is carried back to the left atrium and from there back into the left ventricle. Thus blood makes a complete trip through both circulations; see Figure 2.1.

The systemic and pulmonary circulations exhibit significant differences in terms of blood pressure and blood volume. The pressure in each portion of the circulation is shown in Figure 2.2. In addition, the figure shows the approximate minima, maxima, and average pressures obtained during each cardiac cycle, at different locations in the cardiovascular system. The maximum and minimum pressures in the systemic aorta are approximately 120 mmHg and 80 mmHg, respectively. The corresponding pressures in the pulmonary arteries are 30 and 10 mmHg. The veins exhibit the lowest pressures. In the venules the pressure oscillates around 10 mmHg. Oscillations in the pulmonary veins are more profound. The volume of blood in the pulmonary circulation (in the pulmonary arteries and veins) makes up approximately 14% of the total blood volume. The volume in the systemic circulation is 74% of the total volume. At any given time approximately 54% of the blood is in the veins, 20% is in the arteries, and 12% is in the heart; see Table 2.1.

Table 2.1. *Volume distribution in the cardiovascular system relative to the total volume (Ganong, 1975; Noordergraaf, 1978).*

Location	Volume
Systemic arteries	20%
Systemic veins	54%
Pulmonary circuit	14%
Heart	12%

2.3 The Heart

The heart is considered to be the only source of energy that moves blood in the circulatory system. However, the heart is not an independent pump but a complex organ affected by the rest of the cardiovascular system. Consideration of the detailed structure of the heart may further one's understanding of this interaction. Therefore, this section offers

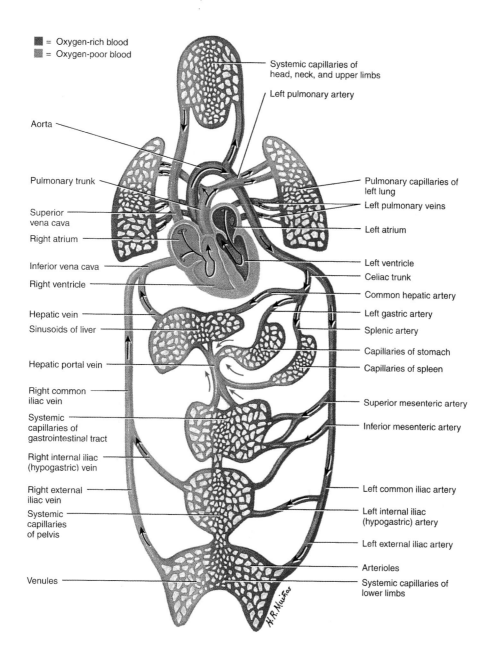

Figure 2.1. *The circulatory system. From* Principles of Anatomy and Physiology. *Copyright 1986. This material used by permission of John Wiley & Sons, Inc.*

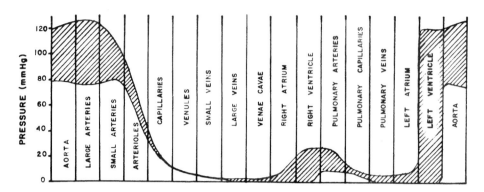

Figure 2.2. *Pressure distribution in the human cardiovascular system. From Chapter 1 in* Circulatory System Dynamics *by Abraham Noordergraaf, copyright* 1978, *Elsevier Science (USA), reproduced with permission from the publisher.*

an introduction to the anatomy of the heart, its conduction system, and the physiology of its muscles.

2.3.1   The Cardiac Cycle

As shown in Figure 2.3, the heart consists of four chambers split between two sides. Each side has a ventricle and an atrium. The left ventricle and left atrium are connected via the mitral valve, which during normal conditions prevents flow from the ventricle into the atrium. The right ventricle and the right atrium are connected via the tricuspid valve. In addition to the valves between the heart chambers, each chamber has a valve connecting the arteries and veins to the heart. The aortic valve is positioned at the outflow from the left ventricle. At the outflow from the right ventricle is the pulmonary valve. A more detailed description of the internal anatomy is given in section 2.3.2.

The time course of the cardiac cycle can be divided into an active phase and a relaxed phase. At the onset of the active phase, electrical stimulation causes the ventricular muscles to contract. As a result, ventricular pressure increases isovolumically (i.e., with no inflow and nonejecting) until the ventricular pressure equals the arterial pressure. At this point the aortic valve opens (onset of systole) and blood flows into the aorta. During systole, ventricular pressure rises and falls as dictated by muscle contraction and prevailing conditions in the vasculature. The aortic valve closes (ending systole) when the ventricular pressure drops below the arterial pressure and initiates an isovolumical relaxation phase. When the ventricular pressure falls below the atrial pressure, the mitral valve opens and blood flows from the atrium into the ventricle. This nonejecting period is denoted *diastole*. The left atrium follows a similar track. The left atrium is filled from the pulmonary circulation during systole and supplies the ventricle actively with blood during diastole. However, only 30% of the ventricular filling (for both the right and the left ventricles) is due to the atrial contraction. The remaining 70% of the filling occurs during diastole. The amount of blood ejected by each ventricle, per stroke, is about 70 ml. The ejection fraction, the percentage of the ventricular volume ejected at each stroke, is about 65%. The duration of the cardiac

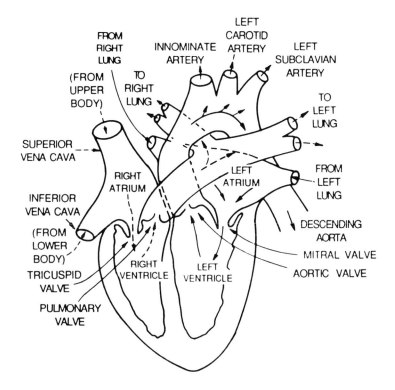

Figure 2.3. *The four-chambered heart is divided into two separate parts, the left and the right sides. Both parts consist of a ventricle and an atrium. The left side of the heart is anatomically larger than the right side; see section 2.3.2. From* Mathematical and Computer Modelling of Physiological Systems, *by V.C. Rideout, Prentice–Hall Biophysics and Bioengineering Series, 1991. Used by permission from the author.*

cycle is approximately 0.8 s during rest. The amount of blood ejected from the left ventricle into the ascending aorta is called cardiac output. In a resting adult the cardiac output is approximately 5 l/min (70 ml × 72 beats/min). Figure 2.4 shows the ventricular pressure and flow curves representative of a normal left ventricle.

2.3.2 Internal Anatomy

As described above, the heart has four chambers (two atria and two ventricles) divided between two sides (left and right). The left and right sides of the heart are separated by a septum. The atria are divided by the interatrial septum, and the ventricles are separated by the interventricular septum. The atrioventricular openings connect the atria and ventricles. This opening is surrounded by a ring of fibrous tissue, the annulus fibrosus. The lower part of the heart is called the apex, while the base refers to the opposite, broad, end of the heart; see Figure 2.5.

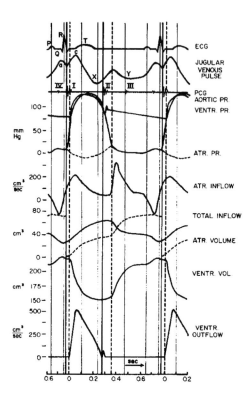

Figure 2.4. *Electrocardiogram (ECG) showing venous pressure, phonocardio-gram, root aortic pressure, ventricular pressure, atrial pressure, arterial inflow, total inflow to the heart, atrial volume, ventricular volume, and ventricular outflow. The solid lines and the dashed lines indicate valve closure and opening, respectively. From Chapter* 1 *in* Circulatory System Dynamics *by Abraham Noordergraaf, copyright 1978, Elsevier Science (USA), reproduced by permission from the publisher.*

The heart walls consist of muscle tissue called the myocardium. The thickness of the myocardium varies over the heart and increases with workload. The atria do less work than the ventricles. Consequently, the atrial wall is thin, approximately 2 mm. The right ventricle is exposed to the pressure of the pulmonary circulation and has a greater wall thickness, approximately 5 mm. The greatest amount of work and hence the greatest wall thickness, approximately 15 mm, belongs to the left ventricle, which ejects blood into the systemic circulation.

Due to the two sets of valves (atrioventricular and semilunar), blood flow in the heart is unidirectional. The atrioventricular (AV) valves are located between each atrium and ventricle. The semilunar valves are found in the opening between each ventricle and artery. Valves are not found at the venous inlets into the atria; see Figure 2.5.

The atrioventricular valves consist of leaflets (cusps) with a triangular shape. The right heart chamber contains three leaflets, and consequently its valve is denoted the tricuspid

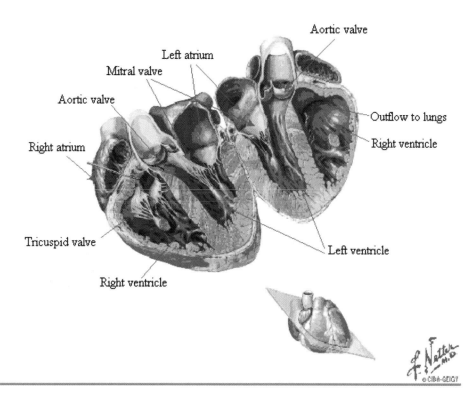

Figure 2.5. *The heart split along a plane perpendicular to the interventricular septum. Adapted from Netter (1991). Copyright 1989. Icon Learning Systems, LLC, a subsidiary of MediMedia USA Inc. Reprinted with permission from ICON Learning Systems, LLC, illustrated by Frank H. Netter, M.D. All rights reserved.*

valve. The valve in the left chamber contains two leaflets and is thus referred to as the bicuspid valve (or the mitral valve). Leaflet tendons or chordae tendineae are found under each valve. They originate from the papillary muscles located on the inner surface of each ventricle. When the atrium contracts, the leaflets hang slack into the ventricle, but when the ventricle contracts, the leaflets are pushed together, closing the atrioventricular opening. The chordae tendineae secure the leaflets, preventing them from moving into the atrium during the contraction of the ventricle.

The semilunar valves prevent blood from flowing back into the heart after ejection into the arteries. Each of the semilunar valves consists of three crescent-shaped leaflets. These leaflets have no chordae tendineae. They are dilated with blood and close the opening when the pressure in the artery exceeds the pressure in the ventricle. The right semilunar valve is also called the pulmonary valve. The corresponding valve on the left side is called the aortic valve.

2.3.3 Conduction System of the Heart

Contraction of the heart is initiated by electrical stimulation. Electrical impulses are spread to all parts of the myocardium by the conduction system. The conduction system consists of the sinoatrial (SA) node, the AV node, the AV bundle (or bundle of His), the bundle branches, and conducting fibers called Purkinje fibers; see Figure 2.6. All parts of the conduction system are able to send out periodic impulses without neural stimulation. This lack of reliance on outside control is called automatism.

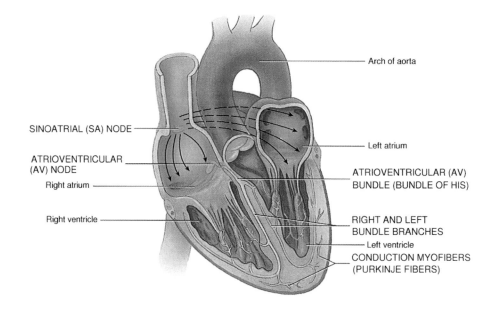

Figure 2.6. *The conduction system of the heart. From G.J. Tortora,* Principles of Human Anatomy, 9*th ed., copyright* 2002. *This material is used by permission of John Wiley & Sons, Inc.*

The SA node is a small mass of muscular fibers located in the myocardium of the right atrium near the inflow of the superior vena cava; see Figure 2.6. The SA node generates impulses faster than the other parts of the conduction system and thus initiates each heartbeat and controls the frequency of the heart rate.

The impulses from the SA node reach the atria first, causing them to contract. Much later the impulses reach the AV node, which is located in the interatrial septum; see Figure 2.6. The AV node is slowly conducting and thus the impulses are delayed there. This allows for the atria to empty their blood (by contraction) into the ventricles before the ventricles start to contract. From the AV node the impulses are sent through the bundle of His, which is the only electrical connection between the atria and ventricles. The bundle of His branches into two: one that goes to the right ventricle and one that goes to the left ventricle. Both branches go in the direction of the apex of the heart and subdivide into a complex network of Purkinje fibers (see Figure 2.6). These fibers stimulate ventricular contraction, which starts at the apex and spreads superiorly toward the base of the heart.

2.3.4 Muscle Physiology

The heart pumps due to the contraction of its muscle fibers. Muscle fibers receive energy from biochemical processes and develop a force that manifests itself as an increasing cavity pressure.

Muscles in the heart lie in a complicated pattern that may be understood via the band concept introduced by Guasp (1980). The cardiac muscle itself consists of individual muscle fibers, or myocytes, that are the smallest functional units in the structure. Each muscle fiber is approximately 40 to 100 μm long with a diameter of 10 to 20 μm.

Each muscle fiber contains a number of fibrils placed in parallel; see Figure 2.7. The fibrils have a characteristic striated pattern. This pattern results from the parallel bundles of interdigitating thick and thin filaments lying between the Z-lines; see Figure 2.8. Filaments lie along the fibrils divided into approximately fifty 2 μm blocks. A block is called a sarcomere and consists of approximately 1000 single thick-thin units of the type in Figure 2.8.

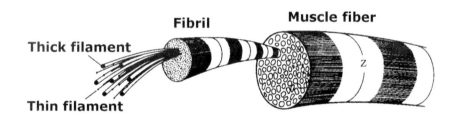

Figure 2.7. *Muscle fibers contain approximately 100 fibrils in parallel. The fibrils have a characteristic striated pattern that stems from the interdigitating thick and thin filaments, as indicated. The striated pattern is also shown in Figure 2.8. From Warberg (1995); used by permission of Polyteknisk Press, Lynaby, Denmark.*

Figure 2.9 shows the sarcoplasmic reticulum between the fibrils in the muscle fiber. The sarcoplasmic reticulum contains a Ca^{2+} reservoir in the terminal cisternae, which is essential during contraction. In addition, the muscle fibers contain T-tubules vital for conduction of action potentials to the sarcoplasmic reticulum, which release Ca^{2+} to the fibrils. The contraction of muscle fibers is explained by the sliding filament theory. This is the most widely embraced theory and is based on a mechanical concept. But this theory is not the only one to explain muscle contraction. Rather than mechanical descriptions, other theories propose a field approach (Spencer and Worthington, 1960; Elliott, Rome, and Spencer, 1970). The field approach has not enjoyed the same popularity since evidence exists in favor of a mechanical type (Noordergraaf, 1978). Accordingly, we will describe the fundamentals of the sliding filament theory.

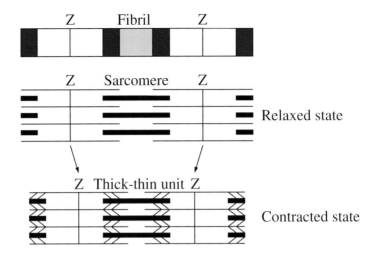

Figure 2.8. *Section of a fibril.   The striated pattern follows from the bundle of interdigitating thick and thin filaments shown in the middle panel in the relaxed state. A sarcomere is also indicated. The lower panel shows a single thick-thin unit with crossbridge bonds between thick and thin filaments. As crossbridge bonds attach between thick and thin filaments, force develops with a concomitant increased overlap between the filament and diminished distance between neighboring Z-lines, as indicated in the lower panel.*

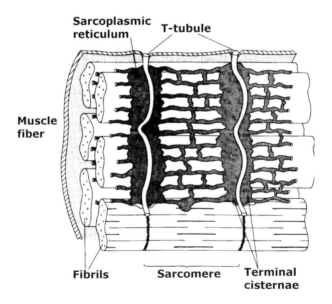

Figure 2.9. *Part of a single muscle fiber with the sarcoplasmic reticulum. The T-tubule runs through the muscle fiber in a transverse direction. Ca^{2+} is released from the terminal cisternae and diffuses into the fibrils. From Warberg (1995); used by permission of Polyteknisk Press, Lynaby, Denmark.*

Sliding Filament Theory

In the 1950s it was observed that the thick filaments remained in a stable suspension during contraction, whereas the distance between neighboring Z-lines diminished. This observation led to the sliding filament theory, which states that the thick and thin filaments slide during contraction, changing the overlap between them. Sliding is accomplished by mechanical connections between thick and thin filaments, which induce thin filaments to move with respect to thick filaments. The connections are called crossbridge bonds. During contraction, crossbridge bonds attach from the thick and thin filaments. According to the sliding filament theory, crossbridge bonds continue to attach, push off, and reattach. However, not all filaments continue to push off and reattach, some simply stay attached. As a result of attachment, force is developed and the distances between neighboring Z-lines are diminished with a concomitant increased overlap between the filaments, as shown in Figure 2.8.

The formation of crossbridge bonds yields a rise in biochemical energy, which in turn results in the development of a force. The force continues to increase until the crossbridge bonds detach in sufficiently high numbers. At this point, the force decreases and the muscle starts to relax. After a bond has detached, it can attach again during the same contraction (cycling of bonds). At the chamber level, ventricular pressure increases during formation of bonds between thick and thin elements and decreases during detachment of bonds.

Biochemical Energy

In the sliding filament theory force is assumed to develop from a combination of mechanical and biochemical processes. In short this sequence of events can be described as follows. The available biochemical energy is closely related to the amount of Ca^{2+}, while the actual release of energy stems from interaction between the proteins in the filaments. Thick filaments consist mainly of the protein myosin, and thin filaments of the protein actin. During contraction, the fibrils are electrically activated in the direction of the axes and in the radial direction by the T-tubule. This electrical activation promotes release of Ca^{2+} near the Z-lines from the sarcoplasmic reticulum. After the release, Ca^{2+} binds to the protein troponin. Troponin sits on tropomyosin, which wraps around actin molecules in the thin filaments. Troponin inhibits reaction between myosin and actin, but the Ca^{2+} binding promotes structural changes, which release this inhibition. Subsequently, myosin interacts with actin via the crossbridge bonds and releases energy. Shortly after this, Ca^{2+} is pumped back into the sarcoplasmic reticulum, which requires energy and takes time, longer than starting the contraction takes. This enhances inhibition of the actin-myosin interaction and thus formation of new bonds. Eventually force decreases and the muscles relax.

2.4 Systemic Arteries

The systemic arteries are composed of large arteries, small arteries, and arterioles; see Figure 2.10. Their topological pattern forms a vast network of branching vessels. The total cross-sectional area increases from 5 cm^2 at the root of the aorta (the biggest artery) to approximately 400 cm^2 at the entrance to the arterioles; see Table 2.2. These numbers should be seen as orders of magnitude because it is impossible to measure the area of the

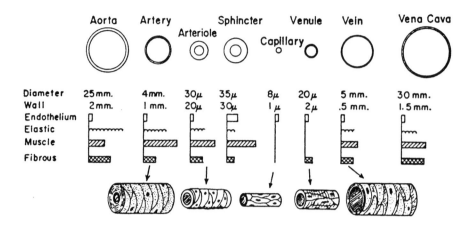

Figure 2.10. *Cross sections of arteries and veins. The vessels have an inner layer of endothelial cells and an outer layer composed of a varying mix of muscle and elastic fibers. The figure shows the relative content for each group of vessels. From Li (1987). Reprinted from Arterial System Dynamics, R.F. Rushmer, Organ Physiology: Structure and Function of the Cardiovascular System, copyright 1972, with permission from Elsevier Science.*

Table 2.2. *Lumen diameter, wall thickness, approximate total cross-sectional area, and percentage of blood contained in the given group of arteries. The total volume of blood is not 100% since the table does not account for the 12% blood in the heart and 18% in the pulmonary circulation (Gregg, 1966).*

	Diameter [mm]	Wall thickness [mm]	Cross-sectional area [cm²]	Percentage of blood volume contained
Aorta	25	2	5	2
Arteries	4	1	20	8
Arterioles	0.03	0.02	400	1
Capillaries	0.008	0.001	4500	5
Venules	0.02	0.002	4000	
Veins	5	0.5	40	54
Vena Cava	30	1.5	18	

arterioles precisely. Consequently, there is a considerable variation in the tabulated values given in different textbooks of physiology (Guyton, 1991; Gregg, 1966; Caro et al., 1978).

The diameters of blood vessels range over several orders of magnitude. This variation may impose a problem for modeling purposes, but that problem can be overcome by dividing the arteries into several groups: large arteries, small arteries, arterioles, and capillaries. This distinction is somewhat arbitrary, but can be justified by the different properties of the vessels as they gradually become smaller.

The large systemic arteries are characterized by strong, highly elastic, vascular walls. The walls of the small arteries are less elastic. The walls of the arterioles and capillaries are almost rigid and contain more smooth muscle than the walls of the aorta, large arteries, and small arteries. This musculature will be described in detail in section 2.4.1. It is the change in wall properties, together with the vast expansion of the network, that enables a significant drop in blood pressure and flow at the arteriolar level. The most important regulation of blood flow is in the arterioles. They act as control valves through which blood is released into the capillaries. For this function, they have strong muscular walls that can constrict the vessels completely or dilate them severalfold. The purpose of varying the cross-sectional area of the arterioles is to alter the blood flow into the capillaries in response to the needs of tissues. The resistance resulting from this behavior is often referred to as the peripheral resistance.

In contrast to the arterioles, the capillaries contain no muscles. Until the capillary level the branching of blood vessels is mostly binary (e.g., in the coronary arteries and arterioles 98% of the bifurcations are binary (Kassab and Fung, 1995)). However, the capillary network extends into a huge "swamp" without a certain (e.g., binary) branching structure (for example, capillaries branch in bifurcations and trifurcations and contain many loops). The region has a high surface area, and the flow is low and no longer pulsatile. The purpose of the capillaries is to exchange oxygen and nutrients between the blood and the interstitial fluid of the cells (or between the blood and alveoli in the case of the pulmonary circuit). This diffusion is achieved through a slow and steady flow and is mediated by the permeability of the capillary walls to small molecular substances. Since flow through the capillaries has to be steady and slow, another important role of the arteries is to damp the waves resulting from the pulsatile flow entering the aorta from the left ventricle. Again, this damping is achieved via a distal increase in the cross-sectional area of the network and the elasticity of the arteries.

In contrast to arteries, veins are low pressure vessels with a low flow, and their vessel walls are thin, with low elastic properties and a low resistance. The latter characteristic makes veins ideal for storage of blood because large volume alterations can be achieved without significant pressure changes. The veins contain muscles that can move blood volume to other parts of the cardiovascular system. Furthermore, none of the arterial pulsations are transmitted into the veins. However, pulsations can be observed in the veins. These pulsations are due either to heart-generated waves passing retrogradely toward the periphery or to respiratory fluctuations (O'Rourke, Kelly, and Avolio, 1992).

The amount of blood ejected from the right ventricle into the pulmonary artery is slightly smaller (1–2%) than the amount ejected into the ascending aorta. The reason for this difference is that the oxygenated blood needed to supply the lung tissues is not returned to the right atrium but continues through the lung into the pulmonary veins. Then this blood continues in the pulmonary vein and enters the left atrium, rather than passing back through the systemic veins into the right atrium.

Seen from a mechanical point of view, this distinction makes perfect sense because blood flow in the arteries is significantly different than that in the arterioles. The difference can be explained in terms of the fluid-mechanical characterization of the flow. If the flow has a Reynolds number significantly larger than one, it is dominated by inertia; if the Reynolds number is much smaller than one, the flow is dominated by viscosity. In the case of blood flow the Reynolds number drops below one when the vessel diameter becomes less than

100 μm. This diameter corresponds to the diameter of the larger arterioles, which range in size from 50 to 100 μm. The diameter of the arterioles decreases by progressive bifurcations down to the smaller arterioles (sometimes called the met-arterioles), where the diameter is approximately 30 μm; see Table 2.2. In addition, there is a functional difference between the two types of arteries. The purpose of the larger arteries is to distribute blood to the different muscles and organs, while the role of the smaller arteries and arterioles is to distribute blood (and, in the case of the arterioles, to control its distribution) within those muscles and organs.

The arteries and larger arterioles make up a complex bifurcating tree. Henceforth we refer to it as the arterial tree. However, the met-arterioles do not have a bifurcating tree structure; instead multiple branches and loops often occur. The order of the arterial tree is large. Assume an arteriolar diameter of 30 μm and a total cross-sectional area of 400 cm^2; see Table 2.2. If we then construct a binary tree consisting of the aorta, the arteries, and the larger arterioles, it will have 26 generations. Even if we neglect the arterioles and consider only the larger and smaller arteries (i.e., vessels with diameters larger than 0.25 cm and 100 μm, respectively), the tree will have as many as 19 generations. Such a tree cannot be depicted, but the tree shown in Figure 2.11, where only the larger and a few of the smaller arteries are shown, is still highly complex.

2.4.1 Arterial Wall

The arterial wall is composed of variable amounts of elastic fibers and smooth muscle, enabling it to dilate when the pulse wave propagates along an artery. It is not purely elastic but exhibits some viscoelastic behavior. As a first approximation, arteries are circular vessels tapering along their length. As mentioned in the previous section, arteries can be subdivided into the following three groups according to their elastic behavior:

- elastic arteries, which are the major distribution vessels, such as the aorta, the common carotid arteries, or the subclavian arteries;

- muscular arteries (see Figure 2.12), which make up the main distributing branches of the arterial tree and include the radial or femoral arteries;

- arterioles; see Figure 2.12.

The transition in structure and function in the arteries is gradual. Generally, the amount of elastic tissue decreases as the vessels become smaller and the smooth muscle component becomes more prominent (Wheater, Burkitt, and Daniels, 1987). Consequently, the arteries become markedly stiffer with increased distance from the heart. The parameters characterizing the elastic properties are Young's modulus, which is greater for arteries farther away from the heart, and relative wall thickness, which is constant for the larger arteries but increases for the smaller arteries and arterioles. The arterial wall is composed of three layers characterized by their predominant structure and cell types:

- The internal layer, the tunica intima, is composed of an endothelial layer and an outer elastic laminar layer.

  The endothelial layer comprises an inner layer consisting of a single layer of endothelial cells and an outer subendothelial layer. The single-cell endothelial layer is present

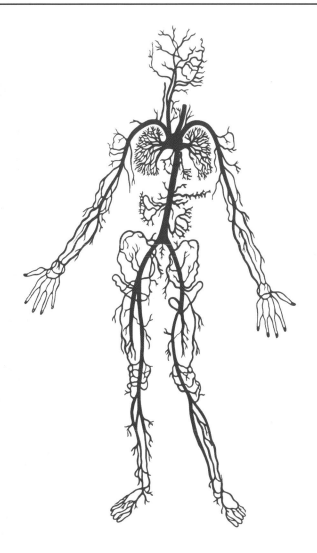

Figure 2.11. *The arterial tree. From* Human Anatomy & Physiology, *2nd ed., by E.P. Solomon, P.R. Smidt, and P.J. Adragna. Copyright 1990. Reprinted with permission of Brooks/Cole, a division of Thomson Learning: www.thomsonrights.com. Fax 800 730-2215.*

as a border to all surfaces that come in contact with the blood. It is rather fragile and easily damaged, e.g., by excessive shear rates. However, it also easily regenerates. The subendothelial layer contains a few collagen-generating cells and collagen fibers.

The elastic laminar layer consists of branching elastic fibers. It is particularly well defined in the smaller arteries, where it forms a clear boundary to the middle layer.

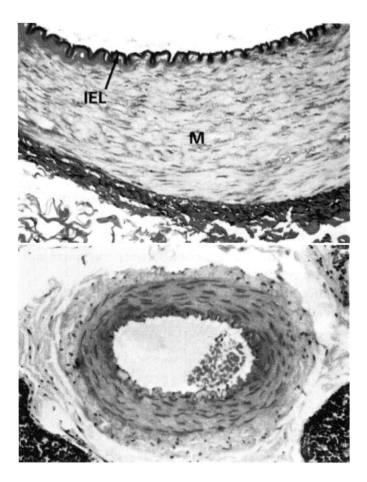

Figure 2.12. *The wall of a muscular artery (top) and a large arteriole (bottom). The muscular artery internal layer consists mainly of a thin elastic sheet (marked with IEL). The middle layer, tunica media (marked with an M), is composed mainly of smooth muscle. The outermost layer, tunica adventitia, is composed of a diffusive external elastic lamina. In addition, collagen fibers are scattered throughout the vessel wall. Arterioles have a thin internal layer that comprises an endothelial lining, little collagenous connective tissue, and a thin, but distinct, internal elastic lamina. The middle layer is almost entirely composed of smooth muscle cells organized in concentric circles. The outermost layer, the tunica adventitia, is thick and merges with the surrounding connective tissue. From Wheater, Burkitt, and Daniels (1987). Reprinted from* Functional Histology, *1st ed., by P.R. Wheater, 1979, Figure 8.8a, page 122, by permission of the publisher Churchill Livingstone.*

- The middle layer, the tunica media, is the thickest layer in the wall. It is also the layer with the greatest variation in structure and properties between different regions of the circulatory system. Transitions in the structure of this layer allow the division of arteries into the elastic and muscular categories. The tunica media of the elastic arteries is made up of multiple concentric layers of elastic tissue separated by thin layers of connective tissue. Collagen fibers and sparse smooth muscle cells are organized in a longitudinal way, forming cross links to the successive elastic layers. More details are given in Caro et al. (1978). In the corresponding layer of the muscular arteries elastic tissue is reduced and the smooth muscle cells are dominant. These cells are oriented circumferentially in spiral structures.

- The external layer, the tunica adventitia, can be as thick as or even thicker than the tunica media. However, it is less prominent microscopically. It is composed of loose connective tissues and relatively sparse elastin and collagen fibers, running in a predominantly longitudinal direction. The boundary with the surrounding tissue is often not well defined.

In Figure 2.10 the various layers of the arteries are shown. Also shown is the size (to order of magnitude) of the various vessels. The walls of arteries larger than about 1 mm in diameter have their own nutrient blood vessels, the vasa vasorum. These vessels originate either from the parent artery or from a neighboring one and break up into capillary networks, which supply the tunica adventitia and part of the tunica media. The tunica intima and the innermost layers of the tunica media are primarily supplied via transport of materials from the arterial lumen. Due to the complex composition of arterial walls, the distensibility or elastic properties of the arteries are nonlinear and therefore not easily described by a mathematical model.

In order to model the mechanical properties of blood flow, an important input parameter to know is the thickness of the arterial wall. It is difficult to describe the wall thickness precisely. The arteries are not loose vessels inserted in the body but are attached to the surrounding tissue. And the outer layer of the arterial wall, the tunica adventitia, usually merges gradually into the surrounding tissue. We will, however, use the simplifying assumption that the arteries are loose vessels. Generally, the thickness of the arterial wall varies considerably throughout the circulatory system, as is evident from Tables 2.2 and 2.3. Therefore, one often studies the ratio between the wall thickness and the diameter of the vessel. For the larger arteries this ratio is approximately constant, but for smaller arteries it is not constant. Even though the wall thickness decreases as one looks at smaller and smaller arteries, the ratio of wall thickness to radius increases. This increase continues until the smallest of the arterioles, where the external diameter is almost twice that of the lumen, even when the smooth muscle is relaxed. Finally, the thickness of the vessel wall in the capillaries is similar in all mammalian species. This similarity exists because the wall has to be thin and permeable in order for diffusion of molecules to occur. Capillary walls have a fixed size and structure independent of the species in question. However, it should be noted that the wall thickness of the arteries changes significantly with age, as do the elastic properties of the vessels. Aging causes elastic elements in the wall to degenerate. The vessels may become calcified and the collagen fibers increase in number, replacing muscle cells and proliferating in other parts of the wall. The overall effect of aging is that the diameter of the vessel increases, and the wall becomes thick and much less distensible.

Table 2.3. *Physiological data for the various parameters in the circulatory system Caro et al.* (1978).

Normal values for canine cardiovascular parameters. An approximate average value, and then the range, is given where possible.

Site		Ascending aorta	Descending aorta	Abdominal aorta	Femoral artery	Carotid artery	Arteriole	Capillary	Venule	Inferior vena cava	Main pulmonary artery
Internal diameter d_i	cm	1·5 (1·0-2·4)	1·3 (0·8-1·8)	0·9 (0·5-1·2)	0·4 (0·2-0·8)	0·5 (0·2-0·8)	0·005 (0·001-0·008)	0·0006 (0·0004-0·0008)	0·004 (0·001-0·0075)	1·0 (0·6-1·5)	1·7 (1·0-2·0)
Wall thickness h	cm	0·065 (0·05-0·08)		0·05	0·04	0·03 (0·02-0·04)	0·002	0·0001	0·0002	0·015 (0·01-0·02)	0·02 (0·01-0·03)
h/d_i		0·07 (0·055-0·084)		0·06 (0·04-0·09)	0·07 (0·055-0·11)	0·08 (0·053-0·095)	0·4	0·17	0·05	0·015	0·01
Length	cm	5	20	15	10	15 (10-20)	0·15 (0·1-0·2)	0·06 (0·02-0·1)	0·15 (0·1-0·2)	30 (20-40)	3·5 (3-4)
Approximate cross-sectional area	cm²	2	1·3	0·6	0·2	0·2	2×10^{-5}	3×10^{-7}	2×10^{-5}	0·8	2·3
Total vascular cross-sectional area at each level	cm²	2	2	2	3	3	125	600	570	3·0	2·3
Peak blood velocity	cm s⁻¹	120 (40-290)	105 (25-250)	55 (50-60)	100 (100-120)		0·75 (0·5-1·0)	0·07 (0·02-0·17)	0·35 (0·2-0·5)	25 (15-40)	70
Mean blood velocity	cm s⁻¹	20 (10-40)	20 (10-40)	15 (8-20)	10 (10-15)						15 (6-28)
Reynolds number (peak)		4500	3400	1250	1000		0·09	0·001	0·035	700	3000
α (heart rate 2 Hz)		13·2	11·5	8	3·5	4·4	0·04	0·005	0·035	8·8	15
Calculated wave-speed c_0	cm s⁻¹	580		770	840	850				100	350
Measured wave-speed c	cm s⁻¹	500	400-600	700 (600-750)	900 (800-1030)	800 (600-1100)				400 (100-700)	250 (200-330)
Young's modulus E	Nm⁻² ×10⁵	4·8 (3-6)	4-8	10 (9-11)	10 (9-12)	9 (7-11)				0·7 (0·4-1·0)	6 (2-10)

(From C. G. Caro, T. J. Pedley, and W. A. Seed (1974). 'Mechanics of the circulation', Chapter 1 of *Cardiovascular physiology* (ed. A. C. Guyton). Medical and Technical Publishers, London.)

In Table 2.3 typical values for the various physiological parameters are presented. These are based on measurements from dogs, but the human values for most of the parameters are approximately the same.

2.4.2 Blood

Blood consists of plasma with red blood cells (erythrocytes), white blood cells (leuco-cytes), and platelets (thrombocytes) in suspension. The primary function of erythrocytes is to transport oxygen and carbon dioxide. Plasma is composed of 93% water and 3% particles: electrolytes, proteins, gases, nutrients, hormones, and waste products. Leuco-cytes are an important part of the immune system. Thrombocytes are a vital component of the blood-clotting mechanism; they are not cells, but fragments from plasma cells called megakaryocytes. Erythrocytes make up more than 99% of all blood cells and approximately 40 to 45% of the blood (cells plus plasma); this percentage is called the hematocrit. Nor-mally erythrocytes are biconcave discs with a mean diameter of 6 to 8 μm and a maximal thickness of 1.9 μm. The average volume of an erythrocyte cell is approximately 83 $(\mu m)^3$ and the number of erythrocytes per cubic millimeter of blood is approximately 5 to 6×10^6. Leucocytes, which are roughly spherical, are usually larger than the red blood cells, ranging between 6 and 17 μm in diameter. However, their number is small, approximately 7 to 11×10^3 per cubic millimeter in a normal adult. Thrombocytes are much smaller than both erythrocytes and leucocytes. They are rounded or oval and have a mean diameter of approx-imately 2 to 3 μm, so even though there are approximately 2.5×10^5 per cubic millimeter, their total volume is small. Together leucocytes and thrombocytes have a volume concen-tration of only approximately 1% of the total blood volume. Furthermore, all these cells are deformable, with the erythrocytes being the most deformable. Significant deformations occur when the cells are passing through the capillaries. However, the cell membranes do not rupture because each cell has a cytoskeleton that supports its shape.

Therefore, the mechanical properties of blood should be studied by analyzing a liquid containing a suspension of flexible particles. A liquid is said to be Newtonian if the coeffi-cient of viscosity is constant at all rates of shear. This condition exists in most homogeneous liquids, including blood plasma (which, since it consists mostly of water, is Newtonian). But the mechanical behavior of a liquid containing a suspension of particles can vary such that the liquid becomes non-Newtonian. These deviations become particularly significant when the particle size becomes appreciably large in comparison to the dimension of the channel in which the fluid is flowing. This situation happens in the microcirculation (for the small arterioles and capillaries).

Consider a suspension in which the suspending fluid has Newtonian behavior. If the suspended particles are spherical and nonsettling, that is, if they have the same density as the suspending fluid, then for any motion the shear stress will be proportional to the rate of shear and the suspension will behave as a Newtonian fluid. This rule applies as long as the concentration of spheres is low, less than 30%. This rule was arrived at through experiments performed under steady state conditions with suspensions of rigid spheres. These experiments showed that the viscosity of the suspension, defined as the viscosity measured in a particular viscometer under particular conditions, was independent of the shear rate for volume concentrations of suspended spheres up to 30%. However, if the suspended particles are not spherical or are deformable in any way, then the shear stress is not proportional to the shear rate unless the concentration is much less than 30%.

The cells suspended in blood are not rigid spheres, and the volume fraction of erythrocytes is about 40 to 45%. Therefore, one should expect that the behavior of blood is non-Newtonian. But it has been shown that human blood is Newtonian at all rates of shear for hematocrits up to about 12%. In general blood has a higher viscosity than plasma, and as the hematocrit rises, the viscosity of the suspension increases and non-Newtonian behavior is observed, being detectable first at very low rates of shear. Studies with human blood show that viscosity is independent of shear rate when the shear rate is high. With a reduction of shear rate the apparent viscosity increases slowly until a shear rate less than 1 s^{-1}, where it rises extremely steeply (Caro et al., 1978). The shear stresses can be divided into two groups according to the effect of the shear rate:

- At low shear rates, the apparent viscosity increases markedly. The reason for this increase is that at low shear rates a tangled network of aggregated cell structures (Rouleaux) can be formed. If blood is subjected to shear stress less than a critical value, these aggregated structures form even for standing blood (blood that is not flowing). As a result, they exhibit a yield stress. This behavior is, however, only present if the hematocrit is high. If the hematocrit falls below a critical value, there are not enough cells to produce the aggregated structures and no yield stresses will be found.

- At high shear rates, the apparent viscosity in small vessels is lower than it is in larger vessels. The progressive diminution with the size of the vessels is detectable in vessels with an internal diameter less than 1 mm. It is even more pronounced in vessels with a diameter of 100 to 200 μm. The decreased viscosity with vessel size is known as the Fåhraeus and Lindqvist effect. Experiments were performed at high enough shear rates for the erythrocytes not to aggregate. It was found that the viscosity was approximately constant in vessels larger than 0.1 cm, but when the radius dropped below that, there was a substantial decrease in viscosity. This result is shown in Figure 2.13.

In the large vessels it is reasonable to assume blood has a constant viscosity, because the vessel diameters are large compared with the individual cell diameters and because shear rates are high enough for viscosity to be independent of them. Hence in these vessels the non-Newtonian behavior becomes insignificant and blood can be considered to be a Newtonian fluid. Measurements of the apparent viscosity show that it ranges from 0.03 to 0.04 g/(cm·s).

In the microcirculation, it is no longer possible to think of the blood as a homogeneous fluid; it is essential to treat it as a suspension of red cells in plasma. The reason for this is that even the largest vessels of the microcirculation are only approximately 15 cell widths in diameter. Also, as discussed earlier in this chapter, viscosity starts dominating the mechanical behavior, leading to low Reynolds numbers. Typical Reynolds numbers in 100-μm arteries are about 0.5.

In summary, we can conclude that blood is generally a non-Newtonian fluid, but it is reasonable to regard it as a Newtonian fluid when modeling arteries with diameters larger than 100 μm. For very small vessels it is not easy to reach conclusions as to the Newtonian nature of blood because some effects tend to decrease the viscosity and others to increase it. The latter influence on viscosity is due to a small flow, which increases the viscosity

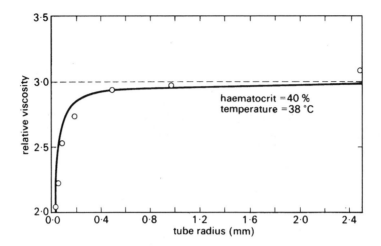

Figure 2.13. *The relationship between the apparent viscosity of blood relative to the plasma viscosity and the diameter of the tube in which the blood is flowing. The tube is assumed to be cylindrical. From Haynes (1960), Am. J. Physiol.* **198**:1193–1200. *Used by permission from the American Physiological Society.*

significantly, as well as to the fact that cells often become stuck at constrictions in small vessels. However, cells become stuck most often in the capillaries. The net effect of all these influences on blood viscosity is that it is reasonable to assume that the overall viscous effects in the small vessels (i.e., vessels with a diameter bigger than 100 μm) are approximately equivalent to those that occur in the larger vessels.

2.5   Cardiovascular Regulation

The regulation of human blood pressure is complex and involves a variety of control mechanisms. The biological function of blood pressure control is to provide adequate blood flow to the various organs connected to the human circulatory system. In essence, pressure control promotes a normal distribution of fluids, hormones, electrolytes, and other agents. The regulatory system consists primarily of two types of control mechanisms: a long-term control and a short-term control. Long-term control provides stabilization of blood pressure over longer periods (minutes, hours, and days), whereas short-term controls are concerned with the immediate and acute circulatory perturbations (seconds and minutes).

Long-term control operates mainly via the renal and hormonal activities. The kidneys increase the output of water and salt in response to an enhanced arterial pressure. These changes in water and salt regulation cause a decrease in blood volume and thus cardiac output. The net effect is a decline in the arterial pressure. A drop in the arterial pressure promotes secretion of renin from the kidneys. The secretion of renin results, among other things, in formation of the hormone angiotensin II, which enhances vessel constriction and thus increases arterial pressure.

Short-term regulation is mediated by autoregulation and by the central nervous system (CNS). The part of the control mediated by the CNS involves baroreceptors, mechanoreceptors, and chemoreceptors. The overall goal of neural control is to redistribute blood flow to the different areas of the body. Nerves communicate the control signals to the heart and the vessels, in response to the needs of the different body regions. Nervous activity in the CNS modifies heart rate, cardiac contractility, and the state of vessel constrictions. Chemoreceptors are sensitive to chemicals in the blood and react to alterations in the concentrations of oxygen, carbon dioxide, and hydrogen ions. A drop in arterial pressure may decrease the concentration of oxygen. The chemoreceptors answer by increasing cardiac strength and vessel constriction. Baroreceptors are stretch receptors, which are sensitive to pressure alterations. The most important receptors are located in high pressure regions such as the carotid sinus and the aortic arch. Mechanoreceptors (or low pressure receptors) are located in low pressure areas such as the atrial and pulmonary veins. Mechanoreceptors are also stretch receptors and provide arterial pressure control by resisting alterations in venous volume. Baroreceptors are the best known and most easily accessible receptors; consequently they have been investigated extensively. Mechanoreceptors are less studied; quantitative experimental data on these receptors are very sparse (Danielsen, 1998).

Autoregulation is a local control mechanism independent of the CNS. Local tissues can control blood flow in response to moderate changes in cardiac output and arterial pressure by dilating or constricting vessels. These responses may be due to a contractile response by the smooth muscles when stretched.

2.6 Pulmonary Physiology

The main purpose of the respiratory system is to transport oxygen and carbon dioxide between the atmosphere and the tissue and organs in the body. Oxygen is a necessity for life, and a human being consumes approximately 260 ml/min at rest (Nunn, 1987). The oxygen is delivered from the atmosphere to the organs and tissue via the lungs and circulatory system. Carbon dioxide is a waste product of oxidative metabolism and is carried by the blood in the opposite direction, from the tissue to the lungs, where it is removed by ventilation. The carbon dioxide elimination rate at rest is about 160 ml/min (Nunn, 1987). Since carbon dioxide dissolved in blood forms carbonic acid, which affects the pH value of the blood, the removal of carbon dioxide plays an important role in the acid-base balance in the blood.

The respiratory cycle starts in the atmosphere outside the body; see Figure 2.14. Oxygen enters the lungs by inspiration, and 21% by volume of atmospheric air consists of oxygen. During inspiration, air enters the lung, where it mixes with the air already in the lung. The upper airways and the lungs form a tree structure, i.e., the pulmonary tree, connecting the atmosphere with the alveoli, which are small air-filled sacs. From the alveoli oxygen diffuses across a membrane into the blood of the pulmonary capillaries; see section 2.6.2. By this diffusion the content of oxygen in the alveoli is reduced, and hence the expiratory air contains only 16% oxygen.

Nearly all respiratory gases are distributed throughout the body by the bloodstream. This transport of gases is much faster than diffusion. Branching of blood vessels into tiny capillaries assures that diffusion lengths are small, both in the lungs and in tissues. Almost all cells in the body are within a few cell diameters of at least one of the smallest branches

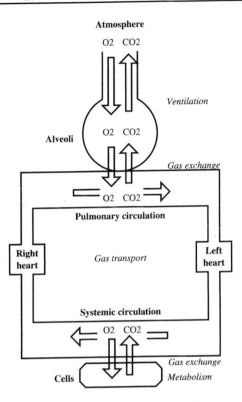

Figure 2.14. *Schematic view of the respiratory system. The figure shows how CO_2 and O_2 are exchanged between the alveoli in the lung (top) and the cardiovascular system (bottom).*

(Vander, Sherman, and Luciano, 1990). When blood flows through the capillaries of the tissues and organs, oxygen leaves the bloodstream by diffusion and enters the cells, where it is used for metabolism. Metabolism produces carbon dioxide, which subsequently diffuses into the blood and is carried to the pulmonary capillaries. From the pulmonary capillaries, carbon dioxide diffuses over the lung membrane into the alveoli. From the alveoli it is transported through the airways to the atmosphere during expiration.

2.6.1 Ventilation

Under normal conditions, breathing continuously renews the air in the lungs. During inspiration, the air passes from the mouth and nose, through the tree-like conducting airways into the alveoli; see Figure 2.15. At expiration the air flows the opposite way. In the alveoli the air and blood are separated by only a thin membrane through which oxygen and carbon dioxide transfer can take place. The area of the blood-gas membrane of the three million alveoli of a standard man is about 90 m² (Grodins and Yamashiro, 1978).

The airways are structured as a binary tree, where each new level of branching, or generation, doubles the number of pipes. The first generation, termed generation 0, consists

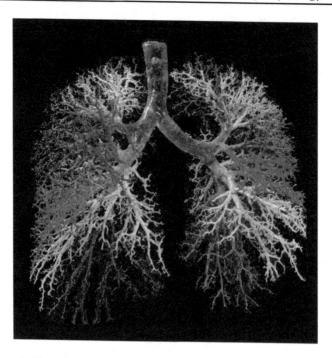

Figure 2.15. *Cast of the airways of a human lung. From* Human Physiology, *4th edition, by Stuart Ira Fox, Wm. C. Brown, 1993. Used by permission from McGraw-Hill.*

of a single pipe, named the trachea. The last generation, denoted generation 23, consists of 8 million pipes. The first generations (0–19) are called the conduction zone. Small alveoli sacs appear on the pipes at the later generations (20–23). Consequently, these generations are termed the respiratory zone.

The lung and airways have no muscles to drive ventilation. Instead, the lungs function like a bellows, with inflow and outflow driven by forces working solely on the outside. Natural ventilation is similar to the normal operation of a bellows, while artificial respiration by a respirator is similar to filling the bellows by blowing into the pipe. Natural ventilation is performed by movement of the pulmonary walls, generating a pressure difference and thus an airflow between the lungs and the surroundings. The alveoli walls contain a fluid, the interpleural fluid, in the interstitial space between the lung and the thorax; see Figure 2.16. This space constitutes a single connected chamber throughout each lung that is "fixed" on the "outside" to the thorax. Consequently, forces acting upon any wall of the interstitial space are transmitted by the fluid to all the rest of the walls by a hydraulic principle. Natural breathing results from rhythmic contraction and relaxation of respiratory muscles. At inspiration the movements of these muscles cause the thorax to enlarge. When the thorax is expanded, the force is transmitted to the lung via the interpleural fluid, forcing each alveolus to enlarge. The expansion causes the pressure within the alveoli to drop to less than atmospheric, and the pressure difference causes an airflow into the alveoli. The ability of the lung to expand is termed elastance, and the inverse of elastance is called compliance.

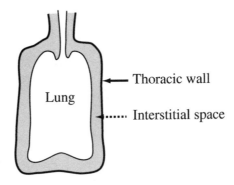

Figure 2.16. *Schematic picture of the lung and its surroundings.*

Pressure from the interstitial space gives the lung an elastic recoil. Normally expiration is caused only by this elastic recoil driving the air in the lungs the opposite way, but active forces contracting the thoracic cage can be applied. At the end of an expiration the interstitial space has a pressure slightly below atmospheric. The force from the interpleural fluid thus prevents collapse of the alveoli.

During artificial ventilation, the driving forces of the respiration muscles are replaced with an externally driven pressure source such as a respirator. Inspiration is obtained by raising the pressure in the ventilatory mask, thus forcing air into the lung. When the pressure is removed, elastic recoil drives expiration.

The volume of air flowing into and out of the lungs during each breath is called the tidal volume; see Figure 2.17. The tidal volume is about 0.5 l at rest. A small amount of air, approximately 0.15 l, reaches only the conducting airways, and will be expired without any exchange with the blood. This amount is called the dead space. Thus approximately 0.35 l per breath participates in gas exchange. At a breathing rate of 15 breaths per minute this gives a volume flow rate of 5.25 l/min.

Ventilation disorders are normally split into two types: obstructive and restrictive. In obstructive ventilation disorders the flow of air through the airways is obstructed. Restrictive disorders are cases in which regions of the lungs are damaged, resulting in lower compliance and possible decreased permeability of the lung membrane.

2.6.2 Gas Exchange between Lungs and Blood

When the inspired air reaches the alveoli, there is only a thin permeable membrane of 0.2 μm separating the air from the blood in the approximately 1800 capillary vessels surrounding each alveolus; see Figure 2.18. Consequently, O_2 and CO_2 are rapidly exchanged (Grodins and Yamashiro, 1978).

Oxygen and carbon dioxide move between the alveoli and blood by simple diffusion. The mechanism of diffusion between air and blood may be understood by considering a small container, half filled with water. The random thermal motions of the gas molecules will let the gas diffuse to areas where the pressure is low, even into the liquid. Since diffusion happens by random movements, a larger number of molecules in one region will result in more molecules diffusing out of that region. Eventually the random movements of

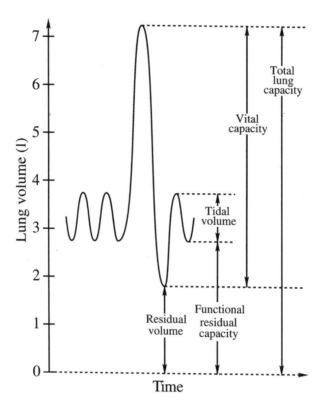

Figure 2.17. *Lung volumes during the breathing cycle. The tidal volume is the normal respiration, but both expiration and inspiration can be increased, yielding the vital capacity.*

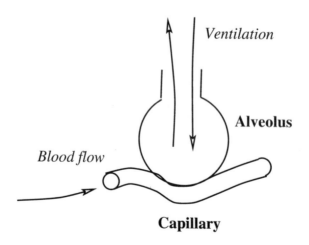

Figure 2.18. *An alveolus and a capillary. Each alveolus is typically surrounded by 1800 capillaries.*

molecules from the gas phase to the liquid phase will equal the random movements in the opposite direction and thus an equilibrium state is reached. At equilibrium the pressure of the gas is uniform throughout the container.

The relationship between the concentration of a gas dissolved in a liquid and the partial pressure expresses the distribution of gas between the two phases. If no chemical reaction takes place, the concentration in the solvent is, to a good approximation, proportional to the pressure. The proportionality factor expresses the solubility of the gas in the liquid. A highly soluble gas will come to an equilibrium with a large number of molecules per volume of the liquid, while a gas with a low solubility will have more gas molecules in the gas phase; see Figure 2.19. The solution of a gas in a liquid may include effects other than random thermal movements, as some gas molecules may react with molecules in the liquid. Yet regardless of how the gas dissociates in the liquid, the gas will adjust toward equal partial pressure in the liquid and gas phases.

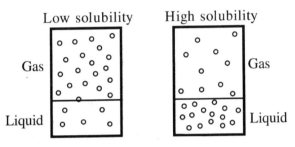

Figure 2.19. *Different gas distributions due to different solubilities.*

In order to distinguish between gas in the gas phase and gas dissolved in liquid, we use the term *pressure* for gas in the gas phase and *tension* for the pressure of the gas dissolved in liquid.

Oxygen is poorly soluble in water. Consequently, the high concentration of oxygen in blood is due to its chemical binding to components in the blood. Without these oxygen-carrying components, a very high partial pressure of oxygen in the alveoli or a much faster blood flow would be required in order to transport the needed 260 ml of oxygen each minute. At normal atmospheric pressure and with the normal amount of blood components, more than 98% of oxygen in the blood is bound to components in the blood. The main carrier of oxygen in the blood is hemoglobin, a protein found in erythrocytes (red blood cells); see section 2.4.2. Oxygen is reversibly bound with hemoglobin. Hemoglobin combined with oxygen is denoted oxyhemoglobin, while hemoglobin not combined with oxygen is termed deoxyhemoglobin or reduced hemoglobin.

Carbon dioxide is much more soluble in blood than oxygen is, but its transport is also improved by chemical reactions. Dissolved carbon dioxide reacts reversibly with water and with hemoglobin. Consequently, some carbon dioxide is transported as bicarbonate or carbamino compounds. When carbon dioxide reacts with water, an acid is formed, and hence there is a close relation between the carbon dioxide level and the acid-base balance in blood. The acid-base balance is expressed by the pH value, which is the negative logarithm of the concentration of hydrogen ions. Fluctuations in the pH value are buffered by the way hydrogen ions participate in certain chemical reactions in blood. Hydrogen ions combine with bicarbonate ions and hemoglobin, and therefore the pH value influences the dissociation of both oxygen and carbon dioxide in blood. Even in this complicated interaction, in which the two gases affect the dissociation of each other through chemical reactions, the gases will each reach an equal pressure between the blood and the surrounding tissue or alveoli as long as the membrane separating the two phases is permeable. This happens because the equilibria of the chemical reactions are shifted when dissolved gas diffuses across the membrane.

In cases of respiratory disorders or hard work, the oxygen supply might become insufficient. In such cases metabolism occurs in the absence of oxygen. However, carbon dioxide is also a waste product of this anaerobic metabolism and is still removed from the tissues by blood. If the elimination of carbon dioxide through the lung is reduced, the consequence will soon be an increased amount of carbon dioxide, and hence a decreased pH value, in the blood.

Acknowledgments

All authors were supported by High Performance Parallel Computing, Software Engineering Applications Eureka Project 1063 (SIMA—SIMulation in Anesthesia). In addition, P.T. Adeler, M. Danielsen, and M. Olufsen were supported by the Danish Academy for Technical Sciences; M. Danielsen was supported by the Danish Heart Foundation (99-1-2-14-22675) and Trinity College, Connecticut; M. Olufsen was supported by the National Science Foundation Group Infrastructure Grant while at Boston University (DMS-9631755).

Chapter 3
Blood Flow in the Heart

P.T. Adeler and J.M. Jacobsen

3.1 Introduction

This chapter presents a two-dimensional (2D) mathematical model of blood flow in the heart. The model has been used to gain insight into the dynamics of the heart and has as such acted as a reference model for the development of the heart model in the SIMA simulator. The model was originally developed by Peskin and was dimensioned to model a dog's heart (Peskin, 1972a, 1972b). Our purpose has been to change the model's dimensions and parameters so it models a human heart and to compare simulated data produced by this modified model with magnetic resonance (MR) measurements of a human heart.

The model represents a 2D section through the left side of the heart. This section lies in a plane, called the long axis plane, that passes through the apex of the heart, bisects the leaflets of the mitral valve, and cuts the aortic outflow tract. The model describes blood flow in the heart and the movement of the left atrium and the left ventricle during a cardiac cycle. The model also includes a description of the movement of the mitral valve, which connects the two chambers.

Venous inflow to the left heart is modeled by a source placed at the center of the atrium; aortic outflow is modeled by a sink at the outflow tract of the ventricle. The model represents the tissues of the heart, such as active muscle or passive valve tissues, as infinitely thin, massless, elastic boundaries totally immersed in blood (i.e., with blood on both sides). These boundaries exert force on the fluid and are themselves moved by the fluid with the local fluid velocity. The beating of the model heart is induced by successive relaxations and contractions of the elastic boundaries representing the heart muscles.

To adapt Peskin's model to the human heart we had to fit parameters for geometry, timing, and muscular contraction. In addition, we introduced modifications based on general physiological considerations regarding tethering and afterload.

Peskin initiated his work on the model with his thesis (Peskin, 1972a). He has since developed and studied both the model and the applied methodology in collaboration with others, in particular McQueen. This work is documented in a series of papers: Peskin (1972b), Peskin (1975), Peskin (1977), Peskin and McQueen (1980), Peskin (1981), McQueen, Peskin, and Yellin (1982), and Peskin and Printz (1993).

Peskin and McQueen's original purpose was to establish a computer tool for studying the function of the mitral valve (Yellin et al., 1981; Meisner et al., 1985). In addition, they wanted to create a tool that could be used to design and test prosthetic mitral valves (McQueen and Peskin, 1983; McQueen and Peskin, 1985). Peskin and colleagues have studied alternative numerical schemes for solving the flow equations and for the treatment of the coupling of boundary forces to the fluid (McCacken and Peskin, 1980; Börgers and Peskin, 1987; Tu and Peskin, 1992; Mayo and Peskin, 1993). These alternative methods are not implemented or discussed in this chapter.

The computational method of the 2D model has since been expanded to three dimensions (Greenberg, McQueen, and Peskin, 1987; Peskin and McQueen, 1989; McQueen and Peskin, 1989), and the work of Peskin and McQueen has culminated in a 3D model of the entire human heart and nearby great vessels (Peskin and McQueen, 1992; Peskin and McQueen, 1993b; Peskin and McQueen, 1993a; Peskin and McQueen, 1994; Peskin and McQueen, 1996; Davis, 1995; McQueen and Peskin, 2000). Due to their high complexity, the computational requirements needed to solve the 3D models make it infeasible to validate the model against data. Model validation requires many simulations to ensure that the geometry and fluid parameters for the model and the experimental data are the same. Therefore, we have chosen to use the 2D model in this paper, the major aim of which is to present validation with data from Magnetic Resonance (MR) imaging. An additional observation supporting our use of the 2D model is that experimental MR results indicate that the flow in the long-axis plane is in fact almost entirely in two dimensions.

In addition to the work by Peskin, the immersed boundary method has been studied and used by a number of people. Beyer and LeVeque (1992) studied the accuracy of various numerical approximations to fluid boundary coupling in a 1D model for the immersed boundary method. In Beyer (1992) the method was used to model cochlea. In Vesier and Yoganathan (1992) the method is validated by a comparison between computational results for a model of flow through a flexible tube and an approximate analytic solution of this flow. A 3D model that uses the immersed boundary method to model blood flow in the left ventricle during part of systole is reported in Yoganathan et al. (1994). This paper also presents some comparisons to experimental results. This model is used in Yoganathan et al. (1995) to study the systolic anterior mitral valve motion syndrome. Stockie and Wetton (1995) presented a stability analysis of the equations for the immersed boundary method.

The immersed boundary method is special in the way it models the interaction between fluid flow and the tissues of the heart. An alternative approach is to let the movement of the boundary be determined in advance, independently of the flow. In this approach, called the method of predetermined boundary motion, only the resulting flow of the fluid has to be computed. This method was carried out in Georgiadis, Wang, and Pasipoularides (1992) for an idealized ventricular geometry in the form of a contracting ellipsoid. In Taylor et al. (1993), Taylor and Yamaguchi (1995b), and Taylor and Yamaguchi (1995a) this method, together with an assumed movement of the ventricular wall, was done for a 3D geometry constructed using a cast of a dog's heart taken in the diastolic phase. Furthermore, Schoephoerster, Silva, and Ray (1993; 1994) used 2D cine-angiographic images to determine the heart wall boundary in both a healthy and a diseased (coronary artery disease) left ventricle, and they subsequently computed the flow during systole. They expanded their method to a simple spherical 3D ventricle in Gonzalez and Schoephoerster (1996), where they also investigated the effect of abnormal wall motion on the flow pattern. In Ding and Schoephoerster (1997) they used cine-angiograms to construct a 3D systolic model of the left ventricle, assuming circular cross sections.

The method of predetermined boundary motion has also been used in combination with MR scanning data. Jones and Metaxas (1998) used MRI SPAMM to give the heart boundary configuration for a flow model of the left ventricle. MRI SPAMM allows tracking of material points on the heart and gives an accurate representation of the left ventricular motion from end-diastole to end-diastole (Park, Metaxas, and Axel, 1996). Saber et al. (2001) used anatomical data obtained by MR scanning to construct a moving meshes model of the left ventricle. Their model simulates the blood flow during a full cardiac cycle, except for late diastole.

Current MR techniques allow the determination of 3D velocity fields for blood flow in the human heart (Houlind et al., 1994; Kim et al., 1995; Walker et al., 1996). In Houlind et al. (1994) and Kim et al. (1995) the three components of the blood velocity field in a single slice (the long-axis plane) were obtained in patients with ischemic heart disease and in normal subjects. Furthermore, Kim et al. (1995) quantified the left ventricle vortex seen during diastole. Walker et al. (1996) measured the 3D blood velocity field in multiple slices in the human heart. In each of these three papers qualitative descriptions of the blood flow are also given. The measurement of the 3D velocity field in the heart should, on the one hand, make possible a better validation of the heart models. On the other hand, it may expand the use of the models, since the flow fields inside the heart may be of clinical importance to the extent that they can be assessed.

3.2 Continuous Formulation

In this model the heart is immersed in a box filled with fluid of the same density and viscosity as blood. The muscular wall and the valve leaflets are treated as massless 1D boundaries immersed in fluid (blood) with the potential to exert force on the fluid. The fluid in the cavities confined by the boundaries represents blood in the heart chambers, while the fluid outside the boundaries represents the surrounding tissue and mass of the heart muscle. The heart model includes a source in the atrium to account for the pulmonary venous inflow and a sink at the aortic outflow tract to account for ejection of blood during systole. Moreover, since the heart expands during diastole and contracts during systole, the model also includes an exterior sink/source, placed along the vertical boundaries of the domain, to make room for the variation in total heart volume during a heartbeat.

The starting point for the continuous formulation of the model is Newton's second law for fluids: Mass times acceleration equals the sum of all forces. The forces acting on the fluid are pressure, viscous forces, and forces from the heart wall; thus

$$\rho \frac{D\mathbf{u}}{Dt} = -\nabla p + \mu \nabla^2 \mathbf{u} + \mathbf{F},$$

where the unknowns are the fluid velocity field $\mathbf{u}(\mathbf{x}, t)$ and the scalar pressure field $p(\mathbf{x}, t)$, with $\mathbf{x}$ denoting the position vector and t denoting time. The constants ρ and μ denote the fluid density and viscosity, respectively. $\mathbf{F}(\mathbf{x}, t)$ denotes the force density (force per unit area) exerted on the fluid by the heart boundary. $\frac{D}{Dt}$ is the material derivative and can also be written as

$$\rho \left(\frac{\partial \mathbf{u}}{\partial t} + \mathbf{u} \cdot \nabla \mathbf{u} \right) = -\nabla p + \mu \nabla^2 \mathbf{u} + \mathbf{F}. \tag{3.1}$$

This is the Navier–Stokes equation for incompressible fluid flow, which together with the continuity equation constitutes the equations of motion for the fluid (Chorin and Marsden, 1998). The continuity equation is given by

$$\nabla \cdot \mathbf{u} = \psi(\mathbf{x}, t), \tag{3.2}$$

where $\psi(\mathbf{x}, t)$ represents the area flow rate density distribution of the sources and sinks. The computational domain Ω is taken to be a square, and the boundary condition is the requirement that the velocity and pressure fields be periodic with this domain equaling one period.

Equations (3.1) and (3.2) are based on assumptions concerning the momentum of the fluid that enters or leaves the model heart at the sources and sinks. Fluid flowing out of the model heart at sinks takes its momentum with it as it leaves, while fluid flowing into the model heart at sources has the same velocity (i.e., same momentum) as the fluid at the point where it enters (Jakobsen and Niss, 2000).

The external fluid force density $\mathbf{F}(\mathbf{x}, t)$ is a δ function layer defined by

$$\mathbf{F}(\mathbf{x}, t) = \int_0^L \mathbf{f}(s, t) \, \delta(\mathbf{x} - \mathbf{X}(s, t)) \, ds, \tag{3.3}$$

where the function $\mathbf{X}(s, t), 0 \leq s \leq L$, parameterizes the heart boundary and $\mathbf{f}(s, t)$ denotes the boundary force density with respect to s; i.e., $\mathbf{f}(s, t)ds$ is the force exerted on the fluid by a piece of the boundary of length ds about s. Figure 3.1 shows the computational domain Ω with the initial heart boundary $\mathbf{X}(s, t_0)$ drawn. The choice of initial geometry will be discussed in section 3.3.5.

The heart boundary is simulated either as passive tissue following the fluid motion or as active muscle tissue. For the passive, elastic parts of the boundary the boundary force density $\mathbf{f}$ depends only on the current configuration $\mathbf{X}$. For the active parts (simulating muscles) it also depends on the history of $\mathbf{X}$ through the variation in time of internal variables (variable resting lengths), which determine the contractile properties of the boundary. The time development is determined from a given function of time, the activation function α, through the interaction of the boundary and the fluid (see sections 3.3.3 and 3.3.4). The particular form of the boundary force density $\mathbf{f}(s, t)$ will be specified in the following sections.

The fluid is subject to the no-slip condition at the heart boundary and therefore moves with the local fluid velocity

$$\frac{\partial \mathbf{X}}{\partial t}(s, t) = \mathbf{u}(\mathbf{X}(s, t), t) = \int_\Omega \mathbf{u}(\mathbf{x}, t) \, \delta(\mathbf{x} - \mathbf{X}(s, t)) \, d\mathbf{x}. \tag{3.4}$$

The motion of the heart boundary is not known in advance, and thus the above equation constitutes an equation of motion for the heart boundary. The fluid velocity is assumed to be continuous across the boundary as a part of the no-slip condition.

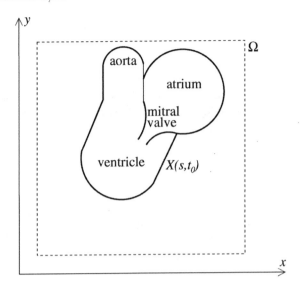

Figure 3.1. *Computational domain Ω and initial heart boundary configuration* $\mathbf{X}(s, t_0)$.

Sources and sinks are incorporated into the model through the density distribution $\psi(\mathbf{x}, t)$ in (3.2). This density distribution is assumed to be of the form

$$\psi(\mathbf{x}, t) = Q_{ex}(t)w_{ex}(\mathbf{x} - X_{ex}(t)) + \sum_{k=1}^{2} Q_k(t)w_k(\mathbf{x} - X_k(t)),$$

where $k = 1$ corresponds to the source in the atrium and $k = 2$ corresponds to the sink in the aortic outflow tract. The index *ex* refers to the exterior source/sink placed outside the heart to accommodate excess inflow or outflow to/from the heart. $Q_k(t)$ and $Q_{ex}(t)$ denote the area flow rates through the source or sink. Conversion from area to volume is defined by specifying a "thickness" of the otherwise 2D domain Ω. w_k and w_{ex} are weight functions (with dimension per unit area) representing the flow distribution across the sources and sinks (assumed to be time and flow independent), and $X_k(t)$ and $X_{ex}(t)$ denote the current positions of the centers of the sources and sinks (which move with the heart).

The mathematical requirement on w_k and w_{ex} is that they be continuous nonnegative functions with compact support satisfying

$$\int_{\Omega} w_k(\mathbf{x}) \, d\mathbf{x} = 1 \quad \text{and} \quad \int_{\Omega} w_{ex}(\mathbf{x}) \, d\mathbf{x} = 1. \tag{3.5}$$

In the model, the weight functions for the interior sources/sinks are chosen to be properly scaled "cosine hat" functions of the form $(1 + \cos x)(1 + \cos y)$ for $(x, y) \in [-\pi, \pi] \times [-\pi, \pi]$, extended to zero outside this domain, while the exterior weight function w_{ex} is a "cosine ridge" of the form $(1 + \cos x)$ for $(x, y) \in [-\pi, \pi] \times [-\pi, \pi]$, extended to zero outside this domain.

Since the velocity field is required to be periodic, there is no net inflow to the domain Ω through the boundaries, so the incompressibility of the fluid implies that the net inflow through the sources must also be zero; i.e.,

$$0 = \int_{\Omega} \nabla \cdot \mathbf{u}\, d\mathbf{x} = \int_{\Omega} \psi(\mathbf{x}, t)\, d\mathbf{x} = Q_{ex}(t) + \sum_{k=1}^{2} Q_k(t).$$

The periodicity of $\mathbf{u}$ implies the first equality above, and the incompressibility is implicit in the assumption that (3.2) represents inflow/outflow. By the above equation we may eliminate Q_{ex},

$$Q_{ex}(t) = -\sum_{k=1}^{2} Q_k(t),$$

so that with the definition

$$\psi_k(\mathbf{x}, t) = w_k(\mathbf{x} - X_k(t)) - w_{ex}(\mathbf{x} - X_{ex}(t))$$

we may write

$$\psi(\mathbf{x}, t) = \sum_{k=1}^{2} Q_k(t)\psi_k(\mathbf{x}, t). \tag{3.6}$$

The flow rate $Q_k(t)$ through the source k is assumed to be a function of the space average $P_k(t)$ of the pressure at the source k. This is defined with respect to the weight function w_k as follows:

$$\begin{aligned}
P_k(t) &= \int_{\Omega} p(\mathbf{x}, t) w_k(\mathbf{x} - X_k(t))\, d\mathbf{x} - \int_{\Omega} p(\mathbf{x}, t) w_{ex}(\mathbf{x} - X_{ex}(t))\, d\mathbf{x} \\
&= \int_{\Omega} p(\mathbf{x}, t)\psi_k(\mathbf{x}, t)\, d\mathbf{x} \\
&= \langle p(t), \psi_k(t) \rangle,
\end{aligned}$$

i.e., with the average pressure over the exterior source/sink $P_{ex}(t)$ chosen as the reference pressure: $P_{ex}(t) = 0$ for all t.

For each source and sink a specific relation between $Q_k(t)$ and $P_k(t)$ is given, making the flow depend upon the pressure $Q_k(p(t), t)$. Hence introducing the relation in (3.6), the source/sink term ψ can now be expressed as a time-varying function $\Psi(p, t)$ of the time-varying pressure field p as follows:

$$\psi(\mathbf{x}, t) = \Psi(p(t), t)(\mathbf{x}) = \sum_{k=1}^{2} Q_k(p(t), t)\psi_k(\mathbf{x}, t).$$

The specific relation between $Q_1(t)$ and $P_1(t)$ for the source in the atrium is chosen to be that of a linear resistance model pumping against a constant pressure reservoir:

$$Q_1(t) = \frac{P_{src} - P_1(t)}{R_{src}}, \tag{3.7}$$

where P_{src} denotes the load pressure at the source and R_{src} denotes the resistance against the flow through the source. In the original model developed by Peskin the aortic sink was also modeled by a linear resistance model pumping against a constant pressure reservoir combined with a diode-like valve. The diode-like valve opened the sink when the pressure at the sink exceeded a given value and closed it again when the pressure fell below that value.

We find that using the linear resistance model with a constant pressure reservoir to represent the relation between the aortic outflow rate and the pressure at the aortic outflow tract gives a rather unrealistic shape to the associated pressure versus time profile. Pressure, under that model, is almost clamped at the prescribed constant pressure in the aorta (see section 3.4). A more realistic curve form can be obtained by combining the diode-like sink with a simple three-element Windkessel model of the arterial impedance. The Windkessel model is also a linear resistance model but with a time-varying pressure reservoir. See Figure 3.2 for the electric analogue of the Windkessel model.

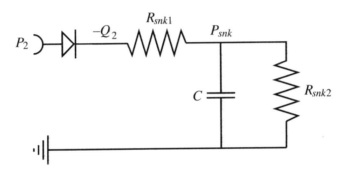

Figure 3.2. *Diagram of a three-element Windkessel model coupled to a diode valve.*

Let $P_2(t)$ denote the pressure at the aortic outflow tract and $Q_2(t)$ the outflow rate from the ventricle. The inflow to the Windkessel model is then $-Q_2(t)$, and we have

$$-Q_2(t) = \begin{cases} \dfrac{P_2(t) - P_{snk}(t)}{R_{snk1}} & \text{if } P_2(t) \geq P_{snk}(t), \\ \\ 0 & \text{if } P_2(t) < P_{snk}(t), \end{cases} \tag{3.8}$$

where R_{snk1} represents the proximal arterial resistance to the flow and where the pressure $P_{snk}(t)$ may be viewed as the pressure at a representative proximal location in the arterial system. $P_{snk}(t)$ is governed by the differential equation

$$C\frac{dP_{snk}}{dt} = -Q_2(t) - \frac{P_{snk}(t)}{R_{snk2}}, \tag{3.9}$$

where R_{snk2} represents the peripheral resistance and C represents the arterial compliance. If the compliance C is allowed to be dependent on the pressure P_{snk}, which makes the impedance model nonlinear, then it is possible to achieve quite good fits to measured pressure-flow relation data with this model; see Cappello, Gnudi, and Lamberti (1995).

Equations (3.1) to (3.4) constitute the continuous equations of motion for the heart model.

3.3 Discrete Formulation

The square domain Ω is divided into a square grid of size N by N, with $h = \Delta x = \Delta y$ denoting the distance between neighboring points. Time is divided into intervals of length Δt, and the parameter interval for the boundary is divided into subintervals of length Δs. Thus the independent variables are restricted to the values $\mathbf{x} = (x, y) = (ih, jh)$, $t = n\Delta t$, and $s = k\Delta s$, where i, j, n, and k are integers.

Hence at each time level $t = n\Delta t$ we have the field quantities $\mathbf{u}^n$, p^n, and $\mathbf{F}^n$ defined on the grid points; i.e.,

$$\mathbf{u}_{ij}^n = \mathbf{u}(ih, jh, n\Delta t),$$
$$p_{ij}^n = p(ih, jh, n\Delta t),$$
$$\mathbf{F}_{ij}^n = \mathbf{F}(ih, jh, n\Delta t)$$

for $i, j = 1, \ldots, N$. The boundary configuration $\mathbf{X}^n$ at time $n\Delta t$ is defined as a finite sequence of points, not necessarily coinciding with the grid points, with associated boundary force densities. Thus

$$\mathbf{X}_k^n = \mathbf{X}(k\Delta s, n\Delta t),$$
$$\mathbf{f}_k^n / \Delta s = \mathbf{f}(k\Delta s, n\Delta t)$$

for $k = 1, \ldots, N_b$, where N_b denotes the number of boundary points in the discretization and we have chosen to let $\mathbf{f}_k^n$ denote a boundary force rather than the force density.

3.3.1 Discretization of Equations of Motion

The differential equations in the continuous formulation (3.1) are discretized by replacing the differential operators with difference operators. The centered, forward, and backward differences in the two space directions, denoted x and y, are used. They are defined for a function ϕ on the domain grid as follows:

$$(D_x^0\phi)_{ij} = \frac{\phi_{i+1,j} - \phi_{i-1,j}}{2h}, \quad (D_x^+\phi)_{ij} = \frac{\phi_{i+1,j} - \phi_{i,j}}{h}, \quad (D_x^-\phi)_{ij} = \frac{\phi_{i,j} - \phi_{i-1,j}}{h},$$

$$(D_y^0\phi)_{ij} = \frac{\phi_{i,j+1} - \phi_{i,j-1}}{2h}, \quad (D_y^+\phi)_{ij} = \frac{\phi_{i,j+1} - \phi_{i,j}}{h}, \quad (D_y^-\phi)_{ij} = \frac{\phi_{i,j} - \phi_{i,j-1}}{h}.$$

In addition to these, a vector difference operator $\mathbf{D} = (D_x, D_y)$, adapted to the chosen approximation to the coupling between the fluid and the boundary (the δ function), will be used. The chosen approximation $\delta_h(\mathbf{x})$ to the δ function is defined as follows (Peskin and Printz, 1993):

$$\delta_h(\mathbf{x}) = \delta_h(x)\,\delta_h(y) \quad \text{for } \mathbf{x} = (x, y), \tag{3.10}$$

where

$$\delta_h(x) = \begin{cases} \dfrac{1}{4h}\left(1 + \cos\dfrac{\pi x}{2h}\right) & \text{if } |x| \leq 2h, \\ 0 & \text{if } |x| \geq 2h. \end{cases} \tag{3.11}$$

Figure 3.3 shows a plot of the δ_h function.

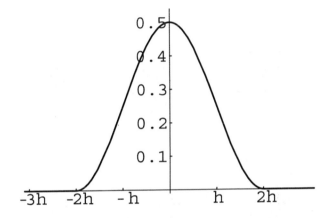

Figure 3.3. *Plot of the function $\delta_h(x)$.*

The virtues of this choice of δ_h function are elucidated in Peskin (1977). One virtue is the invariance of this function's discrete integral under arbitrary (continuous) translation, as discussed below for the source weight functions.

Peskin and Printz (1993) define the vector difference operator $\mathbf{D} = (D_x, D_y)$, where

$$(D_x\phi)(x, y) = \sum_{x',y'} \phi(x', y')\,\gamma(x - x')\,\omega(y - y'), \tag{3.12}$$

$$(D_y\phi)(x, y) = \sum_{x',y'} \phi(x', y')\,\omega(x - x')\,\gamma(y - y'), \tag{3.13}$$

with

$$\gamma(x) = \delta_h(x + X)\big|_{X=-h/2}^{X=h/2}, \quad \omega(x) = \int_{-h/2}^{h/2} \delta_h(x + X)\,dX,$$

$$\gamma(y) = \delta_h(y + Y)\big|_{Y=-h/2}^{Y=h/2}, \quad \omega(y) = \int_{-h/2}^{h/2} \delta_h(y + Y)\,dY.$$

The operator works in the following way:

$$\mathbf{D}p^n = (D_x p^n, D_y p^n), \quad \mathbf{D} \cdot \mathbf{u}^n = D_x u_x^n + D_y u_y^n,$$

where $\mathbf{u}^n = (u_x^n, u_y^n)$.

The Navier–Stokes equations (3.1) and the continuity equation (3.2) are discretized following a scheme adapted from Chorin (1968; 1969); see Peskin (1972b), Peskin (1972a), and Peskin (1977). This scheme introduces the fractional step velocity fields $\mathbf{u}^{n+1.0}$, $\mathbf{u}^{n+1.1}$, and $\mathbf{u}^{n+1.2}$ found for given $\mathbf{u}^n$ and $\mathbf{F}^n$ by solving in succession the equations

$$\rho \frac{\mathbf{u}^{n+1.0} - \mathbf{u}^n}{\Delta t} = \mathbf{F}^n, \tag{3.14}$$

$$\rho \left(\frac{\mathbf{u}^{n+1.1} - \mathbf{u}^{n+1.0}}{\Delta t} + u_x^n D_x^0 \mathbf{u}^{n+1.1} \right) = \mu D_x^+ D_x^- \mathbf{u}^{n+1.1}, \tag{3.15}$$

$$\rho \left(\frac{\mathbf{u}^{n+1.2} - \mathbf{u}^{n+1.1}}{\Delta t} + u_y^n D_y^0 \mathbf{u}^{n+1.2} \right) = \mu D_y^+ D_y^- \mathbf{u}^{n+1.2}. \tag{3.16}$$

Then $\mathbf{u}^{n+1}$ and p^{n+1} are found by solving the system

$$\rho \left(\frac{\mathbf{u}^{n+1} - \mathbf{u}^{n+1.2}}{\Delta t} \right) + \mathbf{D}p^{n+1} = 0, \tag{3.17}$$

$$\mathbf{D} \cdot \mathbf{u}^{n+1} = \Psi^n(p^{n+1}). \tag{3.18}$$

Here $\mathbf{D}$ is the vector difference operator defined above and $\Psi^n(p^{n+1})$ is obtained from a discretization of (3.6):

$$\Psi^n(p^{n+1}) = \sum_{k=1}^{2} Q_k^n(p^{n+1}) \psi_k^n. \tag{3.19}$$

For the source in the atrium we have, according to (3.7),

$$Q_1^n(p^{n+1}) = \frac{P_{src} - P_1^n}{R_{src}} = \frac{P_{src} - \langle p^{n+1}, \psi_1^n \rangle_h}{R_{src}} \tag{3.20}$$

based on the discrete inner product

$$\langle p, \psi \rangle_h = \sum_{ij} h^2 \, p_{ij} \psi_{ij},$$

where the summation is over the grid points of the domain. We note that the source function ψ is defined and continuous everywhere in Ω, while the pressure p is only defined on the grid points. We also note that the choice of weight functions w_k is such that the discrete analogue of the normalization condition (3.5) holds for all translations of w_k. For the treatment of the aortic sink see section 3.3.2.

Equation (3.3) for the coupling of the boundary forces to the fluid is discretized using the approximation to the δ function given by (3.10) and (3.11). The discretized version of (3.3) then reads

$$
\begin{aligned}
\mathbf{F}_{ij}^n &= \sum_k \mathbf{f}^n(k\Delta s)\, \delta_h(\mathbf{x}_{ij} - \mathbf{X}^n(k\Delta s))\Delta s \\
&= \sum_k \mathbf{f}_k^n\, \delta_h(\mathbf{x}_{ij} - \mathbf{X}_k^n). \tag{3.21}
\end{aligned}
$$

The calculation of the boundary forces $\mathbf{f}_k^n$ is discussed in section 3.3.3.

Finally, (3.4), which couples the fluid motion to the boundary motion, is discretized using the same approximation to the δ function; i.e.,

$$
\mathbf{X}_k^{n+1} = \mathbf{X}_k^n + \Delta t \sum_{ij} \mathbf{u}_{ij}^{n+1}\, \delta_h(\mathbf{x}_{ij} - \mathbf{X}_k^n)\, h^2. \tag{3.22}
$$

To solve the discrete equations it is necessary to reduce all lengths and times using a factor of $\gamma = 1/25$ to ensure numerical stability. This scaling reduces the Reynolds number, which is equivalent to raising the viscosity of the fluid by a factor of 25. Another effect of the scaling is that the thickness of the boundary layer of the fluid is raised by a factor of $\sqrt{25} = 5$. The scaling can be avoided through the introduction of a new fluid solver, which is used in the 3D heart model (McQueen and Peskin, 1997) but has not yet been implemented in the 2D heart model.

3.3.2 Discrete Windkessel Model for the Aortic Sink

The relation between flow and pressure for the Windkessel model at the aortic outflow is found by discretizing the equations of this model. Equation (3.9) is discretized according to an implicit Euler method as follows:

$$
\frac{P_{snk}^n - P_{snk}^{n-1}}{\Delta t} = -\frac{Q_2^n}{C} - \frac{P_{snk}^n}{C R_{snk2}}.
$$

We use $n-1$ and n because we need an expression for Q_2^n. This equation may be solved for P_{snk}^n:

$$
P_{snk}^n = \frac{P_{snk}^{n-1}}{1 + \Delta t/(C R_{snk2})} - \frac{\Delta t/C}{1 + \Delta t/(C R_{snk2})} Q_2^n. \tag{3.23}
$$

When C depends on P_{snk}, we will use the approximation $C = C(P_{snk}^{n-1})$, and the method will be only semi-implicit. Inserting relation (3.8) into (3.23) and solving for P_{snk}^n, we get

$$
P_{snk}^n =
\begin{cases}
\dfrac{P_{snk}^{n-1} + \frac{\Delta t}{R_{snk1} C} P_2^n}{1 + \frac{\Delta t}{RC}} & \text{if } P_2^n \geq P_{snk}^n, \\[3ex]
\dfrac{P_{snk}^{n-1}}{1 + \frac{\Delta t}{R_{snk2} C}} & \text{if } P_2^n < P_{snk}^n,
\end{cases}
\tag{3.24}
$$

where $R^{-1} = R_{snk1}^{-1} + R_{snk2}^{-1}$ and $P_2^n = \langle p^{n+1}, \psi_2^n \rangle_h$. From relations (3.8) and (3.23) we deduce that

$$P_2^n \geq P_{snk}^n \qquad \text{if and only if} \qquad P_2^n \geq \frac{P_{snk}^{n-1}}{1 + \frac{\Delta t}{R_{snk2}C}}. \tag{3.25}$$

Inserting (3.24) and (3.25) into (3.8), we obtain the following relation between Q_2^n and P_2^n at a given time level n:

$$-Q_2^n = \begin{cases} \dfrac{P_2^n - \dfrac{P_{snk}^{n-1}}{1 + \frac{\Delta t}{R_{snk2}C}}}{R_{snk1}\dfrac{1 + \frac{\Delta t}{RC}}{1 + \frac{\Delta t}{R_{snk2}C}}} & \text{if } P_2^n \geq \dfrac{P_{snk}^{n-1}}{1 + \frac{\Delta t}{R_{snk2}C}}, \\[2em] 0 & \text{if } P_2^n < \dfrac{P_{snk}^{n-1}}{1 + \frac{\Delta t}{R_{snk2}C}}. \end{cases} \tag{3.26}$$

Hence at each time level n for which the valve is open there is a linear pressure-flow relation between P_2^n and Q_2^n (compare with (3.20) for the source in the atrium). When the compliance is pressure independent, the apparent resistance to the flow (the denominator in the above expression) is constant in time, but the pressure P_{snk}^n against which the heart is pumping varies with time.

When the values of P_2^n and Q_2^n are computed from the interaction with the heart model, then the pressure P_{snk} should be updated according to the following formula:

$$P_{snk}^n = \begin{cases} P_2^n + R_{snk1}Q_2^n & \text{if } P_2^n \geq \dfrac{P_{snk}^{n-1}}{1 + \frac{\Delta t}{R_{snk2}C}}, \\[2em] \dfrac{P_{snk}^{n-1}}{1 + \frac{\Delta t}{R_{snk2}C}} & \text{if } P_2^n < \dfrac{P_{snk}^{n-1}}{1 + \frac{\Delta t}{R_{snk2}C}}. \end{cases} \tag{3.27}$$

The above model for the aortic sink gives rise to the following algorithm for updating pressure and outflow rate from time $n - 1$ to time n and for controlling the opening and closing of the valve:

1. Compute a preliminary update of P_{snk} according to $P_{snk}^n = \dfrac{P_{snk}^{n-1}}{1 + \frac{\Delta t}{R_{snk2}C}}.$

2. Compute the pressure P_2^n from the heart model using relationship (3.26) between pressure and flow in (3.19) for the source/sink term Ψ:

If the valve is open, use

$$-Q_2^n = \frac{P_2^n - P_{snk}^n}{R_{snk1}(1 + \Delta t/RC)/(1 + \Delta t/R_{snk2}C)},$$

otherwise

$$Q_2^n = 0.$$

3. If $P_2^n \geq P_{snk}^n$ and the valve is not open, then the valve should be opened. Thus open the valve by obtaining the flow Q_2^n through recomputing the pressure P_2^n from the heart model using the first of the two pressure-flow relationships in (3.26).

4. If $P_2^n < P_{snk}^n$ and the valve is open, then the valve should be closed. Close the valve by recomputing the pressure P_2^n using the second pressure-flow relationship in (3.26).

5. Compute the correct value of P_{snk}^n according to (3.27):

   If the valve is open, then

   $$P_{snk}^n = P_2^n + R_{snk1} Q_2^n.$$

   Otherwise no further updating is needed.

The actual choice of Windkessel parameters (C, R_{snk1}, and R_{snk2}) will be discussed in section 3.4.

3.3.3 Boundary Forces

Now we turn to the description of the boundary forces at the discrete level.

The Link Formalism

The position of the heart boundary is defined by its configuration vector

$$\mathbf{X} = (\mathbf{X}_1, \ldots, \mathbf{X}_{N_b})$$

consisting of the position vectors $\mathbf{X}_k$ ($k = 1, \ldots, N_b$) for the boundary points. The boundary forces are modeled as arising from elastic springs, or links, connecting these boundary points.

The set of all links is denoted I. Let i denote a link connecting the boundary points k and l with position vectors $\mathbf{X}_k$ and $\mathbf{X}_l$. Then it is a property of the model that the force $\mathbf{f}_k^i$ on the point k due to the link i, at any given time level, is of the form

$$\mathbf{f}_k^i = \mathbf{f}_k^i(\mathbf{X}) = T(|\mathbf{X}_l - \mathbf{X}_k|) \frac{\mathbf{X}_l - \mathbf{X}_k}{|\mathbf{X}_l - \mathbf{X}_k|},$$

where the function T gives the tension in link i versus its length $L = |\mathbf{X}_l - \mathbf{X}_k|$. Note that the superscript i does not denote a time step. The elastic properties of the boundary are thus defined at each time level by a set of length-tension relations $T(L)$, one for each link. The model for these length-tension relations is given below.

The total boundary force on the boundary point k is now given by

$$\mathbf{f}_k = \mathbf{f}_k(\mathbf{X}) = \sum_{\{i \in I_k\}} \mathbf{f}_k^i(\mathbf{X}),$$

where $\{i \in I_k\}$ denotes the set, of all links i, that has the boundary point k as one of its endpoints. More than one link can connect a given pair of points.

Passive Tissue Model

The heart valves and the outflow tract are essentially elastic tissue. These parts of the heart boundary are therefore modeled by linear springs. The length-tension relation in the links modeling the valves is; see Peskin (1977).

$$T(L) = \begin{cases} S_P (L - L_0), & L \geq L_0, \\ 0, & L < L_0, \end{cases}$$

where S_P is the stiffness and L_0 is the resting length. It is seen that the tension is never negative, so the element is slack when its length is less than the resting length. This means that the leaflets can resist tension but not compression (McQueen, Peskin, and Yellin, 1982). In Peskin and McQueen (1980) prosthetic heart valves were modeled by changing the length-tension relation to allow for resistance to compression.

Muscular Model

A link representing a muscular fiber is modeled as two linear springs in parallel, one of which, the active part, has a variable resting length; see Figure 3.4. This resting length is controlled in a specific way by the length of the link L, its tension T, and a given function of time $\alpha(t)$ modeling the activation level of the muscle. Hence the active part is the variable resting length. This is called the contractile element. The spring in the active part is called the series element. The spring in parallel with the active part, and that constitutes the passive part, is called the parallel element.

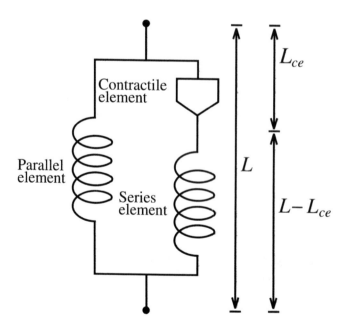

Figure 3.4. *Sketch of the muscular model with two linear springs in parallel, one of which is connected in series to a contractile element.*

The tension T in the muscular link is the sum of a passive and an active part:

$$T = T_p + T_a, \tag{3.28}$$

where

$$T_p = \begin{cases} S_{pe}(L - L_{pe}^0) & \text{if } L \geq L_{pe}^0, \\ 0 & \text{if } L < L_{pe}^0, \end{cases} \tag{3.29}$$

and

$$T_a = \begin{cases} S_{se}(L - L_{ce}) & \text{if } L \geq L_{ce}, \\ 0 & \text{if } L < L_{ce}. \end{cases} \tag{3.30}$$

L is the length of the muscular link, L_{pe}^0 is the resting length of the parallel element (fixed), L_{ce} is the length of the contractile element (i.e., the variable resting length of the active link), S_{pe} is the stiffness of the parallel element, and S_{se} is the stiffness of the series element.

In the continuous formulation of the model the velocity of contraction of the contractile element is a certain function $v(L, T_a, \alpha)$ of the length L of the link, the active tension T_a in the link, and $\alpha = \alpha(t)$, a given function of time called the activation function. For details on the activation function see section 3.3.4. More specifically, the length L_{ce} is governed by the following combination of a differential equation and an algebraic equation:

$$\begin{cases} -\dfrac{dL_{ce}}{dt} = \dfrac{V_{max}}{\alpha(t)} \dfrac{\alpha(t)S_0 L - T_a}{S_0 L + T_a} & \text{if } \alpha(t) > 0, \\[3ex] L_{ce} = L & \text{if } \alpha(t) = 0, \end{cases} \tag{3.31}$$

where V_{max} and S_0 are muscle parameters.

At constant activation α and length L the equilibrium value of the active tension is $T_a = \alpha S_0 L = S_{se}(L - L_{ce})$, so the maximum value of α compatible with a nonnegative resting length L_{ce} is $\alpha = S_{se}/S_0$. Characteristic of the model is that the approach to this equilibrium is faster at smaller activation levels due to the α in the denominator in the expression in (3.31).

Equation (3.31), with its activation dependence, is a generalization of Hill's equation for the velocity of contraction of a tetanized muscle. The model above is known as Hill's three-element model; see Chapters 9 and 10 of Fung (1993).

The resting length L_{ce} and hence the tension are determined by a force balance between the boundary and the fluid.

In the description of the 2D heart model in McQueen, Peskin, and Yellin (1982) the following relations are stipulated among the individual link parameters: $S_{se}/S_{pe} = 5$ (found from physiologically measured values), $S_{se}/S_0 = 5$ (arbitrarily assigned), and $V_{max} = $ (initial length) $\cdot\ 128$ s^{-1} (found by trial and error to give the model output a good match with hemodynamic data).

3.3.4 Activation Function

The activation function $\alpha(t)$ introduced for the muscular link above is the driving force of the heart model. The activation function models the release of Ca^{2+}, which initiates the contraction of the heart muscle, as mentioned in section 2.3.4. Thus the activation function determines the time course in the model, i.e., the contraction and relaxation of the heart muscle.

The time course of the heart cycle is determined by the following parameters: heart rate, onset and duration times of atrial and ventricular systole relative to the cardiac cycle, and excitation and relaxation rate constants (see below). The model sets the initial time of the simulation in the early ventricular diastole during the isovolumic relaxation phase of the ventricle. The activation function $\alpha(t)$ is defined in different ways in diastole and systole and is different for the links in the atrium, the ventricle, and the papillary muscle.

During diastole, the activation function $\alpha(t)$ is defined as a solution to a differential equation of the form

$$\frac{d\alpha}{dt} = -K_r \alpha,$$

where K_r denotes a relaxation rate constant. During systole, $\alpha(t)$ is defined as a solution to a cascaded set of differential equations of the form

$$\frac{d\alpha_*}{dt} = K_e(\alpha_{max} - \alpha_*), \qquad \frac{d\alpha}{dt} = K_e(\alpha_* - \alpha),$$

where K_e denotes an excitation rate constant and α_{max} denotes the limiting value of the activation function. The activation function α in systole is chosen as the solution to a second order equation because doing so gives the function a smooth transition to the systolic phase, whereas the auxiliary function α_* has a more abrupt upstroke. In addition, the activation function is required to be continuous for the entire cardiac cycle. There are different sets of constants for the atrium and the ventricle, but the same set is used for the ventricle and the papillary muscle. The initial values of the activation function will be different for all three types of links.

In Peskin and McQueen's original model the constants for the activation function were set to match a dog's heart. Since our purpose is to make a simulation of the model resemble the MR data of a human, we have changed the timing parameters for the activation function. In a recording of the velocity at a central region in the outflow jet at the level of the aortic valve, the length of the heart cycle, 936 ms, was found as the peak-to-peak time (26 intervals of 36-ms duration each). The MR data do not directly give the timing for the contraction of the heart, but by comparing velocity profiles at the mitral ring and the aortic outflow tract, from the simulation and from the MR recording, the constants can be determined by visual inspection. These are as listed in Table 3.1.

The differential equations have analytic solutions. The solution to the first equation is given by

$$\alpha(t) = \alpha(t_d)e^{-K_r(t-t_d)},$$

where t_d is the initial time, which will be either the starting time for the simulation or the time of onset of diastole. The second set of equations is solved with a common initial value

Table 3.1. *Timings (onset of systole t_s and end time of systole $t_{s,end}$), rate constants, limiting values, and initial values for the activation functions. Initial time is $t = 0$.*

	t_s [ms]	$t_{s,end}$ [ms]	K_r [ms^{-1}]	K_e [ms^{-1}]	α_{max}	$\alpha^0 =$ $\alpha(t = 0)$
Atrium	0.362	0.466	1/55	1/75	1.5	0
Ventricle	0.545	0.851	1/55	1/21	4.9	1.07
Pap. musc.	0.545	0.851	1/55	1/21	4.9	0.107

$\alpha_*(t_s) = \alpha(t_s)$ for the auxiliary function α_* and the activation function α, with t_s denoting the time of onset of systole. The solution is

$$\alpha_*(t) = \alpha_{max} - (\alpha_{max} - \alpha_*(t_s))e^{-K_e(t-t_s)},$$

$$\alpha(t) = \alpha_{max} - (\alpha_{max} - \alpha(t_s))(1 + K_e(t - t_s))e^{-K_e(t-t_s)}.$$

The differential equations are solved numerically using the implicit Euler method. In diastole we have

$$\frac{\alpha^{n+1} - \alpha^n}{\Delta t} = -K_r\alpha^{n+1},$$

$$\alpha_*^{n+1} = \alpha_*^n,$$

and in systole

$$\frac{\alpha_*^{n+1} - \alpha_*^n}{\Delta t} = -K_e(\alpha_{max} - \alpha_*^{n+1}),$$

$$\frac{\alpha^{n+1} - \alpha^n}{\Delta t} = -K_e(\alpha_*^{n+1} - \alpha^{n+1}).$$

Figure 3.5 shows the activation functions for the ventricle, atrium, and papillary muscle.

3.3.5 Topology and Initial Geometry of the Heart

The geometric parameters determine the size and shape of the model heart at the simulation start time. The initial geometry is rather idealized, being composed of straight-line segments and circular arcs. This geometry of course changes during a simulation of the cardiac cycle. The boundary points are connected by links representing muscles and connective tissues. This link structure, the topology of the model heart, is fixed throughout the cycle. In addition to the so-called physical boundary points, the boundary includes one virtual point that is not moved directly by the fluid but is used as an interconnection point between the physical points.

Links connecting boundary points may be either active or passive; see section 3.3.3. A given pair of points may be connected by more than one link. The boundary points are

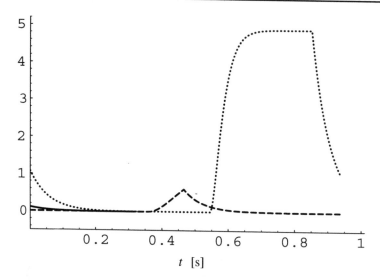

Figure 3.5. *Activation as a function of time for the ventricle (dotted), atrium (dashed), and papillary muscle (solid); the latter is indistinguishable from the ventricular curve in systole.*

connected along the wall and in addition across the wall by cross links. Cross links are needed to simulate circumferential stresses in the heart since the model is only 2D. The links themselves do not interfere with the fluid, except through the forces they exert on their endpoints.

In the original model by Peskin the size of the heart was set to match that of a dog. We have changed the size and shape of the heart to match the MR human data recordings. We used four key parameters for the geometry: the diameter of the mitral ring (2.97 cm), the diameter of the aortic outflow tract (2.58 cm), the angle between these two (45°), and the length of the ventricle (5.42 cm). The diameter values were read off from the widths of the inflow and outflow jets in the MR recordings and then increased by 10% to compensate for the increased thickness of the boundary layer. The increased thickness is due to the scaling of the viscosity mentioned in section 3.3.1 and the size of the region of influence of the boundary on the fluid; cf. section 3.3.1 about the δ function. The angular value, which is a bit on the high side, was chosen to make the simulated inflow jet direction fit the measured direction at the time of early peak inflow. The length of the ventricle was chosen to be large enough that it could withstand contraction until the end of systole. Figure 3.6 shows the final modified initial geometry.

Cross links connect the straight parts of the ventricle to each other and the straight parts of the outflow tract to each other. To make the model correspond better to a realistic human left ventricle we increased the stiffness parameters of the cross links in the straight part of the ventricle and introduced cross links in the circular part of the ventricle; see Figure 3.6.

Cross links extend from a point at the bottom of the ventricle to the virtual point and from the virtual point to the endpoints of the valve leaflets. The link from the ventricle wall to the virtual point represents the two papillary muscles (see Figure 3.6), which in a real

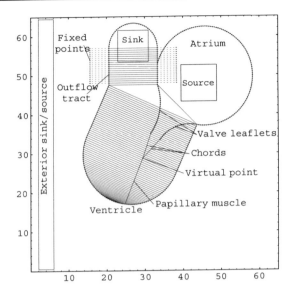

Figure 3.6. *Modified initial heart geometry with cross links and links along heart wall (solid lines) and links tethering the heart to fixed points (dotted lines) shown. The heart is shown in a* 64×64 *grid.*

heart would protrude from the heart wall at points above and below the plane in which the model resides. They are represented in the model by their common projection onto this plane. To control the movement of the posterior ventricle wall we move the point at which the papillary muscle is connected to the ventricle wall. In the original model the papillary muscle was connected to the apex of the heart (Peskin, 1977); we moved it up about 1.3 cm along the posterior wall. The links from the virtual point to the tips of the leaflets represent the chords, or chordae tendineae, which are passive tissue.

All active links are accompanied by a corresponding passive link, but some pairs of points are connected only by a passive link; cf. section 3.3.3 on boundary forces. The links in the outflow tract and the associated cross links are all passive. Also the links in the valve leaflets and the links representing the chords are passive.

In the original model the heart was freely floating in the surrounding domain, which resulted in excessive translational and rotational movements during the ejection phase. To overcome this problem we tethered the heart by linear springs (passive links) to a set of fixed points. After trying different arrangements, we chose to tether the straight part of the outflow tract to fixed points translated horizontally from the initial position of the outflow tract; see Figure 3.6. This change lessened the translational and rotational movement of the heart during a heart cycle considerably, though they still occurred more than in a natural human heart (see section 3.6).

3.4 Evaluation of the Windkessel Model

To investigate the effect of coupling a Windkessel model to the aortic sink, we made a comparison between a simulation with Peskin's original model and a simulation where the

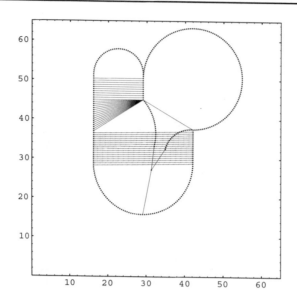

Figure 3.7. *Original initial geometry of heart; compare with Figure 3.6. Note that the heart is not tethered to any fixed points. The heart is shown in a 64 × 64 grid.*

only modification is the replacement of the linear resistance model with the Windkessel model. Here we used the model of a dog's heart; i.e., none of the improvements presented earlier to make the model human were implemented. The length of a heart cycle was taken to be 675 ms. The original initial geometry is shown in Figure 3.7 (compare with Figure 3.6 for the modified model), and the parameters for the activation function are listed in Table 3.2 (compare with Table 3.1).

The remaining figures in this section show the pressure (in mmHg) at three different locations in the model heart: the source in the middle of the atrium, the center of the ventricle, and the sink in the aortic outflow tract. In the figures that show simulation results with the Windkessel model the load pressure P_{snk} is also shown. Furthermore, the inflow rate to the atrium and the outflow rate at the aortic sink are shown (in cm³/s). The volume flow rates are obtained from the simulation's area flow rates by assigning a uniform

Table 3.2. *Timings, rate constants, limiting values, and initial values for the activation functions for the original model (McQueen, Peskin, and Yellin, 1982).*

	t_s [ms]	$t_{s,end}$ [ms]	K_r [ms⁻¹]	K_e [ms⁻¹]	α_{max}	$\alpha^0 = \alpha(t=0)$
Atrium	0.347	0.422	1/60	1/54	2.0	0
Ventricle	0.447	0.547	1/60	1/15	4.9	1.07
Pap. musc.	0.447	0.547	1/60	1/15	4.9	0.107

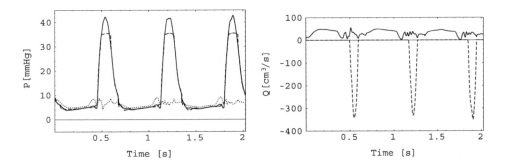

Figure 3.8. *Results for original model. Pressure (left) in mmHg versus time at atrium (dotted curve), ventricle (solid curve), and aortic sink (long-dashed curve). Flow rate (right) versus time for flow through atrial source (solid curve) and aortic sink (dashed curve) in cm^3/s.*

thickness of 1/4 to the computational domain of the heart (McQueen, Peskin, and Yellin, 1982).

Figure 3.8 shows pressures and flow rates from a simulation with the unmodified model over three cycles. It is apparent that the model operates at a physiological pressure that is only about 1/3 that of the human case. (Remember, however, that the model parameters are adjusted to the physiology of a dog.) It is also clear that the pressure at the aortic sink is held at an almost constant level when the sink is open. However, the atrial pressure is at approximately physiological level. Note the presence of the atrial systole in the atrial pressure curve and to a lesser degree in the ventricular pressure curve.

Figure 3.9 shows the corresponding curves for a simulation with the Windkessel model (modified arterial impedance) with the following parameters: $R_{snk1} = 0.033$, $R_{snk2} = 0.8$ mmHg·s/cm^3, and $C = 1.5$ cm^3/mmHg. These parameter values are adapted from

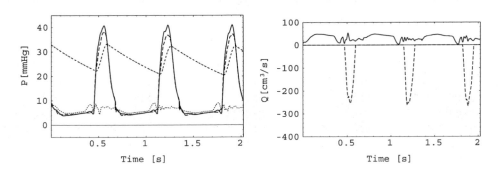

Figure 3.9. *Results for $R_{snk1} = 0.033$, $R_{snk2} = 0.8$ mmHg·s/cm^3, and $C = 1.5$ cm^3/mmHg. Pressure (left) in mmHg versus time at atrium (dotted curve), ventricle (solid curve), aortic sink (long-dashed curve), and the load pressure P_{snk} (short-dashed curve). Flow rate (right) versus time for flow through atrial source (solid curve) and aortic sink (dashed curve) in cm^3/s.*

Cappello, Gnudi, and Lamberti (1995) as follows: Since the model operates at a pressure of 1/3 of the physiological levels used in Cappello, Gnudi, and Lamberti (1995), the values for the resistances are divided by three while the compliance is multiplied by three. One can see that the pressure curve for the aortic sink is more physiologically realistic under these conditions. One can also see that the time interval in which the aortic sink is open is larger with the Windkessel modification and that the corresponding aortic outflow rates are smaller. The stroke volume seems to be roughly the same.

Figure 3.10 shows simulation results for the case of a nonlinear compliance (i.e., pressure dependent compliance). The Windkessel parameters are $R_{snk1} = 0.033$, $R_{snk2} = 0.8$ mmHg·s/cm^3, and $C(P_{snk}) = 10.299 \times \exp(-0.06 P_2)$ cm^3/mmHg. The parameter values and the functional relation between the pressure and the compliance are again adapted from Cappello, Gnudi, and Lamberti (1995). This figure should be compared to Figure 3.9. They are very similar, the biggest difference being that the pressure at which the aortic valve opens is clearly larger with the nonlinear compliance model.

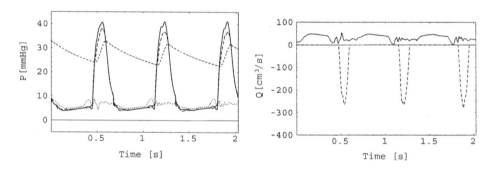

Figure 3.10. *Results for $R_{snk1} = 0.033$, $R_{snk2} = 0.8$ mmHg·s/cm^3, and $C(P_{snk}) = 10.299 \cdot \exp(-0.06 P_2)$ cm^3/mmHg. Pressure (left) in mmHg versus time at atrium (dotted curve), ventricle (solid curve), aortic sink (long-dashed curve), and load pressure P_{snk} (short-dashed curve). Flow rate (right) versus time for flow through atrial source (solid curve) and aortic sink (dashed curve) in cm^3/s.*

The results presented here show that the pressure profile at the aortic outflow tract is considerably improved through the use of a Windkessel model rather than a linear resistance model coupled to a constant pressure reservoir. In addition, the results show that the use of a nonlinear compliance (i.e., a pressure-dependent compliance) does not make a significant difference in the pressure profile. Based on these results we therefore chose to implement a Windkessel model with a constant compliance in the heart model. After implementing all of our modifications in the model, we adjusted the Windkessel parameters based on simulations with the modified model compared with MR data. The parameters of the arterial load model were chosen to be compatible with the specified cardiac output, as were those of the atrial preload. Then some fine-tuning was done to control the shape of the systolic outflow curve. We ended up with the following parameter choices: $R_{snk1} = 0.014$ mmHg·s/cm^3, $R_{snk2} = 0.69$ mmHg·s/cm^3, and $C = 2.63$ cm^3/mmHg.

3.5 MR Data

Evaluation of the entire modified 2D heart model was done by comparing simulation results with MR measurements from a normal human subject. These MR data came from a study previously published by Kim et al. (1995).

The MR investigations were performed with a 1.5 Tesla 15S Gyroscan HP whole-body system (Philips Medical Systems, Best, The Netherlands). The subjects were examined in the supine position. The study was conducted on 26 healthy volunteers: 18 men and 8 women. Their mean age was 25 years (in a range from 21 to 30 years). The study was approved by the Regional Ethical Committee on Human Research, and individual informed consent was obtained according to the Helsinki II declaration. The volunteers had no known cardiac or systemic diseases. One representative dataset was compared with results from our 2D heart model.

In the MR data-collecting process, a conventional spin echo sequence was used for a coronal image of the heart followed by a single angulation through the aortic valve and the apex of the heart to yield a long axis view of the left ventricle. In the original study (Kim et al., 1995) the long axis plane of the left ventricle was chosen because it contains both the aortic outflow and the mitral inflow, as well as the left ventricle and left atrium. The long axis plane is well suited for the comparison between the MR data and the model data, since it is consistent with the computational plane of the 2D model.

Image acquisition was prospectively triggered by the R wave of the electrocardiogram, so the first recorded frame was 8 ms after the R wave. This time corresponds approximately to the time of closure of the mitral valve.

The intracardiac flow velocity measurements were obtained using the standard FLAG (flow adjusted gradients) sequence, which is described in Groen and van Dijk (1987). The recordings were conducted in a square 40 cm field of view with an in-plane resolution of 3.125 mm $\times$ 3.125 mm giving an imaged slice of 128 $\times$ 128 voxels (volume elements). Furthermore, the following parameter settings were used for the flow measurements: 45° flip angle, 9.5 ms echo time, 7.1 ms acquisition time, and 10 mm slice thickness. The velocity sensitivity was set to produce a maximum phase shift at a velocity of ±1.25 to ±1.67 m/s to avoid aliasing. Velocity encoding was performed in all three spatial directions to produce 3D velocity information from each voxel inside the field of view. The time needed for the quantitative velocity acquisitions (using ECG gating to avoid motion artifacts) was 15 to 20 min for each subject, depending on the heart rate. The procedure resulted in the recorded MR data being averaged over 768 heartbeats. For the 26 examined persons the temporal resolution was 28 ± 3.8 ms (mean $\pm$ standard deviation (SD)) with 25 to 32 frames per cardiac cycle. For the specific MR set used in the comparison the temporal resolution was 36 ms with 26 frames per cardiac cycle giving a total length of the heart cycle of 936 ms. The total recording time was 1152 ms, i.e., covering approximately one and one quarter heart cycles.

The accuracy of quantitative flow measurements by the MR phase technique depends on the suppression of artifacts. Therefore, the velocity maps were corrected for possible linear phase errors resulting from eddy currents in the magnetic field. In addition, masking of random noise in the stationary tissue was done. Both procedures were performed by a semi-automated software program, which is described in detail in Walker et al. (1993).

An MR data set consists of the 3D velocity information in each voxel and the modulus image at each time frame. The modulus image contains information on the MR signal amplitude, which depends, e.g., on the specific tissue and the velocity of the flow in the voxel. The modulus image is displayed as a gray-scale picture and can as such be used to roughly determine the internal anatomy and also the magnitude of the flow. For the comparison with the 2D heart model we were interested only in the in-plane components of the velocity. The in-plane velocity components can be displayed as velocity field vectors. Presenting the velocity field plot superimposed on the corresponding modulus image shows both the magnitude and the direction of the blood. It is also possible at each time frame to extract velocity information from one or several voxels, and this information can be used to plot a velocity profile (velocity as function of time) at a specific place in the imaged slice.

3.6   Comparison between Simulation and MR Data

This section presents the comparison between a simulation of the model and the MR data. The simulation was performed over two cardiac cycles, using all the modifications described above, and with the following numerical parameters: a 64×64 grid with a spatial resolution of 2.036 mm and a time step of 0.333 ms. The results presented here are similar to those in Jacobsen et al. (2001). The starting time of the MR data during the cardiac cycle is different from that of the simulation. The natural start of the MR data is at the closure of the mitral valve (see section 3.5), while the simulation starts in early ventricular diastole. We have chosen to show all results according to the time course of the simulation, i.e., starting with early ventricular diastole, and have shifted the velocities from the first cycle of the MR measurements accordingly. The matching was done by synchronizing the time of the early peak velocity (also called the E peak) at the mitral annulus velocity profile. The match placed the R peak in the ECG of the MR data at 550 ms.

The comparison falls into three parts, each represented by its type of data comparison:

- velocity fields: a comparison is done on plots of the velocity field and heart wall boundary; see section 3.6.1.

- velocity profiles: a comparison is carried out of the velocity profiles at the mitral ring and at the aortic outflow tract; see section 3.6.2.

- vortex data: a comparison is carried out on the size and angular velocity of the vortex formed in the left ventricle during a heart cycle; see section 3.6.3.

3.6.1   Velocity Fields

The comparison between the velocity fields from the MR data and from the simulation is presented in this section. In Figure 3.11 the velocity field is shown at eight different times during a cardiac cycle as velocity vectors along with the heart wall boundary. In the model the heart wall boundary is computed and it is therefore easily plotted. It is shown in two gray-scale colors, a dark gray showing the width at half maximum of the δ function and a light gray showing the width at zero height, both centered at the heart wall boundary. The

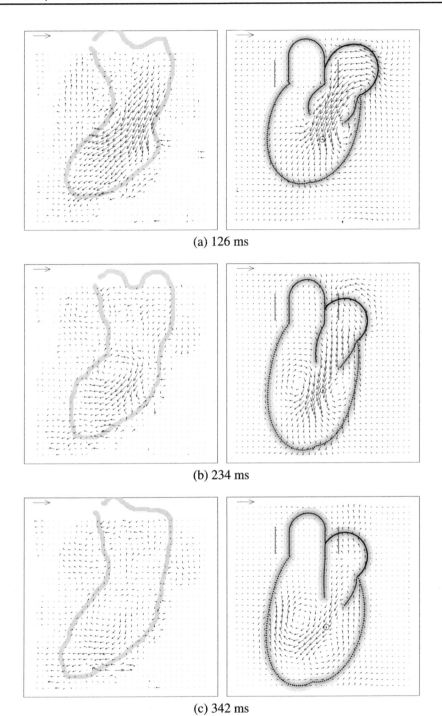

(a) 126 ms

(b) 234 ms

(c) 342 ms

Figure 3.11. *See caption on page* 61.

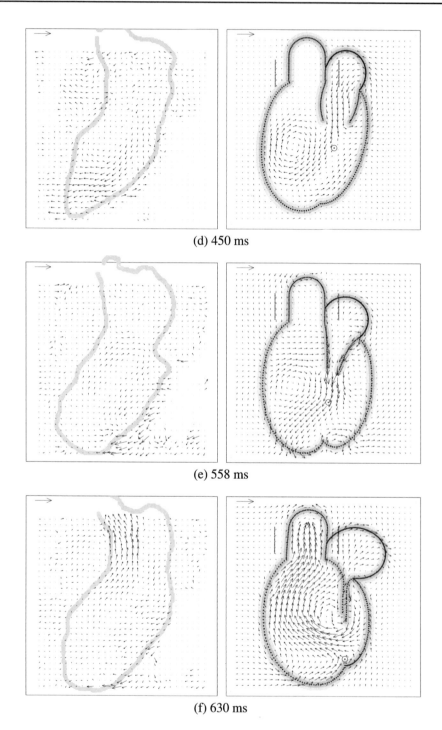

(d) 450 ms

(e) 558 ms

(f) 630 ms

Figure 3.11. *See caption on page* 61.

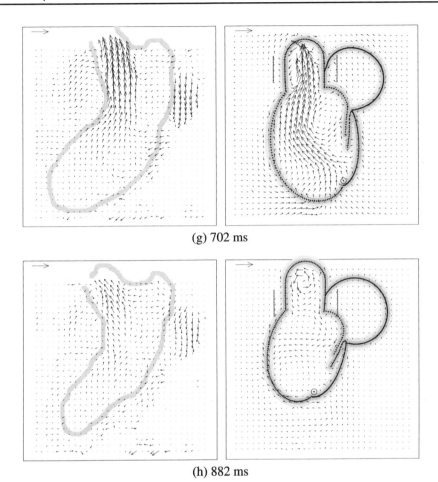

(g) 702 ms

(h) 882 ms

Figure 3.11. *Velocity fields from MR data (left) and simulation (right). The unit size vector in the upper left corner of each picture corresponds to a length of 1.25 cm and a velocity of 1 m/s. Frame* (a) *shows early peak mitral inflow,* (b) *a temporary maximal anterior vortex in the MR data,* (c) *a decline in inflow, but still a clear vortex motion in the ventricle,* (d) *atrial systole in the simulation,* (e) *closure of the mitral valve,* (f) *early aortic outflow,* (g) *peak aortic outflow in the MR data, and* (h) *the time just before closure of the aortic valve in the MR data and just after aortic sink closure in the simulation.*

virtual point (see Figure 3.6) is shown with a circle around it. In the MR data the interior of the heart boundary (not the valve leaflets) was found by manual tracing from plots of the velocity field superimposed on the modulus image; see section 3.5. The boundary is given a width equal to the spatial resolution of the MR measurements, i.e., 3.125 mm. In each picture a single arrow is displayed in the upper left corner representing a length of 1.25 cm and a velocity of 1 m/s. The grid distance in the pictures is 4.072 mm, and the velocity fields shown are interpolated from the original velocity fields for both simulation and MR data.

We begin with a description of the results in Figure 3.11. This description will introduce the time sequence of a cardiac cycle and provide the basis for our comparison.

Early mitral inflow. The results start with the early mitral inflow. The mitral inflow has two peaks during a cardiac cycle, the early peak, also referred to as the E peak, and the late atrial peak, referred to as the A peak. In Figure 3.11a (126 ms), at the time of the E peak, a wide inflow jet is seen in both the MR data and the simulation. In the MR data the jet is wider and reaches deeper into the ventricle compared to that of the simulation. In both the MR data and the simulation, the beginning of the formation of a vortex can be seen at the leaflet tips. (The leaflets are not shown in the MR data.)

Mid-diastole. In the MR data and the simulation at 234 ms (Figure 3.11b) the inflow has decreased and in the simulation partial closure of the mitral valve is seen. Also, the flow pattern in the MR data suggests that the mitral valve is in a semiclosed position. In the MR data a large vortex has developed in the anterior part of the ventricle, while there is only a hint of a posterior vortex. The anterior vortex of the MR data has its clearest appearance at this time. The flow pattern is similar in the simulation: Two vortices are clearly seen, a large anterior vortex and a smaller posterior one. At 342 ms (Figure 3.11c) the inflow has further declined in both the MR data and the simulation, but the vortices are still present at lower angular velocity. The anterior vortex is still the most prominent one, and in the MR data the posterior vortex is hardly detectable.

Late filling. At 450 ms (Figure 3.11d) the inflow through the mitral ring in the simulation has increased compared to Figure 3.11c, and the second peak (the A peak) of the mitral inflow is on its way. In contrast, only a hint of the A peak is seen in the MR data. The presence of the anterior vortex in the MR data is rather vague at this time, while it is still clear in the simulation. No posterior vortex is detectable in the MR data and it is vague in the simulation.

Mitral valve closure. In both the MR data and the simulation, the mitral valve closes at 558 ms (Figure 3.11e). The mitral valve is not shown explicitly in the MR data, but the flow pattern reveals that the valve is closing. At the mitral ring in the simulation a small amount of backflow is observed. In the MR data a rotational flow pattern filling up the ventricle is discernible, but is not a clear vortex, while in the simulation the anterior vortex is still present.

Ventricular systole. The early aortic outflow pattern near the outflow tract at 630 ms (Figure 3.11f) is quite similar in the MR data and the simulation, though the outflow jet is wider in the MR data. Deeper in the ventricle the velocities are higher in the simulation than in the MR data. Furthermore, the flow from most of the ventricle in the simulation converges toward the outflow tract. Based on the position of the heart wall of the ventricle in the simulation (when compared to the previous Figure 3.11e), one can see that ventricular systole has set in. Systole is not clearly indicated by heart wall position in the MR data, but since aortic outflow is present, ventricular systole must have started. In Figure 3.11g at 702 ms aortic outflow velocity has increased in both the MR data and the simulation, the outflow jet being still wider in the MR data. From the velocity profiles shown in the following section it can be seen that this is the time of peak aortic outflow velocity according to the MR data and just after peak velocity according to the simulation. The convergence of the ventricular flow toward the outflow tract from most parts of the ventricle is still more clearly seen in the simulation. In the simulated flow field the artificial termination of the outflow tract causes a backflow along the anterior side of the tract.

End of systole. The flow pattern at 882 ms (Figure 3.11h) shows that the aortic valve is still open in the MR data, but the aortic sink is closed in the simulation. From the velocity profiles in the following section it is seen that this time is just before valve closure according to the MR data and just after valve closure in the simulation. The vortex seen in the outflow tract of the simulation is an artifact of the model.

The above comparison between the velocity fields from the MR measurements and the simulation shows a good agreement in the general flow pattern. A difference in velocity fields is seen in the width of both the inflow jet (from atrium to ventricle) and the outflow jet (at the aortic outflow tract). This difference is caused primarily by the difference in outflow area between the model and the measured heart, but also by a thicker boundary layer for the fluid in the model. The thicker boundary layer is a result of the scaling of time and all lengths by a factor of $\gamma = 1/25$; see section 3.3.1. The flow patterns also differ with respect to the size of the ventricle; the ventricle in the simulation is larger than in the MR data. This is a consequence of the model being 2D. The entire expansion takes place in the considered plane, and to produce the right outflow the ventricle has to expand more. The larger ventricle might be an explanation for the clearer formation of a posterior vortex and the more prominent anterior vortex in the simulation. This will be further discussed in section 3.6.3 on vortex data.

The flow fields from the simulation are affected by the simple aortic valve (sink coupled through a diode to a Windkessel model). Together with the termination of the outflow tract, the aortic sink results in the formation of a vortex at the sink, which of course is not seen in the MR data. Furthermore, tethering the model heart has eliminated the excessive movement of the heart, but has also introduced oscillations in both the atrium and the ventricle, which affects the flow field in an unphysiological way. The oscillations of the atrium can be seen from the plots of the heart wall, while the oscillations of the ventricle are best seen in an animation. In addition, tethering restricts the model to movements along the axis of the outflow tract. Together with the dynamics of the mitral ring (variation in length), the restriction due to tethering results in the ventricle axis being almost aligned with the axis of the outflow tract instead of having the angle we introduced in the initial geometry.

The papillary muscle (modeled by a link between a virtual point inside the left ventricle and a point on the posterior heart wall) results in an unphysiological shape of the apex of the left ventricle (see Figure 3.11e) and a distorted contraction of the posterior heart wall in the computed data (Figures 3.11f–h). Both of these distortions affect the flow pattern and lessen the agreement between modeled and observed data. These distortions suggest that changing the tethering position of the papillary muscle to the ventricle wall would improve the model, but modifying the model in this way has not led to much success. Moving the tethering point farther down the posterior wall toward the apex (i.e., closer to the position in the original model) gives a less distorted contraction of the posterior wall, but this change also results in a much wider expansion of the posterior wall. The increased expansion results in a more circular and unphysiological shape of the ventricle and gives room for a larger posterior vortex, which is undesirable. Therefore, it has been decided to accept the more distorted wall contraction to achieve a more correct flow pattern inside the ventricle.

Generally, we have obtained a geometry that more closely mimics the human left ventricle than did the original model, but the geometry is still simplified. We have achieved a correct length scale of the left ventricle by changing the diameter of the mitral ring and the diameter of the outflow tract.

The following section compares the timing and the velocity profiles of the simulation and the MR data.

3.6.2   Velocity Profiles

From the velocity fields of both the MR data and the simulation we can extract velocities at chosen spatial positions. A sequence of these velocities at a single position is called a velocity profile (velocity versus time). For the MR data we have 26 different time points over a cardiac cycle at which we can extract velocity information; for the simulation we have 2808 different time points. We chose to extract information at 26 different time points for the MR data and at 468 and 104 different time points, respectively, for the center and mean velocities from the simulation.

As well as for comparing the velocities at chosen spatial positions, the velocity profiles can be used to evaluate the timing of the simulation compared to MR measurements. We have no exact measurements of the time of onset and the duration of ventricular and atrial systole (see section 3.3.4) due to the limited quality of the ECG during the MR scan. But by comparing the velocity profiles the timing parameters can be evaluated.

In addition to the velocity profiles, we report for both the MR data and the simulation the early peak (E peak) and the late peak (A peak) mitral velocity across the diameter of the mitral annulus by finding the one voxel (or point in the simulation) at the mitral annulus where the velocity is largest. The peak outflow velocity across the diameter of the outflow tract is found in a similar manner. The peak velocities are generally larger than the maximum values of the velocity profiles.

The in- and outflow velocity profiles from the MR data are obtained by averaging the velocities over two fixed voxels at the center of the left ventricular outflow tract at the level of the aortic valve and over a square of four fixed voxels near the center of the mitral annulus. At the outflow tract the vertical component is recorded (see reference frame in Figure 3.11), while at the mitral annulus the component making an angle of 30° with the vertical is recorded. Because of the small movement of the heart wall, no attempt is made to trace the movements of the heart in choosing these voxels, and thus the velocities are computed with respect to a reference frame fixed in space. The extracted velocities are *mean center* velocities, which are a combination of a mean velocity over and a center velocity at the relevant position. It would be difficult to compute the mean velocity because the posterior wall is not known for the aortic outflow tract and the anterior border is not known for the mitral ring. It is also difficult to report the center velocity (at one voxel) because the velocity oscillates from one voxel to the next over the aortic outflow tract. These are our reasons for reporting a mean center velocity for the MR data.

For the simulation we report both a mean and a center velocity to be able to compare the mean center velocity of the MR data with both types of velocities. We obtain mean and center velocities at positions that follow the heart motion. The mean velocities are computed relative to the moving annulus, i.e., in a reference frame moving with the mitral annulus (because of the relatively large movement of the heart wall), while the center velocities are reported relative to the computational grid, i.e., in a reference frame fixed in space (for ease of comparison with the MR data). The mean and center velocities for the aortic outflow tract are calculated similarly.

Figure 3.12 shows the time course of the inflow velocity at the mitral annulus, extracted from both the MR measurements and the simulation. Outflow velocities at the aortic outflow tract are shown in Figure 3.13. The match of the simulation and the MR data placed the R peak in the ECG of the MR data at 550 ms (as mentioned earlier in this section), and this time is indicated by a vertical dotted line in Figures 3.12 and 3.13.

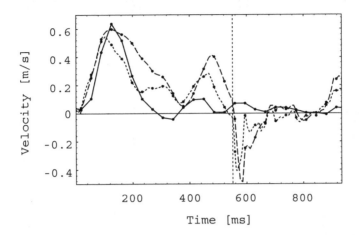

Figure 3.12. *Velocity versus time at the mitral annulus for MR data (solid curve), center velocity in simulation (long-dashed curve), and mean velocity in simulation (short-dashed curve). From Jacobsen et al. (2001), Cardiovasc. Eng.* **1**(2):59–76, *Kluwer Academic/Plenum Publishers. Used by permission.*

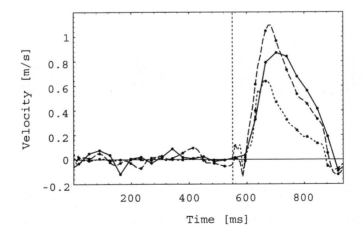

Figure 3.13. *Velocity versus time at the aortic outflow tract for MR data (solid curve), center velocity in simulation (long-dashed curve), and mean velocity in simulation (short-dashed curve). From Jacobsen et al. (2001), Cardiovasc. Eng.* **1**(2):59–76, *Kluwer Academic/Plenum Publishers. Used by permission.*

Figure 3.12 shows that at the mitral annulus there is fairly good agreement between the MR data and simulation results for both the center and the mean velocity. The agreement between time of occurrence of the early maximum (E) velocity is of course due to the matching of the simulation and the MR data at time $t = 126$ ms for center velocity. The E velocity is slightly higher in the MR data than in the simulation (for both mean and center values). The upstroke of the E peak of the simulation results is very similar, while the downstroke is faster for the mean than for the center velocity. The upstroke of the MR data starts off more slowly than that of the simulation, but after 36 ms it is as fast as the simulation. The downstroke of the MR data is closer in slope to the downstroke of the mean velocity of the simulation than to the center velocity. The MR data show negative flow velocity between the E peak and the late (A) peak, and this negative velocity is not seen in the simulation. This difference could be due to an incorrect movement of the mitral valve leaflets in the simulation, but it is difficult to be sure that this is the cause since the mitral valve motion is not detected in the MR measurements. Alternatively, the difference could be caused by an incorrect activation level of the muscle model (see section 3.3.4 on the activation function).

The A peak in the mitral inflow is caused by the onset of atrial systole. Atrial systole can be hard to detect in the MR measurements, because it occurs at the end of each MR recording. This part will be smeared out since the final measurement is an average over 768 heartbeats, which might vary in length. This distortion can explain why the A peak velocity is much higher in the simulation than in the MR data. As seen in Figure 3.12, the simulation shows a clear atrial systole. Although the measurements show only a very vague atrial systole, the atrial systole occurs at approximately the same time in the simulation and MR measurements, confirming the placement of the start time of the atrial systole in the model. The MR data show no backflow at the time of closure of the mitral valve (between 558 and 594 ms), as is seen in both simulation results. This could be a consequence of the averaging of the MR data, or it could be caused by the coarse time resolution in the MR measurements.

Figure 3.13 shows that the outflow through the aortic tract starts at approximately the same time in the MR data and the simulation (for both velocities reported), and the velocity profiles in the simulation have the same slope at the beginning of outflow as the MR data. Thus we can conclude that the onset of ventricular systole in the simulation is correct. The form of the center velocity profile in the simulation and the MR velocity profile are comparable in shape during outflow, even though the maximum center velocity is larger in the simulation and occurs approximately 24 ms earlier. With respect to mean velocity, the agreement between the MR data and the simulation is not quite as good. The upstrokes of the outflow are very similar, but the maximum value of the mean velocity is too low and occurs too early, and the downstrokes differ a great deal. The small difference between the MR data and the simulation with respect to the time of maximum outflow velocity suggests that the onset of ventricular diastole could be further improved.

A primary effect of the use of the Windkessel model is that we obtain a simulated aortic outflow profile that reproduces a characteristic feature of the measured one, namely, the upstroke that has a duration of about one half that of the downstroke. As mentioned above, the actual shape of the simulated outflow curve does not exactly match the measured one as it has a too narrow top and upper part. Further adjustment of the load parameters might achieve a better agreement, but the muscular model behind the heart model affects the simulated profile and therefore should be involved in the tuning process as well. It was also found that the shape of the outflow profile was sensitive to the strength and geometry of the tethering.

Table 3.3 displays the peak velocities found at one specific voxel (as explained above) at the mitral annulus and aortic diameter from (1) the MR data set, (2) the entire MR study (all 26 persons, taken from Kim et al. (1995)), and (3) the simulation. The "1 series" in the MR data is the one used for the detailed comparison.

Table 3.3. *Peak velocities at the mitral annulus and left ventricular outflow tract from model and MR data. The MR data show the specific data series used in the comparison and the mean values and ranges for the entire set of MR data (consisting of 26 series) taken from Kim et al. (1995) For the mitral inflow both the early peak (E) and the late atrial filling peak (A) velocity are reported.*

| | | Mitral annulus | | Outflow tract |
		E [m/s]	A [m/s]	[m/s]
Simulation		0.75	0.41	1.14
MR	1 series	0.66	0.23	1.03
	Mean value	0.64	0.27	1.32
	Range	0.45–0.82	0.18–0.44	0.81–1.84

The comparison of the peak velocities in Table 3.3 reveals almost the same picture as that described above for the velocity profiles. The E peak velocity is slightly higher in the simulation than it is in both the 1 data series and the mean of the 26 data series. This is different from what we have seen for the mitral velocity profile. The E peak velocity for the simulation is well within the range of the 26 data series and falls in the upper part of the range. The A peak velocity in the simulation is twice the value of that of the MR data and falls in the top of the range of the 26 data series. This result is consistent with the velocity profiles and, as mentioned earlier, is a consequence of averaging the MR data. The peak outflow velocity is also slightly higher in the simulation than in the MR data (as seen from the center velocity profile). But it is lower than the mean value of the 26 MR data series and it falls in the lower part of the range of the 26 series.

3.6.3 Vortex Data

We are interested in investigating the vortex pattern in the left ventricle in more detail than just a qualitative comparison of velocity field plots. So we compute a typical radius of the vortex and an average angular velocity of the entire vortex.

We have based our computations on the method presented in Kim et al. (1995). In Kim et al. (1995) vortex data based on 3D velocity components from the MR measurements were reported, though the method for computing radius and angular velocity was described for both 2D and 3D velocity components. Regarding the plots of the velocity field, we have chosen to compute the vortex data based only on the in-plane components of the velocity field, using the method described on pages 227–229 of Kim et al. (1995).

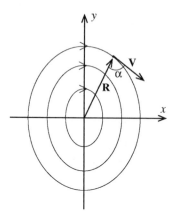

Figure 3.14. *Schematic illustration of clockwise vortex motion. The origin of the coordinate system is placed at a manually found grid point at the center of the vortex.*

Clockwise vortex motion is defined as a flow field for which streamlines are concentric circles, where the oriented angle α measured from the velocity vector **V** at each point to the radius vector **R** of that point lies in the range $90° \pm 45°$. On the average, $\sin \alpha \geq 0.7$; see Figure 3.14.

The computation of radius and angular velocity proceeds as follows: For each velocity field in which a vortex can be distinguished, a grid point appearing to be the center of the vortex and a region of interest are selected manually. For each velocity vector in this region the radius vector **R** and the angle α are computed. The radius R of the vortex is then chosen to be the largest of the consecutive radii for which the average of $\sin \alpha$ over the grid points at this distance from the center is greater than or equal to 0.7. For the MR data this criterion is relaxed for the smaller radii to compensate for the noise in the data. The radius R is easily found by plotting the average of $\sin \alpha$ versus radius R and then determining where the curve falls below 0.7. After the radius R is determined, the average angular velocity over the circular disk with radius R is computed.

In the MR data an anterior vortex was clearly detectable, while the posterior vortex was very vague. The simulation produced both an anterior and a posterior vortex in the ventricle, though the anterior was the most prominent one. Only the anterior vortex is quantified because the posterior vortex was not consistently present in the MR data.

The radius of the anterior vortex from both the MR data and the simulation is shown versus time in Figure 3.15. Only the time period where the vortex is present is shown. Note that at $t = 306$ ms it is difficult to compute vortex parameters from the MR data because of noise in the flow field at the center of the vortex. Figure 3.16 shows the absolute value of the average angular velocity of the entire vortex as a function of time.

The figures show that the vortex is created at the same time, $t = 162$ ms, in both the simulation and the MR data. It is present for 396 ms in the simulation, while it is present for only 288 ms in the MR measurements.

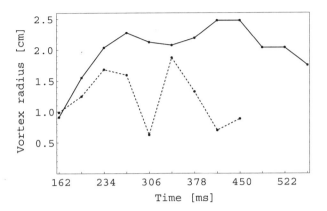

Figure 3.15. *Vortex radius in cm versus time for simulation (solid curve) and MR data (dashed curve). Adapted from Jacobsen et al.* (2001), Cardiovasc. Eng. **1**(2):59–76, *Kluwer Academic/Plenum Publishers. Used by permission.*

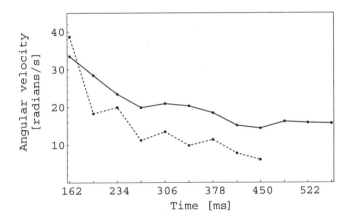

Figure 3.16. *Absolute value of average angular velocity in vortex in radians per second versus time for simulation (solid curve) and MR data (dashed curve). Adapted from Jacobsen et al.* (2001), Cardiovasc. Eng. **1**(2):59–76, *Kluwer Academic/Plenum Publishers. Used by permission.*

Figure 3.15 shows that the radius of the vortex is generally larger in the simulation than in the MR data. Furthermore, in the simulation the radius of the vortex is held at a level of approximately 2.2 to 2.3 cm once it has reached full size, whereas the radius varies more in the MR measurements.

The average angular velocity is also larger in the simulation than in the MR data (Figure 3.16), but here the time variation is quite similar. The maximal angular velocity is found at the time of formation of the vortex in both the simulation and the MR data. Initially the angular velocity decreases quickly but thereafter falls slowly to an almost constant value. (This behavior is most prominent in the simulation.)

From the velocity field plots (Figures 3.11a–h) we saw that the vortex is created at the mitral valve leaflets and then shed down the left ventricle. The size of the vortex depends heavily on the shape and size of the left ventricle and on the movement of the mitral valve. This might be part of the explanation for the difference between the radius of the vortex found in the simulation and in the MR measurements. The size of the ventricle is larger in the simulation because the entire expansion takes place in the considered plane, which makes room for a larger vortex. But the most important reason for the difference in radius is the fact that the model is 2D, while the MR measurements are of a full 3D velocity field. It is to be expected that the vortex motion in the left ventricle is essentially a 3D phenomenon and cannot be accurately described in 2D.

3.7 Conclusions

We have presented a 2D computational model of blood flow in the heart. This model was obtained by modifying an already existing model, constructed to match a dog's heart, to more closely represent the heart of a human. The modified model has been compared with MR measurements performed on a human heart.

The starting point of the study was a 2D dog's heart model using the immersed boundary method developed by Peskin. This model had a simple outflow condition at the aorta and floated freely in the surrounding domain. These conditions influenced the internal flow pattern and made it unsuitable for comparison with MR measurements of humans. Therefore, modifications were introduced to the model. The modifications were based on general physiological knowledge and on comparisons with MR measurements of a human heart.

First, we changed the initial geometry and topology of the model. Through comparison with modulus images from an MR measurement, the initial size and shape were changed to match those of a human heart. In addition, the topology of the cross links was changed. Cross links were introduced in the circular part of the ventricle (i.e., the part at the apex). The stiffness parameters of the cross links in the straight part were increased, and the connection point of the papillary muscle (link between virtual point and ventricle wall) was moved up along the posterior ventricular wall. In the original model the heart floated freely in the domain, which resulted in unphysiological movement of the heart, especially during ejection. Therefore, a tethering of the heart was introduced. This tethering was achieved by (passive) links from the boundary points on the straight part of the outflow tract to fixed points in the domain.

In the original model the outflow through the aorta was modeled by a linear resistance model pumping against a constant pressure reservoir combined with a diode-like valve. This mechanism resulted in an unrealistic shape of the pressure curve and was therefore changed to a three-element Windkessel model coupled to a diode-like valve. This modification improved the pressure curve and also the aortic outflow profile.

Finally, the timing of the model was adjusted to match, as well as possible, a specific MR measurement. The timing was primarily changed through the parameters determining the activation function $\alpha(t)$, but indirectly the in- and outflow conditions also affected the time development of the model. The times of onset of both atrial and ventricular systole were adjusted along with the durations of atrial and ventricular systole. In addition, the relaxation and excitation rate constants were modified. The parameter adjustment was done by comparing velocity profiles at the mitral ring and aortic outflow tract from simulation and MR measurements.

The dependence of the simulated results on various parameters in the model is very complex. We cannot be certain that a further adjustment of the parameters would not lead to an even better agreement between simulated and measured velocity data. Parameters of particular importance in this regard are duration of ventricular systole, excitation and relaxation parameters, and the Windkessel and muscle parameters.

The comparison between the simulation results and the MR data showed a reasonable agreement. In particular, the timing and the intraventricular flow pattern matched well. The flow pattern showed the various phases of the heart cycle: mitral inflow (early and late peak velocity), the formation of vortices in the ventricle, and the ejection phase with aortic outflow. The velocity profiles at the mitral ring and aortic outflow tract from the simulation and MR measurements matched reasonably well. The timing of the velocity profiles was in good agreement, while the shape and peak values of the velocity curves showed some differences.

Furthermore, the time evolution of the large anterior vortex in the simulation coincided reasonably well with the measured time history with respect to angular velocity and size. The anterior vortex is created at the same time in the simulation and MR measurements, though it is present for a longer time in the simulation than in the MR measurements. One clear difference between the simulation results and the MR data concerns the relative sizes of the vortices shed by the two leaflets of the mitral valve. In the simulation both leaflets shed prominent vortices. These were unequal in size, with the anterior vortex being larger. In the MR data only the anterior vortex was clearly seen. Although the MR velocity field contained a hint of a small posterior vortex, it was much smaller than in the simulation results.

The work reported in this chapter shows that, despite the geometrical constraints posed by the 2D modeling of a 3D phenomenon, a fair agreement is obtained between simulated and measured results. Small improvements in the comparison could possibly be achieved by adjusting the parameters of the model further. But to obtain a much better agreement it is probably necessary to use a full 3D heart model, with a more correct and detailed geometry. Implementing such a model should improve the internal flow pattern and, in particular, the vortex motion in the left ventricle.

Acknowledgments

Both authors were supported by High Performance Parallel Computing, Software Engineering Applications Eureka Project 1063 (SIMA—SIMulation in Anesthesia). In addition, P.T. Adeler was supported by Math-Tech, UNI-C, and the Danish Academy for Technical Sciences. In addition, the authors want to acknowledge the scientific support from Dr. Charles Peskin, Courant Institute for Mathematical Sciences, New York University.

Chapter 4

The Ejection Effect of the Pumping Heart

M. Danielsen

4.1 Introduction

In Chapter 3 we presented a detailed 2D model describing the flow in the left ventricle. This model used sources and sinks to describe the flow into and out of the ventricle. The inflow from the right atrium was described using passive resistance, and the outflow into the aorta was described using a simple Windkessel model. The model in Chapter 3 did not take into account that the ventricular contraction depends on outflow; this is analogous to tissue mechanics, where muscle contraction depends on the velocity of the muscle shortening. In this chapter, we focus on deriving a model that can be used to analyze and describe the function of the left ventricle during ejection. To obtain a model that is physiologically correct, the ventricular pressure generation or contraction is described as a function of outflow among other factors, as discussed previously.

The heart is the central organ and the pump of the cardiovascular system. However, it is not an independent energy source but interacts strongly with the rest of the cardiovascular system. This interaction constitutes one of our key interests in this chapter. In this chapter we describe the impact of blood ejection on ventricular contraction and introduce a new approach to modeling the heart, which considers the heart to be a pressure source dependent on time, volume, and flow. Studying the impact of ejection, or properties of the ejecting heart, demands that these properties be isolated from the properties of the heart when it is not ejecting. This isolation is possible through the use of the modeling approach presented in this chapter. This approach separates isovolumic (no ejection of blood) and ejecting heart properties. The latter will be represented by the so-called ejection effect, which will be carefully described in section 4.3. The modeling approach consists of two steps. The first step describes the isovolumic beating heart, viewing the ventricle as a pressure source dependent on time and volume. This analytical description was first established by Mulier (1994) based on experiments on dog hearts and is presented in section 4.2. In the second step the ejection effect, which describes the ejecting heart properties, is added. The ejection effect consists of two phenomena, deactivation and hyperactivation, which represent negative and positive effects, respectively, on developed ventricular pressure during ejection. The reason that the ejection effect is needed is because isovolumic heart properties alone cannot fully describe the function of the heart during blood ejection (Danielsen and Ottesen, 2001; Danielsen, Palladino, and Noordergraaf, 2000a; Danielsen, 1998). In simple terms, a model based on isovolumic heart properties alone generates higher than expected ventricular pressure in early ejection and lower than expected ventricular pressure later. In essence, the ejection effect lowers the blood pressure during early ejection (deactivation) and increases the blood pressure later (hyperactivation). This behavior pattern of the ejection effect is directly related to the dynamic of the muscle fibers and is discussed carefully in section 4.3. That section also discusses possible causes of the ejection effect, such as reflected waves and inertial effects. The final mathematical formulation of the ejection effect and the heart model are presented in section 4.4 together with computed pressure and flow curves for a normal human being. In section 4.5 we give a summary and discussion.

The modeling approach presented in this chapter breaks with the concept behind the classical time-varying elastance description of the pumping heart. That description is used, for example, in the model of the cardiovascular system in Chapter 6, section 6.3.1. When we consider pressure and flow curves as functions of time, the time-varying elastance concept is the most widely embraced description of the heart and is adopted in several recent cardiovascular models (Martin et al., 1986; Tham, 1988; Rideout, 1991; Neumann, 1996; Stergiopulos, Meister, and Westerhof, 1996; Danielsen and Ottesen, 1997; Ursino, 2000). Mathematical models based on this elastance concept suffer from a number of deficiencies. These deficiencies may be related to the model's predetermined time course making it immune to different conditions in the cardiovascular system (Danielsen and Ottesen, 2001). The models fail, for instance, to cover both isovolumic (no ejection of blood) and ejecting heartbeats. Moreover, and perhaps more importantly for this chapter, the time-varying elastance function is determined from pressure and flow curves taken during blood ejection. These data consequently contain information about both isovolumic and ejecting heart properties. These properties are subsequently difficult to distinguish. They may, for instance, be hidden in the same parameters. In order to gain better insight into isovolumic versus ejecting heart properties, a modeling approach separating these properties is warranted (Danielsen and Ottesen, 2001).

4.2 Model of the Isovolumic Ventricle

In the isovolumic contracting ventricle the volume remains constant during the entire heartbeat. The ventricular pressure becomes extremely high and increases with volume. As volume (or end-systolic pressure) increases, the ventricular pressure will reach an upper physiological limit. This is known as the Frank mechanism, and the result was included in Otto Frank's first study on the heart, "On the dynamics of cardiac muscle," published in *Zeitschrift fuer Biologie* in 1895. The Frank mechanism is the key element in the experimentally based ventricular model by Mulier (1994) that views the isovolumic ventricular pressure as a function of time t and volume V_v contained in the ventricle. The model assumes that ventricular isovolumic pressure p_v equals a passive (diastolic) $p_d(V_v)$ plus an active (systolic) $p_s(V_v)g(t)$ component:

$$p_v(V_v, t) = p_d(V_v) + p_s(V_v)g(t), \tag{4.1}$$

where

$$g(t) = \begin{cases} \left(1 - e^{-\left(\frac{t}{\tau_c}\right)^\alpha}\right), & 0 \le t \le t_b, \\[3mm] \left(1 - e^{-\left(\frac{t}{\tau_c}\right)^\alpha}\right) e^{-\left(\frac{t-t_b}{\tau_r}\right)^\alpha}, & t_b < t < t_h. \end{cases} \tag{4.2}$$

The second term on the right-hand side of (4.1) describes the active contractile processes in systolic pressure generation. The time constants τ_c and τ_r in (4.2) characterize the contraction (pressure increase) and relaxation (pressure decrease) processes, respectively, while α is a measure of the overall rate of onset of these processes. The constant t_h denotes the heart period. The constant t_b is the time when the relaxation process begins to evolve and is directly related to τ_c, τ_r, and α and to the time of peak generated pressure, t_p, by

$$t_b = t_p \left\{ 1 - \left(\frac{\tau_r}{\tau_c}\right)^{\frac{\alpha}{\alpha-1}} \left[\frac{e^{-\left(\frac{t_p}{\tau_r}\right)^\alpha}}{1 - e^{-\left(\frac{t_p}{\tau_c}\right)^\alpha}} \right]^{\frac{1}{\alpha-1}} \right\}. \tag{4.3}$$

The function g is the product of two factors, as shown in Figure 4.1. The first describes the buildup of pressure resulting from crossbridge formation in the constituent muscle fibers composing the ventricle. The second describes the relaxation of pressure resulting from bond detachment.

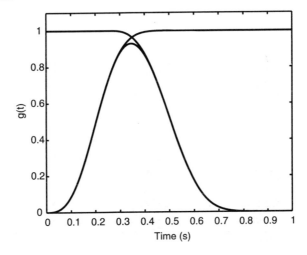

Figure 4.1. *The function $g(t)$ consists of two factors. The first factor describes the buildup of ventricular pressure during contraction via muscle crossbridge bond attachment (increasing curve). The second factor describes the relaxation following bond detachment (decreasing curve).*

Explicit expressions for $p_d(V_v)$ and $p_s(V_v)$ in (4.1) follow from a number of experiments by Mulier (1994) done on dog hearts. These experiments showed that the peak developed pressure is best described by a linear relation

$$p_s(V_v)g(t_p) = p_v(V_v, t_p) - p_d(V_v) = cV_v - d, \tag{4.4}$$

where c and d are directly related to the volume-dependent and volume-independent components, respectively, of developed pressure. Experiments also revealed that the passive component $p_d(V_v)$ best follows a quadratic relationship between ventricular pressure p_v and volume V_v:

$$p_d(V_v) = a(V_v - b)^2, \tag{4.5}$$

where a is a measure of diastolic ventricular elastance and b corresponds to diastolic volume for zero diastolic pressure (Mulier, 1994).

Using (4.1) to (4.5), ventricular pressure in the isovolumic beating heart depends on ventricular volume V_v and time t as

$$p_v(V_v, t) = a(V_v - b)^2 + (cV_v - d)f(t), \tag{4.6}$$

where f is the normalization of g:

$$f(t) = \begin{cases} \dfrac{\left(1 - e^{-\left(\frac{t}{\tau_c}\right)^\alpha}\right)}{\left(1 - e^{-\left(\frac{t_p}{\tau_c}\right)^\alpha}\right)e^{-\left(\frac{t_p - t_b}{\tau_r}\right)^\alpha}}, & 0 \leq t \leq t_b, \\[4mm] \dfrac{\left(1 - e^{-\left(\frac{t}{\tau_c}\right)^\alpha}\right)e^{-\left(\frac{t - t_b}{\tau_r}\right)^\alpha}}{\left(1 - e^{-\left(\frac{t_p}{\tau_c}\right)^\alpha}\right)e^{-\left(\frac{t_p - t_b}{\tau_r}\right)^\alpha}}, & t_b < t < t_h. \end{cases} \tag{4.7}$$

The isovolumic pressure model (4.6) contains nine parameters $a, b, c, d, t_h, t_p, \tau_c, \tau_r$, and α, which are derived from experimental data. The last three parameters were obtained using curve fitting. All parameters are given a physiological interpretation. The parameter values can be determined from at least two isovolumic contractions with different volumes. Further details are available in Mulier (1994).

The isovolumic pressure model (4.6) describes the ventricular pressure in the isolated heart. But the model embraces the major features of an ejecting ventricle when coupled to a description of the human cardiovascular system (Danielsen, 1998; Palladino et al., 1997; Palladino, Mulier, and Noordergraaf, 1997; Palladino, Ribeiro, and Noordergraaf, 2000; Danielsen, Palladino, and Noordergraaf, 2000a). For the purpose of permitting the ventricle to eject, the isovolumic pressure model (4.6) is coupled to a three-element modified Windkessel arterial load, as shown in Figure 4.2. The input impedance of the Windkessel then relates root aortic pressure p_{ao} to ventricular volume V_v as

$$\dot{p}_s = -\frac{R_s + R_0}{R_0 R_s C_s} p_s + \frac{1}{R_0 C_s} p_v(t, V_v), \tag{4.8}$$

$$\dot{V}_v = \frac{1}{R_0} p_s - \frac{1}{R_0} p_v(t, V_v), \tag{4.9}$$

$$p_{ao} = p_s - R_0 \dot{V}_v, \tag{4.10}$$

where R_0 is the characteristic aortic impedance, R_s is the total peripheral resistance, and C_s is the total arterial compliance. Parameter values for the arterial load can be found in Table 4.1. Representative computed ventricular pressure p_v, root aortic pressure p_{ao}, and ventricular outflow Q_v $(= -\frac{dV_v}{dt})$ for the human left ventricle are shown in Figure 4.3 together with the isovolumic pressure p_{iso} for the same end-diastolic volume. The ventricle

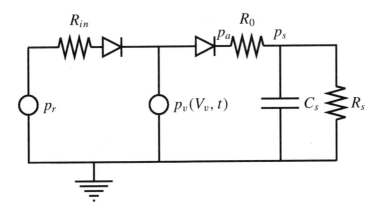

Figure 4.2. *The isovolumic pressure model (4.6) coupled to a three-element modified Windkessel arterial load and a venous pressure reservoir p_r. End-diastolic volume is 125 ml (filling pressure $p_r = 10$ mmHg). Parameter values for the arterial load and the constant pressure reservoir p_r can be found in Table 4.1.*

Table 4.1. *Parameter values for the three-element modified Windkessel model and the constant pressure reservoir p_r in Figure 4.2.*

Parameter	Value	Units
R_0	0.08	mmHg·s/ml
C_s	2.75	ml/mmHg
R_s	1.0	mmHg·s/ml
R_{in}	0.001	mmHg·s/ml
p_r	10	mmHg

Table 4.2. *Parameter values for the isovolumic pressure model (4.6).*

Parameter	Value	Units
a	0.0007	mmHg/ml
b	5	ml
c	1.6	mmHg/ml
d	1	mmHg
τ_c	0.15	s
τ_r	0.175	s
t_p	0.3	s

is defined by the parameter values in Table 4.2. The large pressure difference between p_v and p_{iso} is caused by the changing volume V_v during ejection. Of special interest is that Starling's law emerges. Consequently, Starling's law follows from the Frank mechanism. In this sense, Starling's law is reduced to the Frank mechanism and thus disqualified as an independent law of the ventricle (Danielsen, 1998).

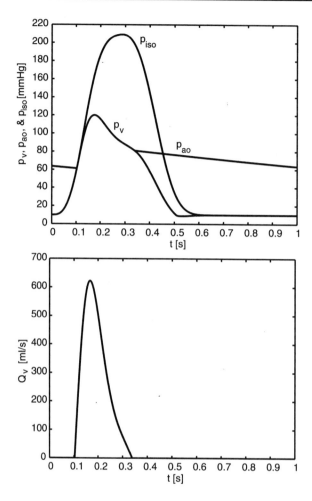

Figure 4.3. *The top panel shows isovolumic ventricular pressure p_{iso}, ventricular pressure p_v, and root aortic pressure p_{ao} (all in mmHg) as functions of time. The bottom panel displays the ventricular outflow Q_v. All curves are computed for a normal human ventricle ejecting into the arterial load in Figure 4.2. End-diastolic volume is 125 ml, stroke volume 72 ml, ejection fraction 0.58, and heart rate $H = 1$ Hz. Ventricular parameters are given in Table 4.2.*

4.3 The Ejection Effect

Despite the fact that the isovolumic pressure model (4.6) embraces the major features of an ejecting ventricle, the Frank mechanism alone cannot fully describe an ejecting heartbeat. Close inspection reveals a number of discrepancies between model-predicted results and their measured counterparts. The waveforms of the ventricular pressure and outflow curves differ from the profiles reported in Guyton (1991) and Noordergraaf (1978). The descending part of the computed ventricular outflow in Figure 4.3 tends to be concave, though it often appears convex in nature. Also, the top descending part of the ventricular pressure p_v is concave. In addition, the outflow Q_v is too narrow. The most significant difference is found between measured and model-predicted ventricular pressures, as shown in Figure 4.4 (Mulier, 1994). The measured ventricular pressure p_{mes} in the early phase of ejection, when ventricular outflow is large, is lower than the model-predicted pressure p_v when using the isovolumic pressure model (4.6), which is termed *deactivation*. Later the model-predicted pressure is somewhat higher and is denoted *hyperactivation*. The combination of deactivation and hyperactivation is denoted the ejection effect (Danielsen, 1998; Danielsen, Palladino, and Noordergraaf, 2000a; Danielsen and Ottesen, 2001). Thus the ejection effect is observed as the difference between experimentally observed ventricular pressure curves during ejection and the corresponding predicted pressure curve computed using the isovolumic heart model. In the following, the ejection effect is also used to denote both the underlying physiological mechanisms and the experimentally observed differences in the pressure curves. However, the meaning will be clear from the context.

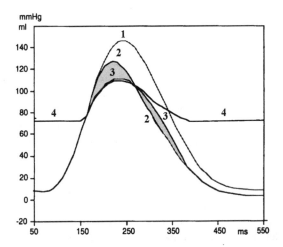

Figure 4.4. *Ventricular pressures measured on one isolated dog heart. Curve 1 is the isovolumic pressure at fixed end-diastolic volume, and curve 2 is the computed pressure p_v computed from the isovolumic pressure model (4.6). Curve 3 is the measured pressure for an ejecting beat. Comparing curves 2 and 3, the measured ventricular pressure is lower in early systole (deactivation) and higher later (hyperactivation). The first shaded area from the left represents deactivation, and the second shaded area represents hyperactivation. Adapted from Mulier (1994).*

At the muscle level, deactivation may be qualitatively related to *Hill's force-velocity relation* stating that the force of muscle contraction is more reduced the higher the velocity of muscle shortening (Hill, 1938; Palladino, Mulier, and Noordergraaf, 2000). The act of muscle shortening forces crossbridge bonds to detach and results in diminished force. On the chamber level this is equivalent to a reduction in ventricular pressure with increased early flow (Mulier, 1994; Palladino, 1990). Hyperactivation follows deactivation and is manifested as an increase in ventricular pressure in late ejection. Hyperactivation may be related to formation of new crossbridge bonds between thick and thin filaments. The strength of hyperactivation is subsequently conditioned by the available biochemical energy. In support of the above De Tombe and Little (1994) concluded on experimental grounds that these positive and negative inotropic effects of ejection are myocardial properties.

Other effects, such as inertial effects and reflected waves, may contribute to the ejection effect. The inertial effects of blood in the ventricle tend to lower root aortic pressure in early systole (similar to deactivation) and increase pressure in late systole (akin to hyperactivation). Moreover, these initial effects may improve the bending of the ventricular outflow curve. In order to test whether inertial effects can explain the ejection effect, the isovolumic pressure model (4.6) was coupled to the arterial load in Figure 4.2 with an inductance L_v placed before the aortic valve. Thus root aortic pressure p_{ao} and ventricular outflow Q_v relate as

$$\dot{p}_s = -\frac{1}{R_s C_s} p_s + \frac{1}{C_s} Q_v, \tag{4.11}$$

$$\dot{Q}_v = -\frac{1}{L_v} p_s - \frac{R_0}{L_v} Q_v + \frac{1}{L_v} p_v(t, V), \tag{4.12}$$

$$p_{ao} = p_s + R_0 Q_v. \tag{4.13}$$

Figure 4.5 shows that inertial effects of ventricular blood provide a phase shift between ventricular and root aortic pressure as observed experimentally (Pasipoularides et al., 1987). However, the results show no significant improvement in the shape of ventricular pressure p_v and outflow Q_v for physiological values of the inductance. Also, interaction with reflected waves was eliminated by making the compliance equal to $100C_s$ (Berger et al., 1993). Previous computations have shown that reflected waves play only a minor role in the profiles. Although the inertial effect of ventricular blood contributes in the right direction, it, and reflection, cannot explain the ejection effect (Danielsen, 1998).

Positive and negative effects of ventricular ejection have been observed previously but not with the same definition. In order to make the discussion about positive and negative effects of ejection clear, it is essential that one precisely define the two phenomena. On the chamber level, we define deactivation and hyperactivation with the two definitions listed below. (These definitions are closely related to the ones given by Mulier (1994).)

Definition 1: Deactivation. *If $p_v < \tilde{p}_v$, where p_v is the actual ventricular pressure during ejection and $\tilde{p}_v$ is the corresponding ventricular pressure predicted from isovolumic heart properties alone, deactivation is defined as the negative pressure difference $\Delta p = p_v - \tilde{p}_v$.*

Definition 2: Hyperactivation. *If $p_v > \tilde{p}$, where p_v is the actual ventricular pressure during ejection and $\tilde{p}_v$ is the corresponding ventricular pressure predicted from isovolumic heart properties alone, hyperactivation is defined as the positive pressure difference $\Delta p = p_v - \tilde{p}_v$.*

Thus deactivation is present when the pressure difference $\Delta p = p_v - \tilde{p}_v$ becomes negative during ejection, and hyperactivation occurs when the pressure difference Δp becomes positive.

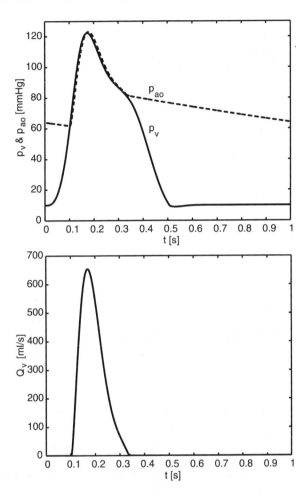

Figure 4.5. *The top panel shows ventricular pressure p_v in mmHg (solid line) and root aortic pressure p_{ao} in mmHg (dashed line) when the ventricle ejects into the modified Windkessel arterial load of Figure 4.2 obtained by using (4.12) and (4.13) with an inductance $L_v = 0.000416$ mmHg·s²/ml. The bottom panel shows the corresponding ventricular outflow Q_v in cm³/s.*

4.4 Formulation of the Ejection Effect

The ejection effect can be introduced into the isovolumic pressure model (4.6) by modifying the function f, which describes the actual force generation such that

$$p_v(t, V_v, Q_v) = a(V_v - b)^2 + (cV_v - d)F(t, Q_v), \qquad (4.14)$$

where Q_v is ventricular outflow and $F(t, Q_v)$ is given by

$$F(t, Q_v) = f(t) - k_1 Q_v(t) + k_2 Q_v^2(t - \tau), \quad \tau = \kappa t. \qquad (4.15)$$

The positive parameters k_1 and k_2 characterize the strength of deactivation and hyperactivation, respectively, and τ represents a time delay. The second term on the right-hand side of (4.15), $k_1 Q_v(t)$, describes deactivation since increased outflow $Q_v(t)$ results in a drop in ventricular pressure. The term $k_2 Q_v^2(t - \tau)$ is responsible for hyperactivation and becomes active τ later in time than deactivation. The time delay $\tau = \kappa t$ allows almost immediate cycling of crossbridges in early ejection and a slower formation of bonds later. The parameter κ represents the change in the rate of new bond formation with time. Further details of this and alternative models of the ejection effect are available (Danielsen and Ottesen, 2001; Danielsen, Palladino, and Noordergraaf, 2000a; Danielsen, 1998).

The extended model (4.14), termed the heart model, was coupled to a three-element modified Windkessel arterial load and filled from a constant pressure reservoir p_r, as in Figure 4.2. The computed ventricular pressure p_v and outflow Q_v are shown in Figure 4.6 using the parameter values for the heart model (4.14) given in Table 4.3. The ventricular pressure p_v and outflow Q_v curves are far more representative of the normal human ventricle and resemble the ones found in the literature (Guyton, 1991; Noordergraaf, 1978). In particular, the descending part of the ventricular outflow Q_v in Figure 4.6 exhibits a clearly convex shape. Also, the top descending part of the ventricular pressure is convex. Table 4.4 compares model results with typical values of the human ventricle adopted from Noordergraaf (1978). Figure 4.7 demonstrates that both deactivation and hyperactivation are introduced by the model of the ejection effect.

Both deactivation and hyperactivation are influenced by interaction with the vasculature. According to Mulier (1994), deactivation increases with higher ventricular outflow Q_v. This is shown in Figure 4.8, where deactivation is slightly higher when ventricular outflow Q_v is increased. (Ventricular outflow is increased when peripheral resistance is reduced from 1 mmHg·s/ml to 0.5 mmHg·s/ml.) This figure also shows that deactivation

Table 4.3. *Parameter values of the heart model* (4.14) *when ejecting into a normal arterial load defined by the parameter values in Table* 4.1.

Case	k_1	k_2	κ
R_{ps}	0.0004	0.0000015	0.45

Table 4.4. *Typical values of the human ventricle (Noordergraaf, 1978) and corresponding model-computed values. The time interval Δt_{pres} is measured between the 20-mmHg levels of the ventricular pressure p_v. Time duration of the systole is Δt_s, and Δt_p denotes time to peak developed flow Q_v.*

Type	Human values	Model values	Units
Δt_{pres}	400	450	ms
Δt_s	275	265	ms
Δt_{peak}	70	84	ms

is diminished by low ventricular outflow. (Ventricular outflow is lowered when peripheral resistance is raised from 1 mmHg to 2 mmHg.)

Maximum hyperactivation is almost unaltered when the peripheral resistance is lowered to 0.5 mmHg·s/ml, which may be an attempt by the heart to compensate for the reduced ventricular pressure p_v. When ventricular outflow Q_v is reduced, hyperactivation is diminished, which may be related to the corresponding increased ventricular pressure. Also, variations in the filling pressure or end-diastolic volume alter the ejection effect pattern, as depicted in Figure 4.9. When end-diastolic volume is lowered to 75 ml from 125 ml, both

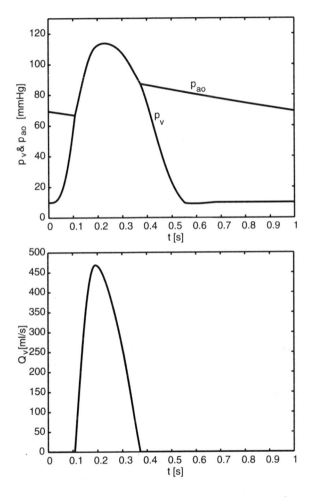

Figure 4.6. *The top panel shows ventricular pressure p_v and root aortic pressure p_{ao} in mmHg computed for a normal human ventricle using the heart model (4.14) ejecting into the arterial load of Figure 4.2. The bottom panel displays the corresponding ventricular outflow Q_v in cm³/s. The parameter values of the ejection effect and the ventricle can be found in Tables 4.2 and 4.3. End-diastolic volume is 125 ml, stroke volume 78 ml, ejection fraction 0.62, and heart rate $H = 1$ Hz.*

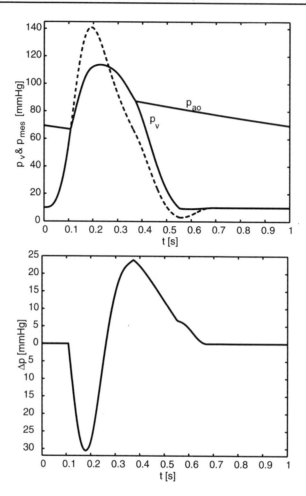

Figure 4.7. *The top panel displays ventricular pressure p_v and root aortic pressure p_{ao} from Figure 4.6 computed using the heart model (4.14) (solid line) and predicted pressure $\tilde{p}_v$ obtained using the isovolumic pressure model (4.6) without ejection (dashed line). The bottom panel depicts the corresponding pressure difference $\Delta p = p_v - \tilde{p}_v$ in mmHg.*

deactivation and hyperactivation are reduced. This reduction may be related to the lower number of attached bonds in diastole. When end-diastolic volume is elevated to 175 ml from 125 ml, an increased number of bonds form, resulting in higher amounts of both deactivation and hyperactivation.

Figure 4.10 illustrates that hyperactivation is diminished compared to the normal situation in Figure 4.6 when the ventricle ejects into a reflection-free afterload (the subsystem following the ventricle; in practice this is obtained by increasing C_s significantly). But hyperactivation is still significantly present, and previous computations show that physiological curves cannot be obtained without hyperactivation (Danielsen, 1998).

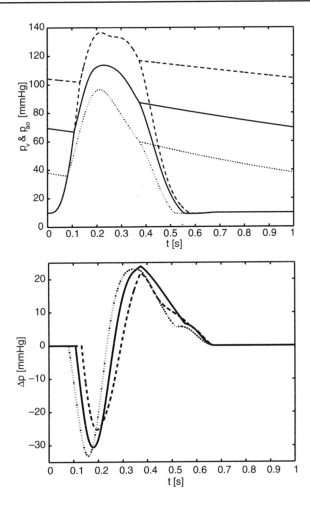

Figure 4.8. *The top panel shows ventricular pressure p_v and root aortic pressure p_{ao} in the normal human ventricle (solid line). Also shown is the result of raising the peripheral resistance from 1.0 mmHg·s/cm³ to 2 mmHg·s/cm³ (dashed line) and of reducing it to 0.5 mmHg·s/cm³ (dotted line). The corresponding variation in $\Delta p = p_v - \tilde{p}_v$, where $\tilde{p}_v$ is computed using the isovolumic pressure model (4.6), is shown in the bottom panel.*

4.5 Summary and Discussion

In this chapter we identified and described the ejection effect of the heart, which consists of deactivation and hyperactivation. These two phenomena are observed as discrepancies between measured and computed ventricular pressures when using a model based solely on isovolumic heart properties. During early ejection, predicted pressure is higher than its measured counterpart, termed deactivation, and later, computed pressure becomes lower than measured ventricular pressure, denoted hyperactivation. Deactivation may be seen as a consequence of muscle shortening forcing crossbridge bonds to detach, while hyper-

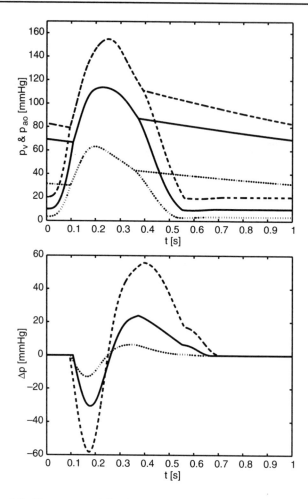

Figure 4.9. *The top panel displays the computed ventricular pressure p_v and root aortic pressure p_{ao} in mmHg in the normal human ventricle (solid line) and when the end-diastolic volume is raised to 175 ml from 125 ml (dashed line) and reduced to 75 ml (dotted line). The bottom panel displays the corresponding variation in $\Delta p = p_v - \tilde{p}_v$ in mmHg, where $\tilde{p}_v$ in mmHg is computed using the isovolumic pressure model (4.6).*

activation may result from formation of new bonds. By modifying the function f in the isovolumic pressure model (4.6), directly related to the actual force generation, we expanded the model to include the ejection effect. This resulting heart model (4.14) generates ventricular pressure and flow curves far more representative of the human ventricle and thus broadens the range of the description from isovolumic contractions to also cover ejecting heartbeats. The computed results suggest that ventricular ejection directly changes the underlying muscle contraction process and that the ventricle is influenced by alterations in the vascular parameters. The ejection effect cannot be explained only by factors such as inertial effects of ventricular blood and vascular reflections, although these effects change ventricular pressure in the right direction.

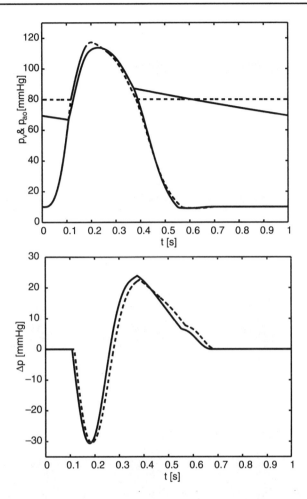

Figure 4.10. *The top panel depicts the ventricular pressure p_v and root aortic pressure p_{ao} in mmHg (dashed line) when the ventricle ejects into a reflection-free after-load (compliance C_s in ml/mmHg is increased by a factor of* 100*), which is superimposed on curves for a normal human from Figure 4.6 (solid line). The corresponding variation in $\Delta p = p_v - \tilde{p}_v$ in mmHg, where $\tilde{p}_v$ in mmHg is computed using the isovolumic pressure model* (4.6), *is shown in the bottom panel.*

The modeling approach behind the heart model presented here separates isovolumic and ejection heart properties from each other. This separation is key when considering the ejection effect and allows isovolumic and ejecting heart properties to be studied individually. The result is achieved by a procedure that first describes the isovolumic contracting ventricle and subsequently expands this description with a formulation of the ejection effect. Each parameter of the isovolumic model describes only properties of the isovolumic contraction, and parameters of the ejection effect model cover only ejecting heart properties. This modeling approach, however, has broader implications. Using this heart model, it can be shown that the popular time-varying elastance concept is disqualified as an independent description of the heart in the sense that it follows from isovolumic heart properties (the Frank mechanism) and the ejection effect (Danielsen and Ottesen, 2001). In addition, the heart model presented here covers most of the features of a human heart during normal and altered conditions in the cardiovascular system.

The model of the ejection effect presented here is the last one in a series of descriptions. Mulier (1994) offered the first attempt to include deactivation mathematically in the isovolumic pressure model (4.6). He considered deactivation to be an independent mechanism modeled by

$$p_d = -\alpha_d Q_v^2 \tag{4.16}$$

such that ventricular pressure is given by

$$p_v(V_v, Q_v, t) = p_v(V_v, t) - \alpha_d Q_v^2,$$

where the positive parameter α_d represents the strength of deactivation. Hyperactivation was assumed to be caused by reflections. According to Mulier (1994), deactivation is then related to flow alone. Mulier (1994) claimed that deactivation does not alter the fundamental parameters of the isovolumic pressure model (4.6). Later Danielsen (1998) and Danielsen, Palladino, and Noordergraaf (2000a) proposed a modification of the c parameter in the isovolumic pressure model (4.6) related to the contractile properties of the ventricle. He made it depend on the time integral of ejected flow, or instantaneous flow. Rather than alter the fundamental parameters of the isovolumic pressure model, Danielsen, Palladino, and Noordergraaf (2000b) proposed a modification of the function f, closer to the actual force generation, such that

$$F(t, Q_v) = f(t) - k_1 Q_v(t) + k_2 Q_v(t - \tau), \quad \tau = \kappa t, \tag{4.17}$$

in the heart model (4.14). This includes a linear version of the term describing hyperactivation in the heart model (4.14) and generates less convexity in both ventricular pressure and flow. The change from a linear to a quadratic expression alters the time course of crossbridge bonds attaching in the myofilaments. A more detailed discussion of these models of the ejection effect is available (Danielsen and Ottesen, 2001; Danielsen, 1998; Ottesen and Danielsen, 2003).

Previous studies have also observed both deactivation and hyperactivation. In these studies the two phenomena are denoted negative and positive effects of ejection, respectively (Hunter, 1989; De Tombe and Little, 1994; Mulier, 1994; Burkhoff, De Tombe, and Hunter, 1993). Burkhoff, De Tombe, and Hunter (1993) adopted a mathematical description of the isovolumic beating heart different from the isovolumic pressure model (4.6) and allowed it to eject into an appropriate model of the vasculature. They observed experimentally both deactivation and hyperactivation during ejection. Actually, they always found both positive and negative effects of ejection. However, no mathematical formulation of the ejection effect was attempted, and the isovolumic model was not analytic. Negative effects of ejection, or deactivation, were studied separately by Hunter et al. (1983) and Vaartjes and Herman (1987) and are included in several ventricular models in the form of a series resistance (Hunter et al., 1983; Campbell et al., 1986). Hyperactivation, although observed previously, has not been appreciated to the same degree as deactivation. This may be related to the absence of an available isovolumic ventricular model. Ducas et al. (1985) seem to be the first who observed positive effects of ejection by lowering the resistive afterload. Hunter (1989) recorded measurements showing that end-systolic ventricular pressure exceeded the isovolumic pressure in the ventricle when the volume was equal to the end-systolic volume from the previous ejecting beat. This was observed under conditions in which the ejection fraction was 0.3, with the positive effect varying between 1 and 17 mmHg. The positive effect of ejection was shown to disappear during even lower or higher ejection fractions. He related these observations to underlying cardiac properties. Burkhoff, De Tombe, and Hunter (1993) took a broader view, considering the whole heartbeat instead of only one point during ejection. Baan (1992) concluded that the positive effects of ejection and homeometric autoregulation stem from the same mechanism, thus relating these positive effects to biochemical processes. Common among these previous observations is that they relied on an unclear definition of positive and negative effects of ejection. In order to structure this discussion, we provided a definition of the ejection effect in section 4.3. In addition, most of the previous work used the time-varying elastance concept.

The reason that our heart model from this chapter is not used in the cardiovascular model in Chapter 6 relates to the problems we encounter if we extend the heart model (4.14) to real-world applications. First, the heart rate is not easily changed, which is essential when used, e.g., in a simulator. This restriction seems to be related to the choice of the exponential functions in the isovolumic pressure model (4.6). This particular isovolumic model is not essential for the modeling approach. Any isovolumic model may apply. A model based on polynomial functions may provide a perfectly suitable and simple alternative, as suggested by Ottesen and Danielsen. Second, at this stage we cannot determine without further experimental data how, for instance, the baroreceptor mechanism influences the ejection effect. However, this practical view is not the goal of this chapter. The modeling efforts are directed toward investigating the physiological mechanisms behind the ejection effect. Identification of the mechanisms involved may emerge in the future from new animal experiments.

Acknowledgments

The author was supported by the Danish Heart Foundation (99-1-2-14-22675) and Trinity College, Connecticut. In addition, the author wants to acknowledge the scientific support from Dr. Abraham Noordergraf, Cardiovascular Studies Unit, University of Pennsylvania, Philadlephia.

Chapter 5

Modeling Flow and Pressure in the Systemic Arteries

M.S. Olufsen

5.1 Introduction

In the two previous chapters we have studied two different models for the blood flow in the heart. In this section, we focus our attention on the systemic arteries and continue with a spatial model similar to the model described in Chapter 3. Because of the cylindrical shape of the arteries, we have restricted the model presented in this chapter to include one spatial dimension (the length of the arteries).

Flow and pressure in the systemic arteries can be modeled using a broad spectrum of approaches. The approach chosen for a given model should depend on the questions that are asked. In this chapter the main purpose is to develop a model that can faithfully reproduce the pulse wave anywhere in the systemic arteries. It is not possible to develop a mathematical model that can be solved analytically or numerically and can predict all details of flow and pressure in the arteries. As a result, it is necessary to introduce simplifications. While it is possible to develop 3D models, based on fundamental physical principles, of parts of the arteries, it would be very difficult to derive and implement a 3D model of the systemic arterial tree. It is possible to develop a model of the systemic arteries that describes the system's behavior in one spatial dimension, and this chapter describes such a 1D model.

Having a model that includes just one spatial dimension lets one study how flow and pressure waves change as they propagate along the arteries. The form of the waves changes as a result of the arteries' changing geometry and structure (Nichols and O'Rourke, 1998; Peskin, 1976). The main change in the waveform is due to reflections that arise from the tapering of the vessels, from bifurcations, and from the resistance of the arterial bed. The waveform is interesting since studying how it changes can have clinical implications. For example, the dicrotic wave (the second oscillation observed during the cardiac cycle; see arterial pressure in Figure 5.15) is diminished in some people suffering from diabetes or vascular diseases such as atherosclerosis (Feinberg and Lax, 1958; Lax and Feinberg, 1959; Lax, Feinberg, and Cohen, 1956; Nichols and O'Rourke, 1998; O'Rourke, Kelly, and Avolio, 1992). The dicrotic wave appears as a result of wave reflection and because a small amount of blood flows retrogradely into the left ventricle before the heart valves are completely closed (Guyton, 1991). People with stiff arteries have a less pronounced dicrotic wave but an increased systolic pressure (Kannel et al., 1981; Avolio, 1992; Nichols and O'Rourke, 1998; O'Rourke, Kelly, and Avolio, 1992). Therefore, studies of the dicrotic wave and comparison of flow and pressure profiles at different positions could possibly be used for diagnostic purposes, e.g., to locate stenosis (Raines, 1972). It has also been observed that people with stiffer arteries have a less pronounced dicrotic wave but increased diastolic and systolic pressures (Avolio, 1992; Kannel et al., 1981; Nichols and O'Rourke, 1998). The profiles of the blood flow and pressure waves vary significantly even among healthy people (Guyton, 1991). In fact, being able to construct a model based on measured geometry and a single noninvasive flow measurement in the ascending aorta will enable us to study the waveforms for any given person. Computed flow and pressure profiles could be used as part of a diagnostic tool. For example, for a given person, measured pathologic flow profiles could be compared with computed healthy flow profiles. Studies of how the model parameters must change for the model to reproduce the measured pathologic flow profiles might lead to a better understanding of the pathologic condition. In addition, computation of flow and pressure profiles could be used in connection with simulators, e.g., surgical or anesthesia simulators. For such simulators the models should be expanded to include the full cardiovascular system. The model presented in this chapter comprises only the systemic arteries, but it could easily be expanded to comprise the full cardiovascular system by adding a heart, systemic veins, and arteries and veins of the pulmonary circulation using the same principles as the ones explained in this chapter. Another constraint would be that a model that can be used in an anesthesia simulator must run in real time, and the 1D model presented in this chapter does not meet this requirement. In order to obtain sufficient speed for a model to run in real time, it is necessary to develop a zero-dimensional model. As described in Chapter 6, such a model can be developed using electrical circuit analogues consisting of inductors, capacitors, resistors, and diodes, which represent inertia, compliance, resistance, and valves, respectively. Even though zero-dimensional models are based on the underlying physiology, they are relatively crude. For example, these models do not represent the cardiovascular system as a continuum but as a collection of discrete points, so they are not able to predict realistic flow and pressure profiles simultaneously at a number of locations throughout the cardiovascular system. They also cannot include wave propagation effects in the part of the system that they model. A zero-dimensional model can produce correct overall wave shapes if good initial conditions and good values for the total resistance and compliance are applied. Total resistance and compliance are physiologic quantities, but they are not easy to measure, and the pulse profiles are sensitive to the value of these parameters. Furthermore, it is hard to avoid artificial reflections in a system that uses outflow conditions based on a zero-dimensional model. A zero-dimensional model can to some extent predict realistic levels for flow, systolic pressure, and diastolic pressure at certain points along the cardiovascular system. In order to achieve more precise flow and pressure profiles, it is necessary to introduce another layer of "models" that superimpose a correct-looking waveform on top of the ones obtained by the zero-dimensional model. Such waveforms can be obtained either from recordings or from a more complex model such as the 1D model described in this chapter. The cardiovascular model in the SIMA simulator is based on a zero-dimensional model combined with a curve-fitting model that adds correct-looking flow and pressure profiles to the diastolic and systolic values obtained by the zero-dimensional model.

The curve-fitting model could use either curves predicted from measurements or curves generated by an extended version of the 1D model discussed in this chapter. Using the 1D model to predict results that can be used in combination with a zero-dimensional model makes the 1D model a *reference model*. On its own the 1D model is inappropriate for use in an anesthesia or surgical simulator, but it could provide detailed information needed to obtain a correct output from a simulator. In addition, the reference model can, as discussed above, help researchers gain a better understanding of the underlying physiology.

As mentioned earlier, the 1D model presented in this paper comprises only the systemic arteries. In fact it mainly comprises the large arteries, including the first two–three generations of vessels in the arterial tree. The first generation comprises the aorta and the iliac arteries, and the second and third generations contain the vessels branching from these main conduit vessels; see Figure 5.1. In order to take the flow through the arterial bed (the small arteries, arterioles, and capillaries) into account, a cruder model is added at the boundaries of the pruned tree. As a result, the full model of the arterial tree consists of two parts, one for large arteries and one for small arteries.

These models are coupled at the termination points of the large arteries; see Figure 5.1. The combined model, including both the large and the small arteries, is able to reproduce the most prominent characteristic features observed in measured flow and pressure profiles:

- The maximum pressure of the large arteries increases away from the heart toward the periphery because of tapering of the vessels.

- The mean pressure in the arteries drops toward the periphery according to the distribution of flow and impedance of the vascular bed (Noordergraaf, 1978).

- The steepness of the incoming pressure profile increases toward the periphery. This increase is a result of the pressure dependence of the wave propagation velocity, which states that the part of the wave with higher pressure travels faster than that with lower pressure.

- The dicrotic wave separates from the incoming wave and is more prominent at peripheral locations than at proximal locations.

- Wave reflections affect the shape of the pulse wave. These appear as a result of the tapering and branching of the vessels and as a result of the peripheral resistance, which mainly originates from the small arteries and arterioles.

These features are results of a characteristically shaped flow (see Figure 5.8) entering from the left ventricle and propagating along the branching elastic arteries. Having a model that is able to reproduce the above characteristic features has satisfied our overall goal: to develop a model that can faithfully reproduce the pulse wave anywhere in the systemic arteries.

The arterial tree has a complex structure, and for the small branches it is not possible to measure the geometry and wall properties of each individual vessel. Furthermore, it is not practical to compute flow and pressure in all these branches even with a 1D model. Consequently, it is necessary to truncate the arterial tree and include outflow conditions at the truncation points. As a result, the outflow conditions should represent the flow and pressure in the small arteries and arterioles.

There exist several 1D models that can predict flow and pressure in the systemic arterial tree. But, to our knowledge, none of these models includes physiologically based approaches to modeling flow and pressure in both the large and the small arteries. Most studies concentrate on either modeling flow and pressure in the large arteries using outflow conditions based on zero-dimensional models, such as a pure resistance or a Windkessel model (Anliker, Rockwell, and Ogden, 1971; Raines, Jaffrin, and Shapiro, 1974; Avolio, 1980; Stergiopulos, Young, and Rogge, 1992) or modeling the small arteries focusing on the geometric and structural properties of the vessels (Zamir, Langille, and Wonnacott, 1984; Schreiner and Buxbaum, 1993; Kassab and Fung, 1995; Stergiopulos, Young, and Rogge, 1992; Canic and Mikelic, 2002; Canic and Mikelic, 2003; Canic and Kim, 2003). More recent studies use either periodical boundary conditions (Ottesen, 2003), or have used a boundary condition similar to the one we present in this chapter (Wan et al., 2002). Therefore, we propose a model that combines the two types of models mentioned above: For the large arteries our model departs from the early 1D models, and for the small arteries our model departs from the models based on geometry and structure.

More specifically, the large arteries are modeled using the studies of Barnard et al. (1966) and Peskin (1976) as points of departure; see section 5.4. The geometry of the vessels is modeled using the approach suggested by Stergiopulos, Young, and Rogge (1992) and Segers et al. (1998), and the boundary conditions are modeled based on physiological properties of wave propagation in the tree of small arteries combined with the above-mentioned studies on the geometry of the small arteries. The outflow condition is obtained by computing the input impedance at the root of a structured, binary, and asymmetric tree representing the small arteries.

The reason for using a structured tree is that the role of small arteries is to ensure that all muscles and organs are perfused sufficiently with blood. It has been suggested that the organization of the small arteries and arterioles must follow some structured pattern (Bassingthwaighte, Liebovitch, and West, 1994). Physiological studies show that bifurcations of the small arteries are mainly binary. From studies of angiographies (X-rays of blood vessels) we found that the tree can be modeled such that at each bifurcation the radius of the daughter vessels is scaled with factors α and β, both smaller than one. In order to disperse the propagating waves, and hence damp the oscillations, the structured tree cannot be truncated at a fixed number of generations. Instead the tree should be continued to the point where the radius of the vessels becomes smaller than some given minimum radius. Consequently, the origins of the reflected waves are geometrically spread out. With this tree structure, accentuation, due to superimposing waves at bifurcations, will be less pronounced than it would be in trees with a fixed number of generations. However, since the trees are structured, some accentuation will still be present, depending on the asymmetry ratio.

In the large arteries blood flow is dominated by inertia, and it is important to take tapering of individual blood vessels into account. Therefore, when predicting flow and pressure in the large arteries, it is important to include the nonlinear convection terms in the Navier–Stokes equations. Combined with inflow and outflow conditions (the latter provided through the root impedance of the structured tree), an equation ensuring continuity, and a state equation relating fluid influence on the vessel wall to the elasticity of the vessel walls, flow and pressure in a given geometry are fully described. In contrast, blood flow in the small arteries is dominated by viscosity, and tapering of individual vessels is not significant. It is important to account for viscosity in more detail, while inertia and tapering are less important than in the large arteries. Assuming that terms, including inertia, are small makes it possible to linearize the Navier–Stokes equations and arrive at a wave equation that can be solved analytically. From the analytical solution the impedance at the root of the structured tree has been computed, and thereby an outflow condition for the large arteries, which is inherently based on the underlying wave propagation, has been constructed.

The approach taken here has an advantage over traditional models in which outflow conditions are usually based on a zero-dimensional model such as the three-element Windkessel model by Raines, Jaffrin, and Shapiro (1974) and Stergiopulos, Young, and Rogge (1992); see Figure 5.9. While such a model can be tuned to simulate the overall results of wave propagation, a circuit with three components cannot describe the true wave propagation of the tree of small arteries. Furthermore, a large number of parameters (three for each outflow terminal) must be estimated. Since probably the most important factors determining the pulse wave are the inflow profile and the geometry of the arteries, the results will depend significantly on these factors. If the parameters of the Windkessel model are not adapted to fit individual people, artificial reflections or excessive damping of the system can occur. The outflow condition proposed in this study does not suffer from this deficiency, because it is based on underlying physiological structures of the vessels.

As mentioned earlier, mathematical models predicting flow and pressure are of interest to clinicians because the shape of flow and pressure profiles, and the appearance of the dicrotic wave, are of diagnostic significance, e.g., for diabetes, atherosclerosis, and hypertension. Good examples are the old studies by Lax, Feinberg, and Cohen (1956); Lax and Feinberg (1959); and Dawber, Thomas, and McNamara (1973), which suggest that changes in the shape of the dicrotic wave result from changes in the elasticity of vascular walls and changes in peripheral resistance. This conclusion originates from observations showing that the dicrotic wave is diminished or totally absent in people with diabetes, hypertension, or atherosclerosis, and in elderly people. Furthermore, observations by Jensen (1994–1998) have shown that a proximal stenosis, e.g., one at the iliac bifurcation, can lead to amplification of the dicrotic wave simply because the partly or fully occluded artery behaves as a high peripheral resistance, giving an early and more prominent reflection of the pulse wave. These studies suggest that regular observations of changes in the shape of the pulse wave could be used for early detection of some vascular diseases.

In summary, the purpose of this chapter is to study the shape of the pulse wave using a 1D mathematical model based on physiological principles for blood flow and pressure in the systemic arterial tree. To develop this model, the following model assumptions have been made:

- Blood flow and pressure in the large arteries are predicted from a nonlinear 1D model, based on the incompressible Navier–Stokes equations for a Newtonian fluid in a tapering elastic vessel.

- The inflow boundary condition is based on a velocity profile measured in the ascending aorta, while the outflow condition is predicted from a dynamic impedance applied at all terminals of the large arteries. The impedance is found from a separate model of the small arteries.

- Blood flow and pressure in the small arteries are predicted from a linear 1D viscous model (a wave equation), which is derived from linearization of the incompressible axisymmetric Navier–Stokes equations for Newtonian fluid in a nontapering elastic vessel.

In sections 5.2 and 5.3 the structure and geometry of the large and small arteries, respectively, are described. The fluid-dynamical model of the large arteries is established in section 5.4, which sets up the equations for one vessel. And section 5.5 describes the flow and pressure in the tree of large arteries (see Figure 5.1). The model for the small arteries is described in section 5.6, which sets up the equations for one vessel. Finally, section 5.7 describes how to find the impedance, and section 5.8 presents our results. In this section, the model will be validated by comparison with both data and existing models that base the outflow conditions on other (and simpler) models: a pure resistor model (Anliker, Rockwell, and Ogden, 1971), or a three-element Windkessel model (Raines, Jaffrin, and Shapiro, 1974; Stergiopulos, Young, and Rogge, 1992). Finally, section 5.9 concludes and summarizes this chapter.

5.2 Structure of the Large Arteries

The purpose of the systemic arteries is to supply all parts of the body, i.e., all muscles and organs, with oxygenated blood and nutrients. The arteries are organized in a sophisticated network that covers the entire organism. This network is called the arterial tree and it is characterized by

- geometric properties of the individual vessels: their diameter and length;

- structural properties of the vessels: the wall thickness and Young's modulus.

5.2.1 Geometric Properties of the Large Arteries

The large arteries are organized as a bifurcating tree in which individual vessels taper along their length. At each bifurcation the cross-sectional area at the top of each daughter vessel is smaller than that at the bottom of the parent vessel. However, the sum of the area of the two daughter vessels is larger than the area of the parent vessel. As a result, the total cross-sectional area increases from approximately 4.5 cm^2 in the aorta to approximately 400 cm^2 at the arteriolar level. Several papers have investigated this tapering. It is common in these studies to assume that the taper along the individual vessels follows an exponential curve of the form

$$r(x) = r_{top} \exp(-kx), \qquad (5.1)$$

where r_{top} is the mean proximal radius, k is the tapering factor, and x is the location along the artery (Anliker, Rockwell, and Ogden, 1971; Caflisch et al., 1980; Werff, 1974; Li, 1987). In this work $k = \log(r_{bot}/r_{top})/L$, where r_{bot} is the distal radius and L is the length of the vessel. For each vessel data for these parameters are taken from measured values. Therefore, to describe the full geometry of the tree of large arteries measurements of the proximal and distal radius for each vessel, as well as the length of each vessel segment, are needed.

For a man of average height (175 cm) and weight (70 kg) such data can be found in the literature (Westerhof et al., 1969; Stergiopulos, Young, and Rogge, 1992; Segers et al., 1998). The first two papers present data to be used in a mathematical model, while the third paper presents data to be used in a physical latex tube model. We have chosen to base our simulations on the data collected by Segers et al. (1998) because they are new and similar to the data in numerous older studies. It should be kept in mind that the dimensions of the arteries vary significantly—it is not unusual to see deviations of more than 50% from the mean values (Jensen, 1994–1998). This variation can have a significant effect on the shape of the arterial pulse, so a quantitative validation of the mathematical model should match the measured geometry of the arteries to the geometry used in the model. For the standard model, the large arteries are defined as those shown in Figure 5.1 with the

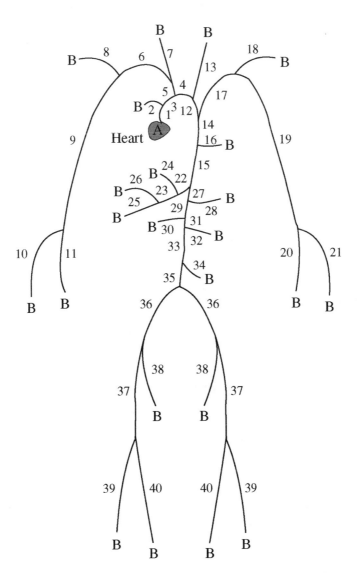

Figure 5.1. *The large arteries. The numbers on the figure refer to Table* 5.1. *Branches marked by the same numbers are identical and thus only modeled once. Blood flows into the arteries at the vessel marked by A and out at the vessels marked by B.*

dimensions presented in Table 5.1. All segment lengths are rounded to units of 0.25 cm. In addition, we assume the following:

- The aorta, and the iliac, femoral, subclavian, and brachial arteries taper continuously with a constant exponential rate.

- The body is symmetric, in the sense that those vessels existing in both the left and the right side of the body have the same dimensions. Some examples are the arm, from the subclavian artery, the renal arteries, and the iliac and femoral arteries. This symmetry has a computational advantage because the inflows into the daughter arteries are identical, $q_p = 2 q_d$, where q_p is the flow in the parent vessel and q_d is the flow in the daughter vessels.

- The coronary arteries, each with cross-sectional area A_1 and length L, can be lumped into one branch of length L and cross-sectional area $A_2 = \sqrt{2}A_1$. This relation is found by letting the flow in the lumped branch be twice the flow in each coronary artery and assuming Poiseuille flow. The same approach is applied to the intercostal arteries. In this case branches are lumped to reduce the number of arterial segments. There are approximately 10 to 15 intercostal arteries, which are all very small. Modeling each of these arteries separately would increase the computational time considerably. An alternative approach to modeling the intercostal arteries would be to include a continuous outflow function along the thoracic aorta; see, e.g., Anliker, Rockwell, and Ogden (1971) or Stettler, Niederer, and Anliker (1981).

- The renal arteries are modeled as two separate branches. They could be lumped together using the same approach used for the coronary arteries because they too are situated close to each other. However, they have been modeled separately because of the magnitude of their outflow. In order for the numerical computations to converge and for the flow to be fully developed, the distance between them has been increased beyond its natural size.

5.2.2 Structural Properties of the Vessel Walls

Mathematically, the arterial wall can be described by the volume compliance or the elasticity of the vessels. The volume compliance C can be approximated by

$$C = \frac{dV}{dp} \approx \frac{3A_0L}{2}\frac{r_0}{Eh},\tag{5.2}$$

where V is the volume of the given segment, p is the pressure, r_0 is the radius, $A_0 = \pi r_0^2$ is the cross-sectional area, L is the length of the artery, E is Young's modulus, and h is the wall thickness. Thus the elastic properties of the vessels can be found either from the compliance or from Young's modulus.

In this study the elasticity is described from estimates of Young's modulus, the radius, and the wall thickness. In fact, it is possible to demonstrate a relation between Young's modulus, the wall thickness, and the vessel radius. This relation is seen in Figure 5.2, where Eh/r_0 is plotted as a function of r_0. The data plotted in the figure are from Stergiopulos,

Table 5.1. *Data for the length and top and bottom radii for the large arteries. The numbering in the left column refers to the numbers shown in Figure* 5.1.

#	Artery	L [cm]	r_{top} [cm]	r_{bot} [cm]
1	Ascending aorta	1.00	1.525	1.502
3	Ascending aorta	3.00	1.502	1.420
4	Aortic arch	3.00	1.420	1.342
12	Aortic arch	4.00	1.342	1.246
14	Thoracic aorta	5.50	1.246	1.124
15	Thoracic aorta	10.50	1.124	0.924
27	Abdominal aorta	5.25	0.924	0.838
29	Abdominal aorta	1.50	0.838	0.814
31	Abdominal aorta	1.50	0.814	0.792
33	Abdominal aorta	12.50	0.792	0.627
35	Abdominal aorta	8.00	0.627	0.550
36	External iliac	5.75	0.400	0.370
37	Femoral	14.50	0.370	0.314
40	Femoral	44.25	0.314	0.200
38	Internal iliac	4.50	0.200	0.200
39	Deep femoral	11.25	0.200	0.200
2	Coronaries	10.00	0.350	0.300
5	Brachiocephalic	3.50	0.950	0.700
6, 17	R. + L. subclavian	3.50	0.425	0.407
9, 19	R. + L. brachial	39.75	0.407	0.250
10, 21	R. + L. radial	22.00	0.175	0.175
11, 20	R. + L. ulner	22.25	0.175	0.175
8, 18	R. + L. vertebral	13.50	0.200	0.200
7	R. com. carotid	16.75	0.525	0.400
13	L. com. carotid	19.25	0.525	0.400
16	Intercostals	7.25	0.630	0.500
28	Superior mesenteric	5.00	0.400	0.350
22	Celiac axis	2.00	0.350	0.300
23	Hepatic	2.00	0.300	0.250
24	Hepatic	6.50	0.275	0.250
25	Gastric	5.75	0.175	0.150
26	Splenic	5.50	0.200	0.200
30, 32	R. + L. renal	3.00	0.275	0.275
34	Inferior mesenteric	3.75	0.200	0.175

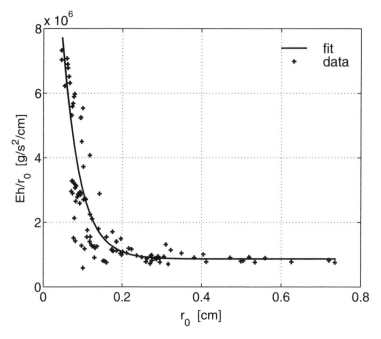

Figure 5.2. *Fit of data for Eh/r_0, where E is Young's modulus, h is the wall thickness, and r_0 is the radius, to the exponential function in (5.3). From M.S. Olufsen (1999), Structured tree outflow condition for blood flow in larger systemic arteries, Am. J. Physiol.* **276**:H257–267. *Used by permission from the American Physiological Society.*

Young, and Rogge (1992). Their data give the volume compliance and are therefore converted to Young's modulus using (5.2). The data for Eh/r_0 are marked with $+$. Through these points a curve of the form

$$\frac{Eh}{r_0} = k_1 \exp(k_2 r_0) + k_3 \tag{5.3}$$

has been fitted using a least squares fit. The parameters for this fit are $k_1 = 2.00 \times 10^7$ g/(s$^2 \cdot$cm), $k_2 = -22.53$ cm^{-1}, and $k_3 = 8.65 \times 10^5$ g/(s$^2 \cdot$cm). The discrepancies between the observations and the fitted curve are due to variations in compliance throughout the body. This variation means that two arteries with the same radius may have different compliance, and hence different values for Eh/r_0, if they belong to different organs.

5.3 Structure of the Small Arteries

The role of the small arteries and arterioles is to distribute blood provided by the large arteries to the capillaries and thus to all muscles and organs. As with the large arteries, the small arteries are arranged in a bifurcating tree. However, for the small arteries tapering of the individual vessels is no longer significant, so each vessel can be modeled as a straight segment.

The total cross-sectional area of the small arterioles in a normal adult is approximately 400 cm^2. This figure should be compared with the total cross-sectional area of the large arteries, which is approximately 20 cm^2, or with that of the aorta, which is approximately 4.5 cm^2. Since the small arterioles have a diameter of approximately 0.003 cm, the arterial tree, ranging from the aorta to the arterioles, will have approximately 26 generations if it is assumed that the tree is binary and symmetric.

The large arteries can be modeled as the tree shown in Figure 5.1. According to this model, the small arteries generally originate 2 to 3 generations from the aorta, iliac, and femoral arteries, e.g., starting at the end of the tibial arteries, the internal or external carotids, or the radial arteries. As a result, they have approximately 24 generations. Dealing with such large trees is impractical unless they are structured in some way. However, there is evidence that the small arteries are distributed in a structured and optimal way; see, e.g., Zamir (1978); Bassingthwaighte, Liebovitch, and West (1994); or Schreiner and Buxbaum (1993).

These studies mostly comprise 2D models of local areas such as the coronary arteries, but Bassingthwaighte, Liebovitch, and West (1994) have suggested that such models could be generalized to be valid for all small arteries. Some of these results have been used to construct a 1D asymmetric structured tree model of the small arteries. In this model the radii of the daughter vessels are scaled linearly, relative to their parent vessel, by factors α and β, so that

$$r_{d_1} = \alpha \, r_p \quad \text{and} \quad r_{d_2} = \beta \, r_p.$$

The subscript p refers to the parent vessel, and the subscripts d_1 and d_2 refer to the two daughter vessels. The structure of the small arteries is shown in Figure 5.3. The coupling of small arteries to large arteries is shown in Figure 5.4.

In order to determine α and β, a relation describing how the geometry (radius or cross-sectional area) changes over a bifurcation is needed. In addition, an area ratio and an asymmetry ratio between the cross-sectional area of the parent and those of its two daughter vessels are needed. As for the large arteries the length L and the compliance for each vessel must also be described. It is not practically possible to determine these parameters individually for each vessel in the structured tree. Thus the possibility of determining these parameters as functions of the vessel radius must be investigated; i.e., data should be found to yield

- a radius relation over the bifurcations,

- area and asymmetry ratios as functions of the vessel radii,

- termination criteria for each of the subtrees,

- the length of the arteries as a function of the vessel radius,

- the wall thickness and Young's modulus as functions of the vessel radius.

Because of the repetitive scaling of daughter arteries with factors α and β, and because all geometric parameters are defined as functions of the vessel radius, the tree is geometrically self-similar.

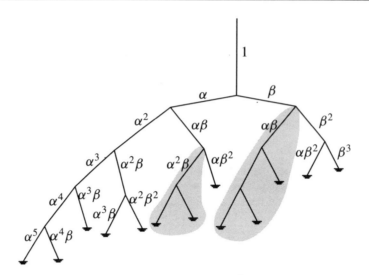

Figure 5.3. *A structured tree. At each bifurcation the radii of the daughter vessels are scaled by a factor of α to the left and β to the right. Because each branch is terminated when the radius is less than some given minimum radius, the tree does not have a fixed number of generations. Shaded areas indicate parts of the tree in which the impedance is already computed. Used with permission of the Biomedical Engineering Society.*

5.3.1 Radius and Asymmetry Relations

The relation determining how the radius changes over an arterial bifurcation was first suggested by Murray (1926a; 1926b) and later generalized by Uylings (1977). They derived a power law based on the principle of minimum work given by

$$r_p^\xi = r_{d_1}^\xi + r_{d_2}^\xi, \tag{5.4}$$

where $\xi \geq 3.0$ corresponds to laminar flow (the power law suggested by Murray used the value $\xi = 3$) and $\xi \leq 2.33$ corresponds to turbulent flow. Based on numerous studies (e.g., Kamiya et al., 1988; Horsfield and Woldenberg, 1989; Pollanen, 1992; Rossitti and Löfgren, 1993; Kassab and Fung, 1995), a value of 2.76 has been chosen. Constructing an asymmetric tree requires some information about the area and asymmetry ratios, i.e., an area ratio relating the cross-sectional area of the sum of the two daughter vessels to the cross-sectional area of the parent vessel and an asymmetry ratio relating the areas of the two daughter vessels. In Zamir (1978) the asymmetry and area ratios are given by

$$\gamma = \left(\frac{r_{d_2}}{r_{d_1}}\right)^2 \quad \text{and} \quad \eta = \frac{r_{d_1}^2 + r_{d_2}^2}{r_p^2} = \frac{1+\gamma}{(1+\gamma^{\xi/2})^{2/\xi}}. \tag{5.5}$$

The above relations give three conditions, (5.4) and (5.5), that characterize the structure of the tree. The parameters ξ, η, and γ are not independent; in (5.5) we find the area ratio

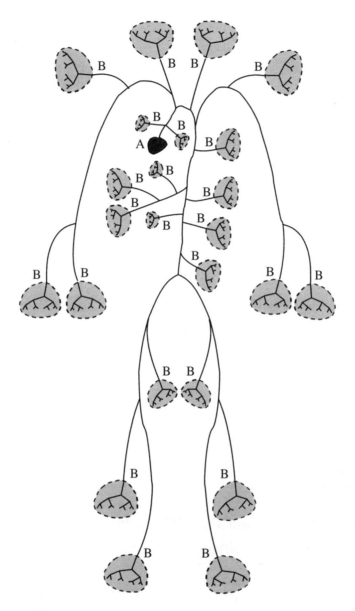

Figure 5.4. *The arterial tree. The tree consisting of the large arteries originates at the heart (marked by A) and terminates at the points marked by B. The structured trees, representing the small arteries, originate at these terminals and provide the main tree with outflow conditions. From M.S. Olufsen (1999), Structured tree outflow condition for blood flow in larger systemic arteries,* Am. J. Physiol. **276**:H257–267. *Used by permission from the American Physiological Society.*

$\eta(\beta, \gamma)$. Hence, the scaling parameters α and β can be determined from ξ and γ:

$$\alpha = \left(1 + \gamma^{\xi/2}\right)^{-1/\xi} \quad \text{and} \quad \beta = \alpha\sqrt{\gamma}. \tag{5.6}$$

Using $\xi = 2.76$ as mentioned above and $\eta = 1.16$ adapted from studies by Uylings (1977), Rossitti and Löfgren (1993), and Papageorgiou et al. (1990) yields an asymmetry ratio $\gamma = 0.41$ and scaling parameters $\alpha = 0.9$ and $\beta = 0.6$.

5.3.2 Order of the Structured Tree

As mentioned earlier, the small arteries and arterioles make up a tree with approximately 24 generations. Including a fixed number of generations with an asymmetric tree would yield a significant variation in the diameters at the terminal branches. This variation does not reflect actual physiology; arterioles for a given muscle or organ have approximately the same diameter in order to ensure an even blood supply. Therefore, the structured tree model is terminated when the radius of the terminal vessels becomes less than some given minimum radius.

5.3.3 Length of Segments

A number of papers offer suggestions for how to estimate the physical lengths of the various arterial segments. In this work the ideas suggested by Iberall (1967) have been followed. Assume that

$$L/d \approx 25 \pm 5, \tag{5.7}$$

where L is the length and d is the diameter of the vessel. This relation is extrapolated from measurements by Suwa et al. (1963), who found that L/d tends to be constant for $d \in [20; 4000]\,\mu\text{m}$, and Patel, DeFreitas, and Greenfield (1963), who found that L/d was constant for $d \in [2; 20]\,\text{mm}$. The constant length-to-diameter (or similarly length-to-radius) relationship is also repeated in West, Bhargava, and Goldberger (1986).

5.3.4 Wall Thickness and Young's Modulus

If the expressions relating the wall thickness h and Young's modulus E to the radius r_0 of the blood vessel can be fixed, it is possible to determine the compliance solely as a function of the radius r_0.

This is exactly what we did for the large arteries; see (5.3) and Figure 5.2. The relation in (5.3) does not necessarily apply to the small arteries because it is not possible to extrapolate the interpolating function outside its domain. The relation in (5.3) is extracted from vessels with a radius that is bigger than 0.05 cm, and as defined earlier, the smallest arterioles have a radius of approximately $100\,\mu\text{m}$, which is an order of magnitude smaller. Furthermore, one has to be careful because the walls of the small arteries and arterioles have a different composition of layers than the large arteries. Very few observations exist regarding the wall thickness and Young's modulus specifically in small arteries. However, this relation has been chosen in spite of these caveats.

5.4 Fluid Dynamic Model of a Large Artery

In this and the following sections the 1D model of blood flow and pressure in the large systemic arteries is derived, i.e., a model in which flow and pressure are determined as functions of time and one spatial component, along the vessels. A model including all fluid dynamic properties of the large arteries should take into account all three spatial dimensions. However, in most clinical situations the pressure is measured only at discrete locations in one dimension, and it is our purpose to derive a model that can be used to predict phenomena seen using such measurements. Furthermore, it can be shown that it is reasonable to simplify the model to one spatial dimension when the flow studied is away from bifurcations and stenosis, i.e., when the measurements are taken away from areas where vortices are created. Finally, restricting the model to one spatial dimension makes it computationally feasible.

A 1D model can be built using three equations describing the motion of fluid in a given vessel (in the longitudinal direction, here in the x direction), the motion of the vessel walls, and the interaction between fluid and the walls. These three equations are the Navier–Stokes equations, which describe flow and pressure of the fluid; a continuity equation, which ensures that what flows in also flows out; and a state equation, which relates the fluid influence of the vessel wall (through the pressure) to the elastic properties of the vessel wall (through the cross-sectional area).

In order to simplify the 3D problem to a 1D model, further assumptions must be made. First, assume that all vessels are circular and that the flow is axisymmetric. Having an axisymmetric flow in a circular vessel eliminates the third dimension from the model, leaving a 2D flow. Second, assume that the velocity profile across the vessels is flat, with a thin boundary layer providing some friction to the system. Specifying the velocity profile enables us to integrate over the cross-sectional area and eliminate the second dimension from the model. As a result, a 1D model is obtained. Finally, we assume that the large arteries consist of a binary tree of compliant and tapering vessels containing an inviscid, incompressible, Newtonian fluid.

The derivation presented in this chapter is based on the integral form of the Navier–Stokes equations. We believe these equations provide more insight into and understanding of the simplifications made in the derivation. In section 5.4.1 the momentum and continuity equations are derived. Then in section 5.4.2 the state equation is discussed.

5.4.1 Momentum and Continuity Equations

The derivation presented in this section is primarily based on the work by Barnard et al. (1966) and Peskin (1976). A blood vessel can be regarded as a rotationally symmetric tubular elastic surface S with end surfaces in the planes $x = 0$ and $x = L$; see Figure 5.5. Furthermore, assume the following:

- S moves with velocity $\mathbf{v} = (v_x, v_r, v_\theta)$, where x is the longitudinal coordinate and the polar coordinates are r and θ.

- S encloses a volume V, which is filled with an incompressible fluid moving with velocity $\mathbf{u} = (u_x, u_r, u_\theta)$. Both u_i and v_i are functions of (x, r, θ, t). Note that the surface S does not in general move with the same velocity as the fluid; i.e., $\mathbf{u}$ and $\mathbf{v}$ are generally different.

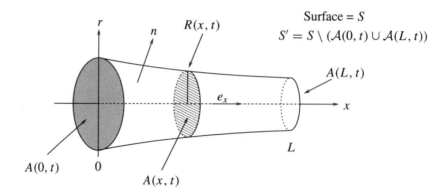

Figure 5.5. *A typical vessel.*

- $\rho = 1.055$ g/cm^3 is the constant density of the fluid.

- $\mu = 0.049$ g/(cm·s) is the constant viscosity of the fluid.

- $p(x, r, \theta, t)$ is the pressure in the fluid.

- $r_0(x)$ is the vessel radius at zero transmural pressure, i.e., when $p = p_0$. As discussed in section 5.2.1, it is assumed that the arteries taper exponentially, as described in (5.1).

- $R(x, t)$ is the radius of the vessel.

- $A(x, t) = \pi R(x, t)^2$ is the cross-sectional area; i.e., the end surfaces of S are $A(0, t)$ and $A(L, t)$, respectively.

- $\mathcal{A}(x, t) = \{(r, \theta) : 0 \leq r \leq R(x, t), 0 \leq \theta \leq 2\pi\}$ is the collection of points in the plane x at time t. The area of $\mathcal{A}(x, t)$ is $A(x, t)$.

- $S' = S \setminus (\mathcal{A}(0, t) \cup \mathcal{A}(L, t))$.

- $\mathbf{v} = 0$ on $\mathcal{A}(0, t)$ and $\mathcal{A}(L, t)$.

- $\mathbf{n}$ is the outward unit normal to the surface S.

- $\mathbf{e}_x, \mathbf{e}_r, \mathbf{e}_\theta$ are the unit vectors in the x, r, θ directions.

Flow through such a vessel can be described by two laws: conservation of volume and conservation of momentum. Conservation of volume ensures that the amount of blood flowing into the vessel during a small period dt flows out of the vessel either at the end of the vessel or through the vessel wall. As a result, it gives rise to the following equation comprising two terms.

Conservation of Volume:

$$\frac{\partial}{\partial t} \iiint_V dV + \iint_S (\mathbf{u} - \mathbf{v}) \cdot \mathbf{n} \, dA = 0$$

$$\Leftrightarrow \frac{\partial}{\partial t} \int_0^L A \, dx + \left[\iint_A u_x \, dA \right]_0^L + \iint_{S'} (\mathbf{u} - \mathbf{v}) \cdot \mathbf{n} \, dA = 0. \tag{5.8}$$

The second law, conservation of momentum (Navier–Stokes equations (5.9)), ensures that Newton's second law (mass times acceleration equals force) is fulfilled. To predict blood flow and pressure along vessels, momentum must be conserved in the longitudinal direction (the x-direction). The resulting equation has two parts. The first part (mass times acceleration) gives rise to the first two integrals in (5.9): transport of mass along the vessel and transport through the vessel wall. The second part of Newton's second law (force) gives rise to the last integral in (5.9). This integral comprises a pressure term describing blood pressure on the vessel wall and a term describing the friction between the vessel wall and the fluid. This ensures the so-called no-slip condition and will be included through a shear tensor. The shear tensor is part of the last surface integral of the momentum equation (5.10).

Conservation of Momentum:

$$\frac{\partial}{\partial t} \iiint_V \rho u_x \, dV + \iint_S \rho u_x (\mathbf{u} - \mathbf{v}) \cdot \mathbf{n} \, dA + \iint_S (p(\mathbf{n} \cdot \mathbf{e}_x) - (\mathbf{dn}) \cdot \mathbf{e}_x) \, dA = 0$$

$$\tag{5.9}$$

$$\Leftrightarrow \frac{\partial}{\partial t} \int_0^L \left(\iint_A \rho u_x \, dA \right) dx + \left[\iint_A \rho u_x^2 \, dA \right]_0^L + \iint_{S'} \rho u_x (\mathbf{u} - \mathbf{v}) \cdot \mathbf{n} \, dA$$

$$+ \int_0^L \left(\iint_A \frac{\partial p}{\partial x} \, dA \right) dx - \int_0^L \left(\int_0^{2\pi} (\mathbf{dn}) \cdot \mathbf{e}_x \, R \, d\theta \right) \sqrt{1 + \left(\frac{\partial R}{\partial x} \right)^2} \, dx = 0. \tag{5.10}$$

Here $\mathbf{d}$ is a tensor representing the shear stresses (the friction), also called the deviatoric stress tensor; see Ockendon and Ockendon (1995). The expansion of the last integral can be explained from Figure 5.6. The surface stress tensor for incompressible flow is given by Ockendon and Ockendon (1995) as

$$\sigma_{ij} = -p\delta_{ij} + d_{ij},$$

where $-p\delta_{ij}$ is the isotropic part, as would exist in an inviscid fluid, and d_{ij} is the deviatoric part, which is due to the viscous forces in the fluid. The isotropic part of the tensor $-p\delta_{ij}$ is already incorporated in the fourth term of (5.10). Assuming that $u_\theta = 0$, i.e., that there is

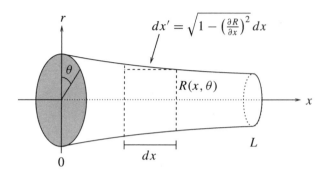

Figure 5.6. *The surface integral in (5.12) can be split into two parts. The first integrates over the circumference with radius R for $\theta \in [0 : 2\pi]$. The second integrates over the length of the vessel, $x \in [0 : L]$. Since the vessel is tapering, each infinitesimal piece is given by dx'.*

no swirl, then, according to Batchelor (1992), the deviatoric part of the surface stress tensor is given by

$$
d = 2\mu \begin{bmatrix} \frac{\partial u_x}{\partial x} & \frac{1}{2}\left(\frac{\partial u_r}{\partial x} + \frac{\partial u_x}{\partial r}\right) & 0 \\[2mm] \frac{1}{2}\left(\frac{\partial u_r}{\partial x} + \frac{\partial u_x}{\partial r}\right) & \frac{\partial u_r}{\partial r} & 0 \\[2mm] 0 & 0 & \frac{u_r}{r} \end{bmatrix}_R
$$

Therefore,

$$
(dn) \cdot \mathbf{e}_x = \left[\mu\left(\frac{\partial u_r}{\partial x} + \frac{\partial u_x}{\partial r}\right) \right]_R \tag{5.11}
$$

If the vessels were not tapering, then $u_r = 0$ and $\partial u_r/\partial x = 0$. Assuming that the tapering is small, this term can be neglected. The last integral in (5.10) can thus be written as

$$
\iint_S (dn) \cdot \mathbf{e}_x \, dA = \int_0^L \left(\int_0^{2\pi} (dn) \cdot \mathbf{e}_x \, R \, d\theta \right) \sqrt{1 + \left(\frac{\partial R}{\partial x}\right)^2} \, dx
$$

$$
= \int_0^L 2\pi \mu R \left[\frac{\partial u_x}{\partial r}\right]_R \sqrt{1 + \left(\frac{\partial R}{\partial x}\right)^2} \, dx. \tag{5.12}
$$

The first equality is explained in Figure 5.6. The assumption of a small tapering factor also justifies neglecting $\partial R/\partial x$. The wall shear stress can be found from

$$
\int_0^L 2\pi \mu R \left[\frac{\partial u_x}{\partial r}\right]_R \, dx. \tag{5.13}
$$

In order to be able to write the equations in a simpler form, $\Psi(x)$ is introduced to be the outflow of volume V, and $\Psi_P(x)$ to be the outflow of momentum (both per unit length); i.e.,

$$\int_0^L \Psi\, dx = \iint_{S'} (\mathbf{u} - \mathbf{v}) \cdot \mathbf{n}\, dA \quad \text{and} \quad \int_0^L \Psi_P\, dx = \iint_{S'} \rho u_x (\mathbf{u} - \mathbf{v}) \cdot \mathbf{n}\, dA.$$

This notation is convenient since both Ψ and Ψ_P vanish when the normal component of the surface velocity equals that of the fluid (when $(\mathbf{u} - \mathbf{v}) \cdot \mathbf{n} = 0$); i.e., all outflow from the system occurs through the end surfaces $A(0)$ and $A(L)$. Ψ and Ψ_P could be used to model outflow through a series of small branches such as the intercostal arteries. Inserting (5.13) in (5.10) and differentiating (5.8) and (5.10) with respect to L, and replacing L with x throughout, gives

$$\frac{\partial A}{\partial t} + \frac{\partial}{\partial x} \iint_A u_x\, dA + \Psi = 0, \tag{5.14}$$

$$\frac{\partial}{\partial t} \iint_A \rho u_x\, dA + \frac{\partial}{\partial x} \iint_A \rho u_x^2\, dA + \iint_A \frac{\partial p}{\partial x}\, dA - 2\pi\mu R\left[\frac{\partial u_x}{\partial r}\right]_R + \Psi_P = 0. \tag{5.15}$$

These equations describe flow in the vessel shown in Figure 5.5. They do not constitute a 1D theory since the velocity distribution over the cross-sectional area appears in the equations.

In order to arrive at the 1D equations, the spatial dependence must be eliminated. To do so assume that p and thus $\partial p/\partial x$ are functions of x and t only, i.e., that pressure is constant over the entire cross-sectional area. Then the average velocity over the cross-sectional area and χ are defined as

$$u = \frac{1}{A} \iint_A u_x\, dA \quad \text{and} \quad \chi = \frac{1}{Au^2} \iint_A u_x^2\, dA.$$

Using these definitions, (5.14) and (5.15) can be written as

$$\Psi_P = \rho u \Psi,$$

$$\frac{\partial A}{\partial t} + \frac{\partial (Au)}{\partial x} + \Psi = 0, \tag{5.16}$$

$$\rho\left(\frac{\partial(Au)}{\partial t} + \frac{\partial(\chi Au^2)}{\partial x}\right) + A\frac{\partial p}{\partial x} - 2\pi\mu R\left[\frac{\partial u_x}{\partial r}\right]_R + \Psi_P = 0. \tag{5.17}$$

These equations constitute a 1D model, but, in order to solve them, a specific relation for u_x is needed. For laminar flow in slightly tapering vessels the velocity profile is rather flat (McDonald, 1974; Pedersen, 1993). Assume

$$u_x = \begin{cases} u & \text{for } r \le R - \delta, \\ u\,(R - r)/\delta & \text{for } R - \delta < r \le R, \end{cases}$$

where δ is the thickness of the boundary layer. According to Lighthill (1975), the thickness of the boundary layer for the large arteries can be estimated from $(\nu/\omega)^{1/2} = (\nu T/(2\pi))^2 \approx 0.01$ cm, where $\nu = \mu/\rho = 0.046$ cm^2/s is the kinematic viscosity, ω is the angular frequency, and $T = 1.25$ s is the length of the cardiac cycle. Measurements show that the velocity profile changes throughout the arterial system from being almost flat in the aorta to a more parabolic shape in the peripheral arteries (McDonald, 1974; Pedersen, 1993; Pedersen et al., 1993). This is a result of the fact that the boundary layer remains constant (1 mm) as the vessels get smaller. Using the above relation gives

$$2\pi\mu R \left[\frac{\partial u_x}{\partial r}\right]_R = -\frac{2\pi\mu u R}{\delta} \quad \text{and} \quad \chi = 1 - \frac{4\delta}{3R} + \frac{\delta^2}{2R^2} \approx 1. \tag{5.18}$$

Note that the last approximation applies only if the boundary layer is thin compared with the vessel radius.

Both of these conditions are derived from an approximation of steady flow. However, blood flow in real arteries is not steady, so the wall shear stresses should be investigated in further detail. In fact there are papers accounting for the wall shear stress in a more detailed way, e.g., Olsen and Shapiro (1967), Wemple and Mockros (1972), Schaaf and Abbrecht (1972), and Stergiopulos, Young, and Rogge (1992). In these, the shear stress is defined as a combination of two terms: one accounting for the steady part of the flow and one for the unsteady part. The first is found from assuming Poiseuille flow and the second from assuming a sinusoidally driven flow in a long straight rigid vessel.

Inserting the conditions arising from the assumption of a flat velocity profile and dividing by ρ gives

$$\Psi_{\tilde{p}} = u\Psi,$$

$$\frac{\partial A}{\partial t} + \frac{\partial(Au)}{\partial x} + \Psi = 0,$$

$$\frac{\partial(Au)}{\partial t} + \frac{\partial(Au^2)}{\partial x} + \frac{A}{\rho}\frac{\partial p}{\partial x} + \frac{2\pi\nu u R}{\delta} + \Psi_{\tilde{p}} = 0.$$

Finally, the equations can be rewritten in terms of the flow $q = Au$, with $\Psi_{\tilde{p}}$ replaced with $q/A\Psi$:

$$\frac{\partial A}{\partial t} + \frac{\partial q}{\partial x} + \Psi = 0, \tag{5.19}$$

$$\frac{\partial q}{\partial t} + \frac{\partial}{\partial x}\left(\frac{q^2}{A}\right) + \frac{A}{\rho}\frac{\partial p}{\partial x} + \frac{2\pi\nu q R}{\delta A} + \Psi\frac{q}{A} = 0. \tag{5.20}$$

These equations cannot be solved analytically, so a numerical method is called for. An easy method to implement is Richtmeyer's version of Lax–Wendroff's two-step method. This method requires that the equations be written in conservation form. For details on the numerical method see Olufsen et al. (2000). In order to rewrite the equations in conservation form, it is an advantage to introduce the quantity B chosen to fulfill

$$B(r_0(x), p(x, t)) = \frac{1}{\rho}\int A\,dp. \tag{5.21}$$

As a result,

$$\frac{\partial B}{\partial x} = \frac{A}{\rho} \frac{\partial p}{\partial x} + \frac{\partial B}{\partial r_0} \frac{dr_0}{dx}.$$

Since the last term does not contain any partial derivatives of p, and hence of A and q, it can be evaluated directly and may therefore be added to both sides of (5.20). The momentum equation (5.20) can thus be rewritten as

$$\frac{\partial q}{\partial t} + \frac{\partial}{\partial x} \left(\frac{q^2}{A} + B \right) = -\frac{2\pi \nu q R}{\delta A} - \Psi \frac{q}{A} + \frac{\partial B}{\partial r_0} \frac{dr_0}{dx}. \tag{5.22}$$

Equations (5.19) and (5.20) or (5.22) constitute the basic equations for the 1D theory for the wave propagation of the arterial pulse. There are two equations and three dependent variables, namely, p, q, and A. Therefore, a third relation is needed, the so-called state equation. This equation is based on compliance of the vessels and gives an equation for pressure as a function of cross-sectional area.

5.4.2 State Equation

The arterial wall is not purely elastic but exhibits a viscoelastic behavior (Caro et al., 1978; McDonald, 1974; Rockwell, Anliker, and Elsner, 1974). However, in order to keep the model simple, viscoelasticity is disregarded and a state equation is derived from the linear theory of elasticity. This is a reasonable assumption because the viscoelastic effects are small within physiological ranges of flow and pressure (Tardy et al., 1991). To develop a relation between the pressure in the vessel and the cross-sectional area, the equilibrium between the internal and external forces acting on a unit element of the wall should be studied. Assume that the arterial vessels are circular, that the walls are thin (i.e., that $h/r \ll 1$ and that the loading and deformation are axisymmetric), and that the vessels are tethered in the longitudinal direction. In this case the external forces are reduced to stresses acting only in the circumferential direction (Atabek, 1968), and from what is often known as Laplace's law the circumferential tensile stress can be obtained:

$$\tau_\theta = \frac{rp_e}{h} = \frac{E_\theta}{1 - \sigma_x \sigma_\theta} \frac{r - r_0}{r_0},$$

where $p_e = p - p_0$ is the excess pressure, i.e., the pressure of the vessel minus the pressure of the surroundings; h is the wall-thickness; $(r - r_0)/r_0$ is the corresponding circumferential strain; E_θ is Young's modulus in the circumferential direction; $\sigma_\theta = \sigma_x = 0.5$ are the Poisson ratios in the circumferential and longitudinal directions; and r_0 is the radius at zero transmural pressure, i.e., at $p = p_0$. Because it was assumed that the vessel is tethered in the longitudinal direction, this is the only contribution obtained when balancing the internal and external forces. And, without loss of generality, it is possible to drop the θ subscript on E. Solving for p_e gives

$$p(x, t) - p_0 = \frac{4}{3} \frac{Eh}{r_0(x)} \left(1 - \sqrt{\frac{A_0(x)}{A(x, t)}} \right), \tag{5.23}$$

where $A_0(x) = \pi r_0(x)^2$ is the cross-sectional area at zero transmural pressure. As shown in Figure 5.2, it is possible to estimate the relation (5.3) between Young's modulus, the wall thickness, and the vessel radius.

5.5   Flow and Pressure in the Tree of Large Arteries

The model derived in the previous sections predicts blood flow and pressure for a single vessel segment. In order to extend this model to the arterial tree shown in Figure 5.1, some appropriate boundary conditions linking the vessels together must be established. Three types of boundaries can be identified: one at the inflow into the aorta from the left ventricle, one at each bifurcation, and one at the terminals.

The system of equations is hyperbolic with a positive wave propagation velocity much larger than the velocity of the blood. Thus, the characteristics will cross and have opposite directions. This means that one boundary condition is needed at each end of the vessel. Consequently, the following boundary conditions for the arterial tree must be derived:

- One equation at the inlet to the arteries, i.e., at the aortic valve. This inlet is marked with an A in Figure 5.1.

- Three equations at the bifurcations: an outflow condition from the parent artery and inflow conditions to both of the daughter arteries.

- One equation specifying the outflow from each of the terminal vessels of the arterial tree. These are marked with B's in Figure 5.1.

Each of these should be specified by an equation for flow, pressure, or a relation between them.

In order to describe these boundary conditions, a tree with only three branches is considered; see Figure 5.7. The inflow appears at In and the outflows at Out, and the bifurcation conditions are applied at Bif. In general, when modeling the arterial tree, the branches will be labeled as shown in Figure 5.1.

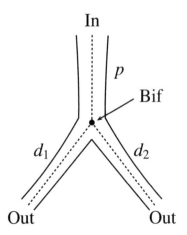

Figure 5.7. *A single binary branch consisting of a parent and two daughters. Even though they are depicted as symmetric they do not need to be so.*

5.5.1 Inflow Condition

At the inflow the flow, the pressure, or a relation between the flow and pressure must be specified. Since the shape of the pulse wave in the upper ascending aorta is generated by the inflow from the aortic valve, the inflow is represented by a periodic extension of a measured flow wave; see Figure 5.8. This curve is measured in the upper ascending aorta using MR (Kim, 1996–1998). The measured curve is modified in a number of ways. First, it is made periodic such that $q(0, 0) = q(0, T)$, where $T = 1.25$ s is the length of the cardiac cycle. Second, it is smoothed to avoid too many high frequency oscillations in our simulated results. The latter is done by fitting a cubic spline through the measured data. Finally, the curve is scaled such that the cardiac output is reduced from 4.14 l/min to 4.03 l/min, a reduction of 3%.

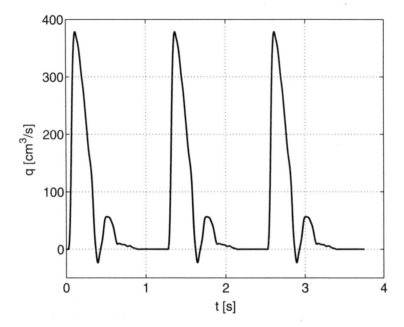

Figure 5.8. *The inflow as a function of time over three cardiac cycles of length 1.25 s. From M.S. Olufsen (1999), Structured tree outflow condition for blood flow in larger systemic arteries,* Am. J. Physiol. **276**:H257–267. *Used by permission from the American Physiological Society.*

This scaling is necessary because not all branches in our arterial model are included, and as a result there is some blood not included in the model.

5.5.2 Bifurcation Conditions

Assuming that the bifurcation takes place at a point (at Bif in Figure 5.7), three conditions are needed to close the system of equations: an outflow condition from the parent vessel (marked by p in Figure 5.7) and an inflow condition to each of the daughter vessels (marked by d_1 and d_2 in Figure 5.7). Assuming that there is no leakage of blood at the bifurcations, i.e., that the in- and outflows are balanced, and that the pressure is continuous over the bifurcation, these conditions give rise to the following equations:

$$q_p = q_{d_1} + q_{d_2} \quad \text{and} \quad p_p = p_{d_1} = p_{d_2}. \tag{5.24}$$

The assumption that the pressure is continuous over the bifurcation can be discussed. If the flow were modeled in detail, this condition would be more complex. If the flow were steady, it would be possible to apply Bernoulli's law, and an exact relation including both flow and pressure could be determined. And at a boundary where the total cross-sectional area decreases as one proceeds downstream one would, according to Bernoulli's law, expect an increase in pressure associated with the increase in velocity. In the arterial system, however, the total cross-sectional area typically increases at junctions (again, proceeding downstream toward the periphery), and hence a decrease in pressure is to be expected. On the other hand, because the flow is unsteady, the change in area at the junction is discontinuous, and, as a result, flow separation and vortex formation are expected just downstream from the bifurcation and Bernoulli's law does not apply. Therefore, in this circumstance, which invokes dissipation of kinetic energy, it seems more appropriate to use pressure continuity. Alternative models of the bifurcations are discussed in Anliker, Rockwell, and Ogden (1971), Stettler, Niederer, and Anliker (1981), Lighthill (1975), Olufsen and Ottesen (1995a), and Olufsen et al. (2000).

5.5.3 Outflow Condition

The outflow condition can be determined in several ways. Three approaches will be discussed in the following: a pure resistance model, a Windkessel model, and our structured tree model.

   The simplest reasonable approach is to let the outflow be proportional to the pressure, i.e., to let the boundary condition be determined by a pure resistive load. This approach has commonly been used in previous studies (Olsen and Shapiro, 1967; Schaaf and Abbrecht, 1972; Anliker, Rockwell, and Ogden, 1971; Streeter, Keitzer, and Bohr, 1963; Forbes, 1981; Stettler, Niederer, and Anliker, 1981).

   It is not obvious how to choose the correct value for the peripheral resistance at points where the large arteries terminate. Furthermore, assuming a constant relation between flow and pressure at the downstream boundary, flow and pressure are forced to be in phase, which is generally not physiologically valid in these relatively large arteries. This is pointed out in Anliker (1977), where it is noted that a boundary condition based on a pure resistance condition applies only if the arteries are sufficiently small. The inconsistency in having a pure resistive load as a boundary condition can be illustrated from the hysteresis curves appearing in phase plots of p versus q, for a fixed x; see Figure 5.14 in section 5.8. The forced in-phase condition propagates back through the vessel, changing the overall slope and narrowing the width of the hysteresis. This change in width means that the phase is disturbed throughout the vessel. Since there are some reflections in the system (just enough to produce the dicrotic wave), a small change in the hysteresis curves is expected, but not as drastic as the one appearing with this boundary condition. In order to avoid these problems, the boundary condition should incorporate a phase shift between p and q.

   Another approach to determining the outflow conditions is to set up a model based on the basic properties of the impedance. This approach was implemented by Wemple and

Mockros (1972), Raines, Jaffrin, and Shapiro (1974), and Stergiopulos, Young, and Rogge (1992), among others. They derive an outflow condition by attaching a three-element Windkessel model at the boundary. Such a model represents the resistance and elasticity of the vessels by an electrical circuit model consisting of a resistance in series with a parallel combination of a resistance and a capacitor; the resistances $R_1 + R_2$ simulate the total resistance and the capacitor C_T simulates the compliance of the vascular bed. This circuit is shown in Figure 5.9. The frequency dependent impedance of the Windkessel model is given by

$$Z(0, \omega) = \frac{R_1 + R_2 + i\omega C_T R_1 R_2}{1 + i\omega C_T R_2},\qquad(5.25)$$

where the parameters C_T, R_1, and R_2 are the volume compliance and resistances shown in Figure 5.9. In order to apply the boundary condition, three parameters must be specified. These are the total peripheral resistance $R_T = R_1 + R_2$, the fraction R_1/R_T, and the total compliance C_T. Such parameter values can be found in Stergiopulos, Young, and Rogge (1992).

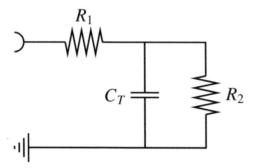

Figure 5.9. *The Windkessel element model used for predicting the impedance at the terminals of the large arteries. The resistances R_1 and R_2 and the capacitance C_T must be estimated for each of the terminal vessels.*

Such a model cannot include wave propagation effects in the part of the arterial system that it models. As a result, it will not be able to capture the phase lag between p and q adequately. Therefore, it is necessary to investigate how the physical domain extends beyond the boundary of the large arteries. As discussed in section 5.3, the small arteries make up a large asymmetric tree with a varying number of generations. One can, as discussed earlier, count approximately 24 generations in this tree before the arteriolar level is reached. After 24 generations of branching (at the arteriolar level), all vessels have approximately the same diameter. Beyond this point loops are formed and the structure becomes too complex to easily describe the geometry. In section 5.3 it was shown that the small arteries can be modeled as an asymmetric structured binary tree, and in section 5.6 we will derive a semianalytical solution for flow and pressure in the structured tree.

It would be too comprehensive to compute the full nonlinear model of such a tree. A more appropriate strategy is to describe flow and pressure in these small arteries with a simpler model that can be solved analytically, e.g., a linear model. From these subtrees of the small arteries it is possible to obtain a boundary condition for the system of nonlinear equations as a time dependent relation between flow and pressure. These subtrees of small arteries are treated as a structured tree of straight vessels in which the corresponding linear equations are solved. The geometry for this model was described in section 5.3; a model for the fluid flow and pressure through these small arteries will be discussed in section 5.6. From these solutions it is possible, using Fourier analysis, to determine a dynamic impedance, which can be used to get a relation between flow and pressure.

The three boundary conditions discussed here, the pure resistance model, the Windkessel model, and the structured tree model, will be compared in section 5.8.2. For any Fourier model, the frequency dependent impedance $Z(x, \omega)$ (e.g., obtained using the structured tree model; see sections 5.6 and 5.7 as well as Appendix B) can be related to flow and pressure by

$$P(x, \omega) = Z(x, \omega) Q(x, \omega), \tag{5.26}$$

where the terminology of electrical networks has been used, with P playing the role of voltage and Q the role of current.

Because the inflow condition is periodic, it is assumed that flow and pressure can be expressed using a complex periodic Fourier series. Then any feature of the system response can be determined separately for each term. Let

$$p(x, t) = \sum_{k=-\infty}^{\infty} P(x, \omega_k) \, e^{i\omega_k t} \quad \text{and} \quad q(x, t) = \sum_{k=-\infty}^{\infty} Q(x, \omega_k) \, e^{i\omega_k t}, \tag{5.27}$$

where $\omega_k = 2\pi k/T$ is the angular frequency and

$$P(x, \omega_k) = \frac{1}{T} \int_{-T/2}^{T/2} p(x, t) \, e^{-i\omega_k t} dt \quad \text{and} \quad Q(x, \omega_k) = \frac{1}{T} \int_{-T/2}^{T/2} q(x, t) \, e^{-i\omega_k t} dt. \tag{5.28}$$

By an inverse Fourier transform the results for $Z(x, \omega)$ can then be transformed to obtain $z(x, t)$. Using the convolution theorem, it is possible to arrive at an analytic relation between p and q:

$$p(x, t) = \int_{t-T}^{t} q(x, \tau) z(x, t - \tau) \, d\tau. \tag{5.29}$$

This is our new outflow condition for the large arteries; see Figure 5.4. These conditions should be evaluated at each of the terminals marked by B in Figure 5.1, i.e., at $x = L_i$, where L_i is the length of the ith terminal segment.

5.6 Fluid Dynamic Model of a Small Artery

In this section the system of equations describing flow and pressure in the small arteries is presented. As with the large arteries, three equations are needed: a momentum equation, a continuity equation, and a state equation. The momentum equation is derived by combining the axisymmetric Navier–Stokes equations for flow in an elastic cylinder with a wall equation balancing the forces of the elastic wall with those acting on the fluid inside. The continuity equation is the same as for the large arteries, but is directly combined with a linearized version of the state equation; see section 5.4.2.

There are two reasons why we cannot directly use the model derived for the large arteries (in sections 5.4 and 5.5). First, solving a nonlinear model for such a large tree is not computationally feasible (see section 5.3.2), and second, it does not model the wall shear stresses in a satisfactory way. Therefore, a model that is linear and at the same time describes the boundary layer in more detail is needed. In the model of the large arteries the wall shear stress was accounted for by assuming a particular velocity profile. This assumption for the wall shear stress gave rise to a friction term on the right-hand side of the equation; see section 5.4.1. For the small arteries viscosity will be included in the momentum equations, and the viscous terms will not be simplified until very late in the derivation. Keeping the viscous terms is important since the wall shear stress is more dominant in the small arteries and thus cannot be dealt with in the approximate way used for the large arteries. Our derivation of a momentum equation for the small arteries will be based on the linearized model originally proposed by Womersley (1957) and later extended by Atabek and Lew (1966), Atabek (1968), Pedley (1980), and Lighthill (1975).

The general model assumptions are similar to those for the large arteries, namely, that the fluid is incompressible, viscous, and Newtonian. In order to set up the momentum equation one must study

- the motion of the fluid,

- the motion of the vessel walls, and

- the interaction between the fluid and the walls.

The derivation is long and tedious and is therefore left out of this section, but the interested reader can find it in Appendix B.

5.6.1 Momentum Equation

For the large arteries the average flow over the cross-sectional area of the vessel must be determined. This is given by

$$Q = \int_0^a w \, 2\pi r \, dr, \tag{5.30}$$

where w_r is the longitudinal velocity. This equation is derived in (B.67) in Appendix B. The longitudinal velocity is found by solving a Bessel equation (established by linearization of the axisymmetric Navier–Stokes equations) combined with equations describing the motion of the vessel walls and the interaction between the fluid and the walls. It is given by

$$w = \frac{p_c k'}{c_0 \rho} \left(1 - \frac{J_0 \left(r w_0 / a \right)}{J_0(w_0)} \right), \tag{5.31}$$

where a is the radius of the vessel and $r \in [0 : a]$, $\rho = 1.055$ g/cm^3 is the density of blood, p_c is an integration constant, $w_0^2 = i^3 w^2$ and $w = a^2 \omega / \nu$ is the Womersley number, ω is the angular frequency, and ν is the kinematic viscosity. $k' = c_0 / c$ is a complex wave propagation velocity, where $c_0^2 = Eh/(2a\rho)$ is known as the Moens–Korteweg wave propagation velocity, and, finally, $J_0(x)$, $J_1(x)$ are the zeroth and first order Bessel functions. Integrating over the cross-sectional area yields

$$Q = \frac{A_0 p_c k'}{c_0 \rho}(1 - F_J), \tag{5.32}$$

where $A_0 = \pi a^2$ is the cross-sectional area of the vessel, and

$$F_J(\mathrm{w}) = \frac{2 J_1(\mathrm{w}_0)}{\mathrm{w}_0 J_0(\mathrm{w}_0)} \approx \begin{cases} 2/(\mathrm{w} i^{1/2}) \left[1 + (2\mathrm{w})^{-1} + \mathcal{O}(\mathrm{w}^{-2}) \right] & \text{for } \mathrm{w} \geq 3, \\ 1 - i(\mathrm{w}^2/8) - (\mathrm{w}^4/48) + \mathcal{O}(\mathrm{w}^6) & \text{for } \mathrm{w} \leq 2. \end{cases}$$

For $2 < \mathrm{w} < 3$ a linear interpolation between the two values is used. This interpolation is needed since the function is derived in Appendix B (B.64) as an approximation for $\mathrm{w} \to 0$ and $\mathrm{w} \to \infty$, so the approximations cannot be made continuous without inserting the linear interpolation. This is a slight modification of the suggestions by Pedley (1980) to use the top approximation for $\mathrm{w} > 4$ and the bottom one for $\mathrm{w} < 4$.

Furthermore, it is known that the pressure gradient (B.68), derived in Appendix B, is

$$\frac{-i\omega p_c}{c} = \frac{\partial P}{\partial x}. \tag{5.33}$$

The final momentum equation can now be obtained by inserting (5.33) into (5.32):

$$i\omega Q = \frac{-A_0}{\rho} \frac{\partial P}{\partial x}(1 - F_J). \tag{5.34}$$

5.6.2   Continuity and State Equations

The 1D continuity equation for the small arteries is the same as for the large arteries, namely,

$$\frac{\partial A}{\partial t} + \frac{\partial q}{\partial x} = 0.$$

Using the Fourier series expansions from (5.27), the continuity equation can be transformed to

$$i\omega C P + \frac{\partial Q}{\partial x} = 0, \tag{5.35}$$

where C is the compliance. The compliance can be approximated by linearizing the state equation (5.23) for the large arteries:

$$C = \frac{dA}{dp} = \frac{3 A_0 a}{2Eh} \left(1 - \frac{3pa}{4Eh} \right)^{-3} \approx \frac{3 A_0 a}{2Eh},$$

which applies since $Eh \gg pa$.

5.6.3 Solution to the Linear Model

For each vessel in the small arteries, (5.34) and (5.35) must be solved. The equations comprise a continuity and a momentum equation determining the flow resulting from an oscillatory pressure gradient in a nontapering vessel where the amplitude and phase depend on the wall distensibility and blood viscosity (through the factor k'). These are

$$i\omega C P + \frac{\partial Q}{\partial x} = 0, \tag{5.36}$$

$$i\omega Q + \frac{A_0(1 - F_J)}{\rho} \frac{\partial P}{\partial x} = 0. \tag{5.37}$$

These equations are periodic with period T (the length of the cardiac cycle) and apply to any vessels of length L. Differentiating (5.36) with respect to x and inserting the result into (5.37) gives

$$\frac{\omega^2}{c^2} Q + \frac{\partial^2 Q}{\partial x^2} = 0 \tag{5.38}$$

where the wave propagation velocity

$$c = \sqrt{\frac{A_0(1 - F_J)}{\rho C}}. \tag{5.39}$$

Solving (5.38) and (5.39) yields

$$Q(x, \omega) = a \cos(\omega x / c) + b \sin(\omega x / c).$$

Inserting this result into (5.37) gives

$$P(x, \omega) = i \sqrt{\frac{\rho}{C A_0(1 - F_J)}} \left(-a \sin(\omega x / c) + b \cos(\omega x / c) \right),$$

where a and b are arbitrary constants of integration. As stated in (5.26), the impedance $Z(x, \omega)$ for any Fourier mode can be defined by the relation

$$P(x, \omega) = Z(x, \omega) \, Q(x, \omega).$$

As a result, the impedance can be found by

$$Z(x, \omega) = \frac{i g^{-1}(b \cos(\omega x / c) - a \sin(\omega x / c))}{a \cos(\omega x / c) + b \sin(\omega x / c)}, \tag{5.40}$$

where $g = \sqrt{C A_0(1 - F_J)/\rho}$. At $x = L$

$$Z(L, \omega) = \frac{i g^{-1}(b \cos(\omega L / c) - a \sin(\omega L / c))}{a \cos(\omega L / c) + b \sin(\omega L / c)},$$

and at $x = 0$

$$Z(0, \omega) = \frac{i}{g} \frac{b}{a}. \tag{5.41}$$

Assuming that $Z(L, \omega)$ is known, b/a can be found using (5.40):

$$\frac{b}{a} = \frac{\sin(\omega L/c) - ig Z(L, \omega) \cos(\omega L/c)}{\cos(\omega L/c) + ig Z(L, \omega) \sin(\omega L/c)}.$$

Then the root impedance for any vessel can be found from (5.41):

$$Z(0, \omega) = \frac{ig^{-1} \sin(\omega L/c) + Z(L, \omega) \cos(\omega L/c)}{\cos(\omega L/c) + ig Z(L, \omega) \sin(\omega L/c)}. \qquad (5.42)$$

5.7 Impedance at the Root of the Structured Tree

Just as for the large arteries, (5.36) and (5.37) could be solved with appropriate boundary conditions to predict the blood flow and pressure at any site along the small arteries. However, we are only interested in the impedance at the root of the structured tree, as explained in section 5.6. Equation (5.42) gives $Z(0, \omega) = f(Z(L, \omega))$ for any of the small arteries. The aim of this section is to find an expression for the root impedance of the structured trees comprising the small arteries. This is done by imposing appropriate boundary conditions, which combine the small arteries into the structured tree. These are

- a bifurcation condition and

- a terminal boundary condition.

5.7.1 Bifurcation Condition

Setting up the bifurcation condition does not require much more than has already been established. For the large arteries it is assumed that pressure is continuous and that no flow is leaking at the bifurcation; see section 5.5.2. Therefore, the bifurcation is analogous to a transmission-line network and the admittances add; i.e.,

$$\frac{1}{Z_p} = \frac{1}{Z_{d_1}} + \frac{1}{Z_{d_2}}. \qquad (5.43)$$

As before, the subscript p refers to the parent vessel and the subscripts d_1, d_2 refer to the daughter vessels.

5.7.2 Outflow Condition

Because the model for the small arteries includes viscosity, the structured tree yields a resistance by itself and it is possible to assume a zero impedance at its leaves. Various parts of the body serve different needs and hence have different peripheral resistances, so when using the approach above, one must choose the minimum radius (r_{min}) applied at the terminals individually for each of the structured trees. While r_{min} may vary for the different structured trees, it is kept constant within each of them. Alternatively, one could choose a common minimum radius and then vary the total impedance by applying a variable terminal impedance at the bottom of the vessels. This terminal impedance would then be propagated back through the tree, together with the induced dynamic impedance.

5.7.3 Root Impedance of the Structured Tree

In order to make the structured tree geometrically self-similar, as discussed in section 5.3, all parameters must be determined as functions of the vessel radius. Both g and the length L for each vessel can be determined as functions of the vessel's radius. Furthermore, the geometric self-similarity makes it very easy to compute the root impedance $Z(0, \omega)$ by solving (5.42) and (5.43) recursively. In order to do so, all basic parameters must be initialized. These parameters are

- the scaling parameters α and β, which are given by (5.6);

- the terminal impedance Z_t (see discussion on page 123);

- the minimum radius r_{min}, at which all vessels of the structured tree are truncated (note that this radius has to be determined separately for each structured tree);

- the root radius r_{root}.

Recall that the root impedance should be used with the outflow condition for the large arteries, i.e., when evaluating the convolution integral in (5.29). Since it was assumed that wave propagation of the large arteries is periodic, the impedance should be determined for all $0 \leq t \leq T$. In the frequency domain this translates to discrete angular frequencies $\omega_k = 2\pi k/T$ for $k = -N/2, \ldots, N/2$, where N is the number of time steps per period. Assuming that $z(x, t)$ is real, then Z must be self-conjugate:

$$Z(0, \omega_{-k}) = \overline{Z(0, \omega_k)}.$$

Therefore, only $Z(0, \omega_k)$ has to be determined for $k = 0, 1, \ldots, N/2$.

Finally, before describing an algorithm for determining the root impedance, we will describe how to take advantage of the fact that the tree of small arteries is structured. A binary tree with N generations would have 2^N terminal leaves. In our case the tree is structured such that the radius of any vessel at the nth generation is scaled with the factor

$$\alpha^k \beta^{n-k} \quad \text{for} \quad k \in [0; n].$$

Each scaling factor appears with frequency

$$\text{Freq} = \frac{n!}{k!(n-k)!}.$$

For a tree with N generations there are

$$\sum_{n=0}^{N} \sum_{k=0}^{n} 1 = \mathcal{O}(N^2)$$

different impedances to be calculated. The scaling ratios are illustrated in Figure 5.3. Modeling the small arteries as an asymmetric tree should be seen in contrast to a general binary tree in which all vessels may have different impedances. In this case the complexity is $\mathcal{O}(2^N)$. Thus the increase in performance achieved by storing each of the impedances once they have been calculated is what makes the algorithm computationally feasible. It should be noted that the estimates above are for a tree with a fixed number of generations. Since the structured trees are truncated when the radius of a given vessel becomes smaller than some minimum radius, the estimates above are not exact. But for most practical situations they are still valid. In any case, the savings obtained from reusing the precomputed values are crucial to this approach.

The recursive algorithm can thus be written as follows.

ALGORITHM 5.1.
Computes the root impedance $Z_0(\omega_k, p_\alpha, p_\beta)$ recursively.

- Determine all parameters for the vessel (as described in section 5.3) and initialize an array computed to zero:

    - The radius $r = \alpha^{p_\alpha} \beta^{p_\beta} r_{root}$.
    - The cross-sectional area $A = \pi r^2$.
    - The elasticity relation $f(r) = Eh/r$; see (5.3).
    - The length $l = r\, l_{rr}$, where l_{rr} is the length-to-radius ratio.
    - The viscosity $\mu = 0.049$ g/(cm·s).

- Determine the wave propagation velocity c and the scalar g. These depend on F_J, defined in (B.64), and thus on the Womersley number w.

- Recursive algorithm:

    - If $r < r_{min}$ then

        ▸ $Z_L(\omega_k, p_\alpha, p_\beta) = $ terminal impedance.

    - else

        ▸ If the root impedance of the left daughter $(p_\alpha + 1, p_\beta)$ has been computed previously then

            ○ $Z_0(\omega_k, p_\alpha + 1, p_\beta) = $ computed $(p_\alpha + 1, p_\beta)$

        ▸ else

            ○ Compute the root impedance of the left daughter by calling $Z_0(\omega_k, p_\alpha + 1, p_\beta)$ recursively.

            ○ If the root impedance of the right daughter $(p_\alpha + 1, p_\beta)$ has been computed previously then

            $Z_0(\omega_k, p_\alpha, p_\beta + 1) = $ computed $(p_\alpha, p_\beta + 1)$

            ○ else

                Compute the root impedance of the right daughter by calling $Z_0(\omega_k, p_\alpha, p_\beta + 1)$ recursively.

    - $Z_L(\omega_k, p_\alpha, p_\beta) = 1/(Z_0^{-1}(\omega_k, p_\alpha + 1, p_\beta) + Z_0^{-1}(\omega_k, p_\alpha, p_\beta + 1))$.

- If $\omega_k \neq 0$ then

  ▪ $Z_0(\omega_k, p_\alpha, p_\beta) = f(Z_L(\omega_k, p_\alpha, p_\beta))$ using (5.42).

- else if $\omega_k = 0$ then

  ▪ $Z_0(\omega_k, p_\alpha, p_\beta) = f(Z_L(\omega_k, p_\alpha, p_\beta))$.

- Update the table of precomputed values:
  computed $(p_\alpha, p_\beta) = Z_0(\omega_k, p_\alpha, p_\beta)$.

For most of our simulations the terminal impedance is taken to be zero and the impedance is predicted solely from the structured tree. It is easy to modify the algorithm to incorporate an additional terminal impedance beyond that provided by the tree itself. Assume, for example, that the total terminal impedance Z_t is given and that it is distributed evenly over all terminals. The terminal impedance for each terminal Z_{t_i} is then

$$Z_{t_i} = N_t Z_t,$$

where N_t is the total number of terminals. This rule applies since the admittances add to give the total admittance, and impedance is the reciprocal of admittance. Because the structured tree has been terminated when the radius $r < r_{min}$, the number of terminals cannot be determined analytically. This number can be counted recursively.

By the inverse Fourier transform of the root impedance $Z(0, \omega)$ the response function $z(0, t)$ in the time domain and hence the impedance that appears in the convolution integral in (5.29) can now be found. Thus the boundary condition for the large arteries can be determined. This closes our model.

Another possibility is to investigate whether the equations can be solved using a self-similar approach. This is not quite possible due to the viscous terms present in the momentum equation. However, this approach gives some insight, which could be used for other purposes.

5.8 Results

This section presents a selection of results that elucidate our hypothesis. Our main aim is to show that the structured tree model is a feasible outflow condition for determining blood flow and pressure in the large systemic arteries. Therefore, we will show that our model captures all essential features characterizing the arterial pulse. In addition, our model is compared with a pure resistance model, a three-element Windkessel model, and measured data.

As discussed in the introduction, the most prominent characteristic features of the arterial pulse are as follows:

- The maximum pressure of the large arteries increases away from the heart toward the periphery because of tapering of the vessels.

- The mean pressure of the arteries drops toward the periphery according to the distribution of flow impedance of the vascular bed (Noordergraaf, 1978).

- The steepness of the incoming pressure profile increases toward the periphery. This increase is a result of the pressure dependence of the wave propagation velocity, which means that the part of the wave with higher pressure travels faster than the part with lower pressure.

- The dicrotic wave separates from the incoming wave and is more prominent at peripheral locations than at proximal locations.

- Wave reflections affect the shape of the pulse wave. These appear as a result of the tapering and branching of the vessels and as a result of the peripheral resistance, which mainly originates from the small arteries and arterioles.

All results shown in this section represent solutions of the nondimensional fluid dynamics equations (5.19) and (5.20) with no leakage, i.e., $\Psi = 0$, state equation (5.23). The inflow profile is the one shown in Figure 5.8, and the outflow condition is given by (5.29). The impedance term in the outflow condition will be determined by the structured tree model, a pure resistance model, or a three-element Windkessel model. It should be noted that the results strongly depend on the parameters used in the computation, especially the inflow profile, the geometry, and the minimum radius of the structured tree, and the total resistance and compliance for the Windkessel and pure resistance models.

5.8.1 Model Problem

The results presented in this section are based on the following choices:

- Geometric data for the large arteries are as described in Table 5.1.

- Parameters for the structured tree follow the choices in section 5.3.

- Terminal resistance for the small arteries is set to zero.

- The minimum radius, $r_{min} = 0.04$.

- Young's modulus times the wall thickness-to-radius ratio is as shown in Figure 5.2.

- Density $\rho = 1.055$ g/cm^3 and viscosity $\mu = 0.049$ g/(cm·s) are kept constant.

The presentation is restricted to flow and pressure profiles for a few representative vessels. These are flow and pressure in the aorta (see Figure 5.10) as well as the subclavian and brachial arteries (see Figure 5.11). All profiles capture the characteristics described in the beginning of this section. It can be difficult to see all of these characteristics in the 3D figures. In particular, it is not easy to see the steepening of the wave front. However, this can be seen in Figure 5.12, where 2D cross sections of flow and pressure profiles are plotted as functions of time at five equidistant locations along the aorta as well as the subclavian and brachial arteries. Furthermore, the 2D plots are more easily compared with measurements that are normally presented as such. For comparison, Figure 5.13 shows pressure profiles at various locations along the aorta and brachial artery. These profiles, recorded from a single person by Remington and Wood (1956), are typical for a healthy adult.

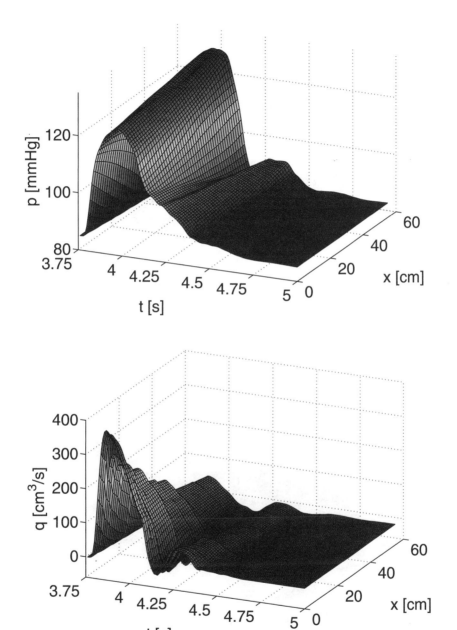

Figure 5.10. *Pressure and flow in the aorta as functions of length x and time t during one cardiac cycle.*

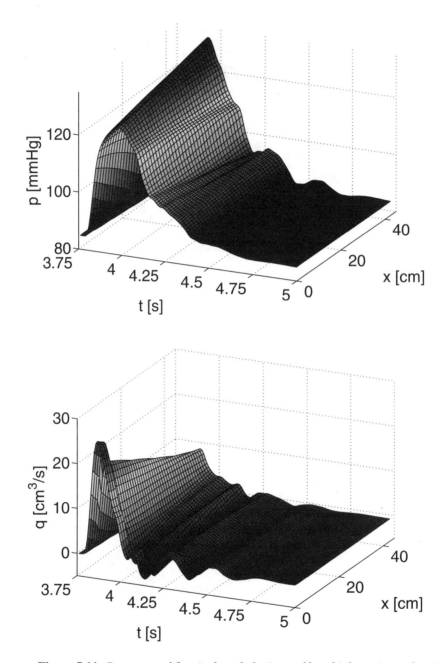

Figure 5.11. *Pressure and flow in the subclavian and brachial arteries as functions of length x and time t during one cardiac cycle.*

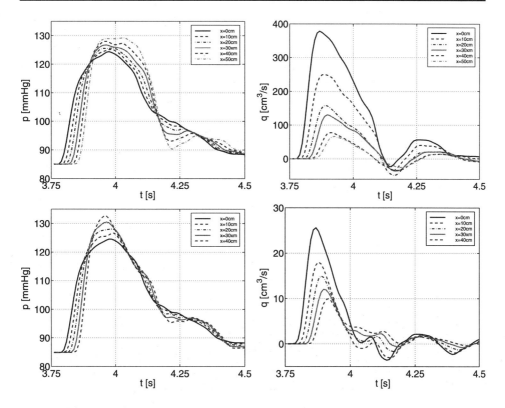

Figure 5.12. *Time dependent pressure and flow in the aorta as well as the subcla-vian and brachial arteries, both during one cardiac cycle. These graphs are cross sections of the 3D plots of Figures 5.10 and 5.11.*

5.8.2 Structured Tree Model, Windkessel Model, Pure Resistance Model, and Measured Data

In order to investigate the performance of the structured tree model, our model is compared with two other models:

1. Pure resistance model:

$$Z_L(\omega) = R_T,$$

    where R_T is a constant representing the peripheral resistance.

2. Three-element Windkessel model: a zero-dimensional model predicting the impedance as a result of a resistive and a compliant behavior of the small arteries; i.e.,

$$Z_L(\omega) = \frac{R_1 + R_2 + i\omega C_T R_1 R_2}{1 + i\omega C_T R_2},$$

    where $R_T = R_1 + R_2$ is the total peripheral resistance and C_T is the total peripheral compliance.

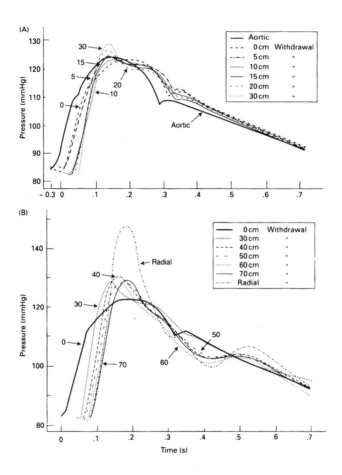

Figure 5.13. *Measured data for the pressure $p(x, t)$ in the aorta and the subclavian and brachial arteries. From J.W. Remington and E.H. Wood (1956), Formation of pulse contour in man,* J. Appl. Physiol. **9**:433–442. *Used by permission from The American Physiological Society.*

The comparison falls into two parts:

- three outflow conditions applied to a single isolated vessel,

- comparison of the Windkessel model, the structured tree model, and measured data for the impedance in people.

Three Outflow Conditions for a Single Isolated Vessel

The advantages of both the pure resistance and the Windkessel models are that they are easy to understand and computationally inexpensive. The disadvantage of the Windkessel and the pure resistance models is that they are not able to capture the wave propagation phenomena in the part of the arterial system that they model. Furthermore, neither the pure resistance model nor the Windkessel model can account for phase lag between flow and pressure. The Windkessel model requires estimates of the total arterial resistance, R_T, and compliance, C_T, for each terminal segment, and the pure resistance model needs the total arterial resistance. Still, when coupled to the nonlinear equations for the large arteries, both models are able to capture the overall behavior of the system.

In order to show the differences between the three models, we have (for simplicity) used a single tapering vessel of length 100 cm, with top radius 0.4 cm and bottom radius 0.25 cm. In order to make the three outflow boundary conditions match as closely as possible, the model parameters have been estimated to match the result obtained by the structured tree model. The parameters have been chosen such that the total resistance R_T (the DC term from the structured tree model) is the same for all three models and the total compliance C_T for the Windkessel model is fitted empirically to match that given by the structured tree model. It should be emphasized that this study is theoretical, and hence the parameters should not be compared with physiological values.

The following plots have been made:

- pressure versus flow; see Figure 5.14;

- impedance (both modulus and phase) versus frequency at the boundary; see Figure 5.16. Since the pure resistance model does not depend on the frequency, this plot comprises only the structured tree and the Windkessel models.

- Pressure as a function of x and t during one cardiac cycle; see Figure 5.15.

The pressure versus flow curves in Figure 5.14 show the phase lag between flow and pressure for a single vessel. The left figure is for the pure resistance model, the middle one for the Windkessel model, and the right one for the structured tree model. Comparing these, the most striking difference is that the pure resistance model affects the overall shape of the curve. The forced in-phase condition at the outflow boundary results in a narrowing of the

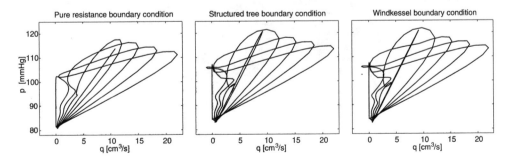

Figure 5.14. *Pressure versus flow in a single vessel at six equidistant locations, i.e., for a vessel of length L, $x = 0, L/4, L/2, 3L/4, L$. Both flow and pressure are plotted as functions of time during one cardiac cycle. The left graph is from the pure resistance model, the middle one from the Windkessel model, and the right one from the structured tree model.*

Pure resistance boundary condition

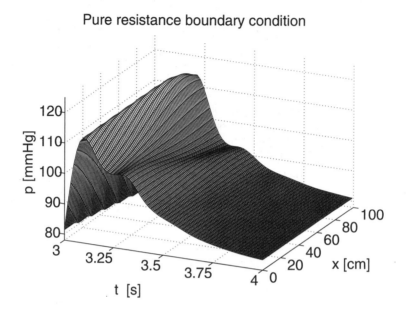

Windkessel boundary condition

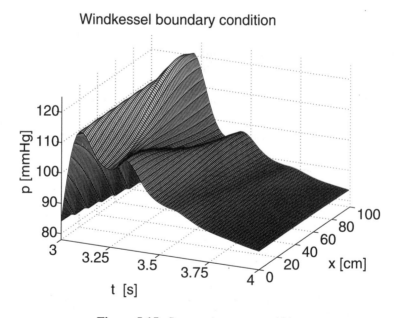

Figure 5.15. *See caption on page* 131.

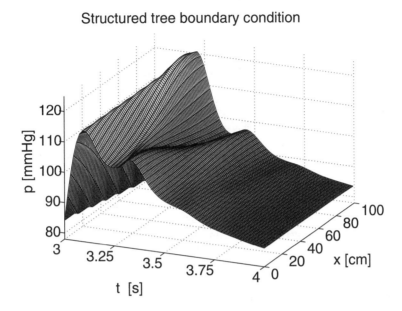

Figure 5.15. *Pressure in a single vessel as a function of length x and time t during one cardiac cycle. The first graph is from the pure resistance model, the second one from the Windkessel model, and the last one from the structured tree model. From M.S. Olufsen (1999), Structured tree outflow condition for blood flow in larger systemic arteries, Am. J. Physiol. **276**:H257–267. Used by permission from the American Physiological Society.*

width of the loop back through the vessel. Furthermore, it is worth noting that the pure resistance model to some extent can be viewed as a special case of the Windkessel model incorporating only the DC resistance. For the Windkessel model the flow and pressure are also nearly in phase, but the narrowing is not reflected back through the vessel. Finally, it is observed that the structured tree model does keep some phase lag between flow and pressure. However, the plots look fairly similar for the Windkessel and the structured tree models. The reason for this is that the local dynamics of the impedance cannot be seen on such a plot. This has to do with the fact that the structured tree model includes wave propagation effects for the entire tree, which the Windkessel model cannot do. As a result, it is likely that the Windkessel model will introduce more artificial reflections than the structured tree model. When the two models are compared (see Figure 5.15), the following differences can be observed: The reflections from the pressure when using the Windkessel model are slightly more pronounced than they are when using the structured tree model. A Bode plot of the impedance at the bottom of the large vessel, $Z_L(\omega)$, versus the angular frequency, ω, i.e., a plot of $\log(|Z_L(\omega)|)$ versus $\log(\omega)$ and a plot of phase(Z) versus $\log(\omega)$, shows the differences between the two models more clearly. The impedances are computed as functions of all frequencies needed to evaluate the outflow condition in (5.26). For example, if the number of time steps in a cardiac cycle of 1 s is $N = 2048$, then $\omega \in [-2\pi N/2, 2\pi N/2]$. First, it is seen that the Windkessel model cannot predict any of the dynamic behavior resulting in oscillations at high frequencies. The question then arises as to whether such oscillations occur in actual data.

Structured Tree Model, Windkessel Model, and Measured Data

In Nichols and O'Rourke (1998) the authors devote a section to discussing the impedance of the large arteries; our results have been compared with those obtained there. The results obtained by Nichols and O'Rourke (1998) for people are mainly from the large arteries. To demonstrate the difference between the Windkessel and the structured tree models the two models have been applied directly as outflow conditions for these large arteries even though our outflow conditions are usually applied further downstream.

The comparisons among the Windkessel, the structured tree, and the measured data are made for the brachiocephalic artery. Since the structured tree model is not designed to be valid for these large arteries, one should not assume that the output from the different models will match perfectly without adjustments of the parameters. The length-to-radius ratio had to be modified to 130. This much larger ratio corresponds with the arteries of the arm. From Table 5.1 it is seen that starting from the subclavian artery, no large side branches occur before the bifurcation between the ulnar and intercostal arteries. The combined length of the subclavian and brachial arteries is 43.25 cm, and the average radius is 0.32 cm. This results in a length-to-radius ratio of 135. Now starting at the brachiocephalic artery there is a major bifurcation after only 3.5 cm resulting in a very short length-to-radius ratio. After the brachiocephalic bifurcation (in the subclavian and carotid arteries), the length-to-radius ratio is fairly long. The very short length-to-radius ratio is not taken into account here. Instead the same length-to-radius ratio as in the subclavian artery has been kept. These large variations confirm our discussion in section 5.3.3, namely, that it is not easy to find a universal length-to-radius ratio for the large arteries. Since it is known that the length-to-radius ratio is smaller for the small arteries, further studies might be able to reveal some functional dependence.

For the brachiocephalic artery an input radius of 0.5 cm and a minimum radius of 0.025 cm have been chosen. An input radius of 0.5 cm is rather small compared with measurements, but in the literature it is not specified exactly where the measurements were taken; i.e., the radius specified does not have to be an input radius. Thus the comparisons should be interpreted with all of these reservations in mind. The results for these comparisons are shown in Figure 5.16.

It becomes clear from all of these figures that the structured tree model is, in fact, able to capture some of the observed oscillatory dynamics in the impedance, dynamics that cannot be captured by the Windkessel model.

5.9   Conclusion

The purpose of this study was to develop a model, based on physiological principles, that can predict flow and pressure in the large systemic arteries. This goal has been achieved by constructing a fluid-dynamical model, including both large and small arteries, where the impedance of the small arteries constitutes a boundary condition for the large arteries. The small arteries were modeled as asymmetric structured trees, while the large arteries were modeled using actual data for the length and top and bottom radii for each of the vessels. Flow and pressure in the large arteries were found by solving the 1D nonlinear Navier–Stokes equation and a continuity equation, combined with a state equation predicting the relation between flow and pressure. For the small arteries only the impedance was computed. It was

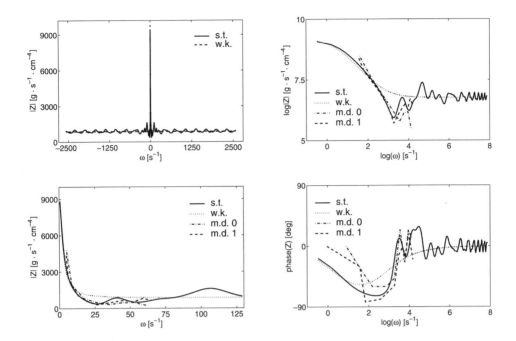

Figure 5.16. *Impedance plots for the brachiocephalic artery compared with measured data. The measured data (m.d.0) and (m.d.1) are from Nichols and O'Rourke (1998). The dotted lines signify results from the Windkessel model (w.k.), and the solid lines signify results from the structured tree (s.t.).*

found by solving a linearization of the 1D Navier–Stokes equation and continuity equation combined with the same state equation that was used for the large arteries.

In section 5.8 it was shown that the resulting flow and pressure profiles all had the correct characteristics:

- The systolic pressure increased away from the heart.

- The mean pressure dropped slowly.

- The steepness of the incoming pressure profile increased toward the periphery.

- The velocity of the wave propagation of the reflected dicrotic wave was slower than that of the main wave, and hence the dicrotic wave separated from the main wave peripherally.

In addition to showing the correct characteristics (see Figures 5.10–5.12), our results also revealed that the structured tree model provided a more dynamical impedance that is much closer to physiological behavior than can be obtained with the Windkessel model; see Figures 5.14–5.16. Furthermore, phase lag between flow and pressure was retained, and correct quantitative as well as qualitative results for both flow and pressure were obtained; see Figure 5.14. For a detailed quantitative comparison, measurements and computations should be carried out in vessels where both geometry and inflow are matched. This has been done in the study by Olufsen et al. (2000), where MR measurements taken at ten locations in the systemic arteries were compared with simulations obtained using a 1D model with structured tree outflow conditions. These results showed an excellent agreement between data and simulations.

The advantage of modeling the small arteries by applying structured trees at all terminals of the large arteries is that it gives rise to a physiological boundary condition that is able to include wave propagation effects such that the impedance shows the right dynamic behavior. Furthermore, this model of small arteries had only one parameter that had to be estimated, namely, the radius at which the structured tree must be terminated. Having the radius as the only parameter required that all other parameters, e.g., vessel length and vessel elasticity, be able to be estimated as functions of the vessel radius. Because of the inclusion of viscosity in the fluid dynamic equations, it was not possible to solve the linearized equations using a self-similar approach. This is because, for these relatively large vessels viscosity is constant and does not scale with the radius of the vessels. As a result, a recursive approach was used to calculate the root impedance of the structured tree. Because no resistance was applied at the terminals of the structured tree, the peripheral resistance of the large arteries was obtained entirely from the solutions of the linearized equations in the asymmetric structured tree. It was shown that it was important to estimate the minimum radii correctly. Furthermore, it was shown that it was possible to estimate the minimum radii by studying the overall terminal resistance of the organ in question.

As mentioned in the section describing the geometry of the structured tree, the diameter of the small arteries varies considerably, and so does the peripheral resistance of those very small vessels with a strong muscular wall. This is consistent with the observations (Guyton, 1991) that the arterioles, and not the capillaries, generate the peripheral resistance. Furthermore, since total peripheral resistance differs from organ to organ, it was important to choose the minimum radius individually for each of the structured trees, i.e., for each of the terminal branches of the large arteries.

In summary, this study has shown the feasibility of limiting the computational domain. Basing parameter estimates and the model foundation entirely on physiological information, a good boundary condition has been derived for the nonlinear model predicting blood flow and pressure in the large systemic arterial tree. And it has been done in such a way that even at the boundary the impedance shows the right characteristics for both low and high frequencies. As a result, the claims stated in our hypothesis have been confirmed.

It is possible to argue that the structured model is too complicated and that the much simpler Windkessel model is adequate because it also provides a dynamic boundary condition. Furthermore, since measurements for both the total resistance and the compliance are available, the Windkessel model provides a simple and yet adequate boundary condition for the large arteries. However, these parameters must be tuned to match the individual studied, and as a result they may vary significantly even among healthy people. For the structured tree model that variation corresponds to changes in the minimum radius, the elastic properties of the vessels, and the length-to-radius ratio. It is our belief that the results using the structured tree model are less sensitive to input variations, but this is a hypothesis that remains to be proved. Generally, the structured tree model gives a more detailed output, but basing the outflow condition on the structured tree model does not yield a decisive advantage over the Windkessel model. Finally, the Windkessel model is local in both time and space, while the structured tree model is periodic in time but local in space. Therefore, replacing the Windkessel model with a structured tree model comes at the cost of replacing an ordinary differential equation with a convolution integral.

Nevertheless, we find that the structured tree model has at least two important advantages. First, it is based on the underlying physiology and includes wave propagation effects. This enables the model to capture the observed high frequency oscillations of the impedance in the part of the arterial system that it models. Second, the structured tree model can predict flow and pressure not just in the large arteries, but also in the small arteries, thus shifting the purpose of the structured tree from being a boundary condition to being a more active part of the model.

The idea of using a structured tree in which a simpler set of equations is solved could also be applied to other modeling problems involving flow in tree-like structures, for example, flow and depth of water in a river delta. The use of the outflow condition stated here is only applicable to phenomena in which there is a scaling law that gives rise to a structured tree.

5.9.1 Perspectives

Several modifications and improvements can be made to the arterial model presented in this chapter. The following sections summarize some of these ideas.

Order of the Structured Tree

It has been shown that the minimum radii of the terminal trees should be varied in order to account for the different resistances of the various organs. It has also been shown that the most significant part of the difference in applying the various terminal minimum radii lies in the magnitude of the average resistance. Therefore, an obvious modeling strategy would be to investigate the possibility of keeping one common minimum radius for all the small vessels and then adjusting the minimum radius for the various organs by attaching a variable pure resistance as a boundary condition at the terminals of the small arteries. In this way the dynamics generated by the structured tree is retained and a pure resistance need not be applied until at the arterioles, where the flow and pressure are almost in phase. Furthermore, having a pure resistance at the bottom makes it easier to adapt the model to various physiological conditions, for example, exercise, when the peripheral resistance of various tissues decreases. Another point that could be interesting to study is the correspondence between the minimum radius predicted for an organ, given a particular root resistance, and the actual physiological radius that can be measured at the arteriolar level for that particular organ.

Young's Modulus and the State Equation

The model in this chapter uses a simple state equation based on linear elasticity theory, but where Eh/r_0 decreases exponentially with an increasing radius. The functional dependence between Young's modulus, the wall thickness, and the radius is an improvement of the basic model. It is still not physiologically correct to use a pure elastic model for the arterial wall. As a result, the wave propagation velocity decreases with increasing pressure, a correlation that should be in the opposite direction. Also, when plotting the cross-sectional area $A(x, t)$ versus pressure $p(x, t)$ for some fixed x, the graph curves the wrong way. This is not a crucial flaw since the curve is almost a straight line. As a result, the model exhibits an overall behavior that is correct. But, as discussed earlier, the model's correspondence with observations may be improved by using an empirically based relation that allows for varying elastic properties for the individual arteries. Constructing such a model would probably be a valuable contribution, since even though the systemic arteries are all composed of the same basic material the elastic properties of the vessels are different depending on their function. In order to develop a better empirical model, a more detailed study must be performed such that the model parameters can be estimated correctly. A good suggestion is the model by Langewouters, Wesseling, and Goedhard (1984). Finally, it should be noted that it is easy to replace the present state equation with some other relation. As a result we highly recommend further investigation of other relations.

5.9.2 Pathological Conditions

In order to get a better idea of the validity of the arterial model, one should use it to study a number of pathological situations. Since our model is based on the assumption of laminar flow, it would not be well suited for studying phenomena related to atherosclerosis, which often gives rise to vortices and turbulence (Nichols and O'Rourke, 1998). It would be useful for studying effects arising from changes in the vascular wall, e.g., aging, diabetes, vasoconstriction, or vasodilation. In the case of vasodilation, $\lim_{r_0 \to 0} Eh/r_0$ would be decreased, which corresponds to stiffening of such small arteries as the radial and femoral, but results in only a slight change in the elasticity of such large arteries as the aorta. This should result in a reduced systolic pressure and a reduction of wave reflections because of a delay in the early reflections (Nichols and O'Rourke, 1998). However, if one wishes to study the implications for specific arteries, i.e., more local changes, a more sophisticated model of the elasticity of the arterial wall should be incorporated.

Acknowledgments

The author was supported by High Performance Parallel Computing, Software Engineering Applications Eureka Project 1063 (SIMA—SIMulation in Anesthesia) and the Danish Academy for Technical Sciences. For scientific support, the author wants to acknowledge Dr. Charles Peskin, Courant Institute of Mathematical Sciences, New York University, and Dr. Jesper Laisen, Math-Tech, Charlottenlund, Denmark.

Chapter 6

A Cardiovascular Model

M. Danielsen and J.T. Ottesen

6.1 Introduction

In contrast to the earlier chapters, the goal of this chapter is to present a model of the overall human circulatory system. We establish a lumped pulsatile cardiovascular model that embraces the principal features of human circulation, based on physiological data available in the literature. Such models are valuable tools in examining and understanding cardiovascular diseases such as hypertension, weak and enlarged hearts, and other life-threatening conditions. Moreover, these models may also facilitate new insights into cardiovascular functions. Interaction between the cardiovascular system and other systems such as the nervous system can also be studied. This is done, for example, in Chapter 7, where the cardiovascular model is coupled to a description of the nervous system represented by the baroreceptor mechanism. New descriptions of the heart, such as the one presented in Chapter 4, can be challenged when connected to such models. However, in this chapter we will establish a traditional model of the human heart based on the elastance concept. The cardiovascular model is implemented in the SIMA simulator, in which it interacts with a number of other models describing different physiological processes. This requires the new model to be stable and reliable: requirements that are difficult for a new heart model to fulfill. The strategy of having simplicity as a guideline is accentuated by the limited knowledge of the explicit interaction between the CNS and the human cardiovascular system that is established in Chapter 7. In addition, the simulator has to run in real time, which generates concerns about computational speed and therefore restricts the complexity of the model. Hence the cardiovascular model must be simple. In essence, a lumped or Windkessel type of model of the cardiovascular system is perfectly suitable.

The human circulation model presented in this chapter provides

- ventricular pressure, root aortic pressure, and ventricular outflow curves that closely resemble observations on humans found in the literature;

- three arterial and two venous pressure and volume levels in both the systemic and the pulmonary circuits, which reflect the pressure distribution in the human circulatory system.

Preceding the description of the cardiovascular model, section 6.2 introduces the major modeling approaches used to describe impedance in the arterial tree and the overall function of the cardiovascular system in terms of pressure and flow. The human circulation model is established in section 6.3. The heart and the vasculature are described in sections 6.3.1 and 6.3.2. The vascular pressure and volume distributions are given in section 6.3.3, which also reveals the strategy used to determine the parameter values in the cardiovascular model. Computed results are given in section 6.3.4. Further details of this model are available in Danielsen (1998). The differential equations making up the cardiovascular model are quantified in section 6.4, and all the parameter values are stated in section 6.5.

6.2   Architecture of Cardiovascular Models

The mathematical formulations of the cardiovascular system are typically divided into two classes of models: distributed and lumped. In the former, arterial pressure and flows may be described by use of partial differential equations. Two examples of such models are the 2D model describing the velocity patterns in the heart (see Chapter 3) and the 1D model describing the flow and pressure in the large systemic arteries (see Chapter 5). Other examples include models by Pedley (1980) and Peskin (1976). Another class of models (lumped models) can be obtained by discretizing the partial differential equations. The descretized equations comprise a system of ordinary differential equations, where each discretized variable defines a segment element (a compartment) (Noordergraaf, 1978; Rideout, 1991). Typically, one element can be described in terms of resistances R, inductors L, and capacitors C. The parameters are computed according to the corresponding vessel radius, viscosity and density of blood, and length of the segment. Calculation of the elastic properties may in addition require Young's modulus of elasticity. The elasticity is represented by the compliance C, which is the reciprocal of the elastance. This class of models can be used to study the input impedance as seen by the heart, wave travel, and wave reflections (Westerhof and Stergiopulos, 1998). Lumped models can also be derived directly without having a particular spatial discretization in mind. For example, models based on random length distribution of the segments, and thus not on the real anatomy, have been shown to exhibit the same wave travel and impedance features as lumped models obtained by discretizing the partial differential equations (Noordergraaf, 1978; Westerhof and Stergiopulos, 1998; Olufsen, Nadim, and Lipsitz, 2001; Olufsen, Tran, and Ottesen, 2003).

An example of a lumped model is the modified Windkessel model in Figure 6.1. Although strikingly simple, the model gives a very good description of the input impedance of the arterial system (Toy, Melbin, and Noordergraaf, 1985; Westerhof and Stergiopulos, 1998; Olufsen, Nadim, and Lipsitz, 2002). Using this type of model, the entire human cardiovascular system may be described as a network of compliances, resistances, and inductances not reflecting anatomical properties. These types of models exhibit major features of their real counterpart, such as higher pressure in response to vessel constrictions, and root aortic pressure and ventricular outflow curves representative of observed values.

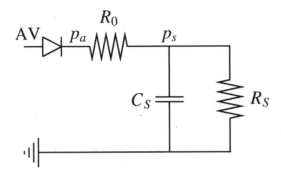

Figure 6.1. *The three-element modified Windkessel arterial load. This model is characterized by a characteristic aortic impedance R_0, a total peripheral resistance R_s, and a total arterial compliance C_s. The aortic valve is indicated (AV).*

In addition, phenomena with low frequency content, such as Mayer waves (Ottesen, 1997a; Ursino, 2000) or autonomic and autoregulation (Olufsen, Tran, and Ottesen, 2003) may be studied with these models. The cardiovascular models can be coupled to mathematical descriptions of nervous control to study effects of the CNS on human circulation. In this case lumped models may be used to study the impact of the venous reservoir during hemorrhages (Ursino, Antonucci, and Belardinelli, 1994; Danielsen and Ottesen, 1997) or the effect of halothane on nervous regulation (Tham, 1988). Knowledge of the complex impact of the CNS on the overall performance of the cardiovascular system is still limited, and thus lumped models may be perfectly suitable for many uses. Lumped cardiovascular models may be divided into two subclasses. The first class contains the nonpulsatile models where the heart is described by some variant of Starling's law. The second class contains the pulsatile models. When pulsatility is included, the heart is typically guided by a classic time-varying elastance function. One exception is Palladino, Ribeiro, and Noordergraaf (2000), where the model of the isovolumic ventricle by Mulier (1994) is applied to a closed circulatory system. In Chapter 4 this heart model was significantly improved with the inclusion of the ejection effect and allowed to eject into a lumped pulsatile arterial model. Otherwise, models following the above lines of thinking include those of Warner (1958), Warner (1959); Grodins (1959); Beneken (1965); Leaning et al. (1983b); Leaning et al. (1983a); Martin et al. (1986); Tham (1988); Rideout (1991); Jin and Qin (1993); Neumann (1996); Sun et al. (1997); Danielsen and Ottesen (1997); Ursino (2000); Palladino, Ribeiro, and Noordergraaf (2000); and Ottesen (2000).

6.3 Cardiovascular Model

The model presented in this chapter consists of a pumping heart coupled to lumped descriptions of the systemic and the pulmonary circulations; see Figures 6.2 and 6.3. The heart consists of four chambers: two ventricles and two atria. The ventricles are guided by a pair of time-varying elastance functions, whereas the two atria la and ra are passive chambers

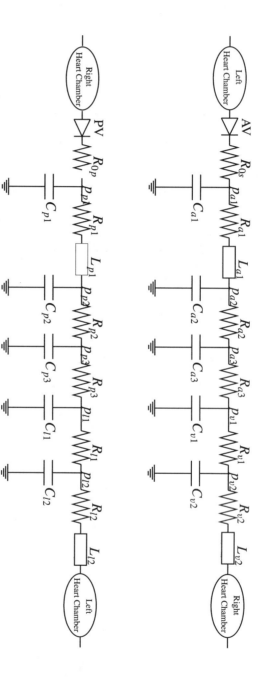

Figure 6.2. *The human circulation model. The cardiovascular model contains a systemic (top part) and a pulmonary (lower part) circuit. The left and right heart chambers are shown in Figure 6.3. The systemic arterial system is described by three sections (marked by subscripts a_1, a_2, and a_3), and the systemic venous system consists of two sections (marked by subscripts v_1, v_2). The pulmonary circulation follows the same architecture. Variables and parameters are explained in the text. Pressure and volume distributions are given in Table 6.1.*

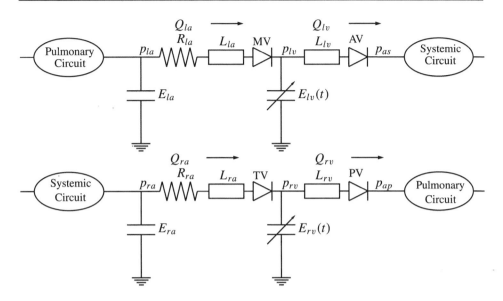

Figure 6.3. *The upper and lower panels show the left and the right heart chambers, respectively, of the cardiovascular model in Figure 6.2. Each side of the heart consists of a passive atrium and an active ventricle. The left atrium is modeled by a constant elastance E_{la}, while the performance of the left ventricle is described by a time-varying ventricular elastance function $E_{lv}(t)$ given by (6.2). The inertial properties of the blood are included by the inductances L_{lv} and L_{la}. The viscous properties of blood are included by the left atrium filling resistance R_{la}. The right heart is described in a similar fashion. The mitral and aortic valves are denoted by MV and AV, respectively. The tricuspid valve and pulmonary valve are denoted by TV and PV, respectively. Flow through each component is marked by Q, and the pressure at each node is marked by p.*

separated from the ventricles via the mitral valve MV and the tricuspic valve TV. The aortic AV and pulmonary PV valves are designed to allow a small amount of volume to flow back into the left lv and right rv ventricles (from the vasculature) before closure is completed, as observed in experiments. When the left ventricular pressure p_{lv} exceeds the root aortic systemic pressure p_{as}, the aortic valve opens and blood flows through the characteristic systemic resistance R_{0s} into the arterial system consisting of three sections. The pressures in the arterial system include the root aortic pressure p_{as} and three pressures p_{a1}, p_{a2}, and p_{a3}. The veins are described by two sections with the pressures p_{v1} and p_{v2}. The veins return the blood to the passive right atrium. When the right atrial pressure p_{ra} exceeds the right ventricular cavity pressure p_{rv}, the tricuspid valve TV opens and the right ventricle is filled with blood flowing through the resistance R_{ra}. Subsequently, blood is ejected into the pulmonary circuit through the pulmonary valve PV and characteristic pulmonary resistance R_{0p}. The architecture of the pulmonary circuit is a replica of the systemic one. The pressure and volume levels in each section and the corresponding physiological locations of the sections can be found in Table 6.1.

6.3.1   Heart

In Chapter 4 the activation function for the heart included an accurate description of the ejection effect. In this chapter we will base our activation function on the simpler elastance model developed by Suga and colleagues (1973, 1976). Nevertheless, despite the failure to encompass the various ejection conditions (see section 4.1), the elastance model provides a good alternative because it is significantly simpler (only three parameters need to be estimated). In addition, a large number of people have used this model (Martin et al., 1986; Tham, 1988; Rideout, 1991; Neumann, 1996; Stergiopulos, Meister, and Westerhof, 1996; Danielsen and Ottesen, 1997; Ursino, 2000), providing several sources for model validation. It should be noted that the heart model presented in this chapter could easily be replaced with the model including the ejection effect.

The elastance model of the pumping heart is shown in Figure 6.3. The cardiac contractile properties of the two ventricles are assumed to be defined by a pair of time-varying elastance functions. The inertia of blood movements in the ventricles is included by inductances that introduce a phase shift between the ventricular pressure and the root aortic pressure. The viscous properties of blood in the two atria are included by ventricular filling resistances, R_{la} and R_{ra}, respectively. The relation between the left ventricular cavity pressure p_{lv} and the ventricular volume V_{lv} is described by

$$p_{lv} = E_{lv}(t)(V_{lv} - V_{d,lv}), \tag{6.1}$$

where $V_{d,lv}$ is the left ventricular volume at zero pressure. The elastance function $E_{lv}(t)$ in (6.1) is given by

$$E_{lv}(t) = E_{min,lv}(1 - \phi(t)) + E_{max,lv}\phi(t), \tag{6.2}$$

where

$$\phi(t) = \begin{cases} a_\phi \sin\left(\frac{\pi t}{t_{ce}}\right) - b_\phi \sin\left(\frac{2\pi t}{t_{ce}}\right) & \text{for } 0 \le t < t_{ce}, \\ 0 & \text{for } t_{ce} \le t \le t_h. \end{cases} \tag{6.3}$$

The parameters $E_{min,lv}$ and $E_{max,lv}$ are minimal diastolic and maximal systolic values of the left ventricular elastance function, respectively. t_h is the heart period and t_{ce} the time for onset of constant elastance.

The relation between heart period t_h and t_{ce} is given by

$$t_{ce} = \kappa_0 + \kappa_1 t_h, \tag{6.4}$$

where κ_0 and κ_1 are constant parameters (Rideout, 1991).

When the aortic valve is open, the root aortic systemic pressure p_{as} relates to ventricular outflow Q_{lv}, left ventricular cavity pressure p_{lv}, and ventricular volume V_{lv} by

$$\frac{dQ_{lv}}{dt} = \frac{1}{L_{lv}}(p_{lv} - p_{as}),$$
$$\frac{dV_{lv}}{dt} = -Q_{lv}.$$

When the aortic valve is closed,

$$Q_{lv} = 0.$$

We allow an amount of volume $V_{lv,b}$ to flow back into the left ventricle before complete closure of the aortic valve. The maximum volume we allow to flow back into the left ventricle is denoted by $\tilde{V}_{lv,b}$. When $V_{lv,b}$ exceeds the volume $\tilde{V}_{lv,b}$, the valve closes. The behavior of the right ventricle is modeled by a similar description. Both the left and right atria follow the architecture above. However, the atria are passive and described by constant elastances, E_{la} and E_{ra}, respectively. In addition, ventricular volume cannot flow from the ventricle to the atria. Detailed mathematical formulations are given in section 6.4. Parameter values for the heart chamber can be found in Tables 6.4 and 6.5.

6.3.2 The Vasculature

The structural pattern of the cardiovascular model in Figure 6.2 follows the architecture of many previous cardiovascular descriptions and may differ from these previous models only in the number of sections it uses to characterize the systemic and pulmonary circuits. The model in Figure 6.2 is divided into ten sections with different pressure and flow levels. The systemic arteries are described by three sections, and the systemic veins are characterized by two. The pulmonary circuit follows the same architecture. Each section may contain up to three components: a viscous loss term R_j, an inertial term L_j, and a term describing the elastic vessel properties represented by the compliance C_j (which is the inverse of the elastance). The index j denotes the particular section. The mathematical relation between vascular pressure and volume in each section amounts to three equations characterizing motion, conservation of mass, and the state of the system. For example, the arterial pressure p_{a1}, flow Q_{a1}, and volume V_{a1} in the first section of the cardiovascular model in Figure 6.2 are described, in the manner of Warner (1959), by the following three equations:

1. an equation of motion:

$$p_{a1} - p_{a2} = R_{a1} Q_{a1} + L_{a1} \frac{dQ_{a1}}{dt};$$ (6.5)

2. an equation for the conservation of mass:

$$\frac{dV_{a1}}{dt} = Q_{lv} - Q_{a1},$$ (6.6)

where Q_{lv} is the left ventricular outflow;

3. a state equation:

$$p_{a1} = \frac{1}{C_{a1}}(V_{a1} - V_{un,a1}).$$ (6.7)

In (6.7) we assume that the elastic vessel properties are independent of the vascular pressure levels in each section. Thus we have adopted a linear relation between pressure and volume.

This may be justified in the high pressure areas such as the arteries, but the relation is non-linear in the low pressure areas such as the veins (Ganong, 1975; Noordergraaf, 1978). For simplicity's sake we have adopted a linear relation in both pressure areas. The parameter $V_{un,a1}$ in (6.7) is the unstressed volume, which is defined as the volume at zero pressure. The vessels collapse when the volume falls below the unstressed volume, and the elastic vessels exhibit highly nonlinear properties (Ganong, 1975; Noordergraaf, 1978; Guyton, 1981).

The entire system of ordinary differential equations making up the human circulation model is given in section 6.4. Parameter values of compliance, resistances, and inductances can be found in Tables 6.3 and 6.6.

6.3.3 Determination of Parameter Values

Our selection of parameter values is guided by data available in the literature. The goal is to obtain realistic average pressure levels and volume distribution in the system. In addition, computed ventricular pressure, root aortic pressure, and outflow curves should closely resemble the corresponding curves found in people.

The strategy we use to achieve the goals may be broken down into two steps:

1. The desired average pressure and volume distribution and cardiac output determine the parameter values of compliances and resistances.

2. Inductances and elastance functions are subsequently assigned values that produce representative pressure and flow pulses.

Iterations between the two steps are inevitable in determining the final parameter values.

We follow this strategy by first determining the pressure and volume distribution in the model. To this end, average pressure and volume levels for each section are shown in Table 6.1. The volume in each section is divided into a stressed and an unstressed component. The ratio between the unstressed and stressed components is called n. The separation between the two types of volume is guided by pressure-volume curves for the vasculature. Since the number of sections in the cardiovascular model is not chosen in accordance with anatomical divisions, it is not surprising that the values of n vary among different lumped cardiovascular models (Beneken, 1965; Tham, 1988; Ursino, Antonucci, and Belardinelli, 1994). We have adopted a set of average values of n based on pressure-volume curves found in Schmidt and Thews (1976) and Guyton (1981).

The parameters of the left ventricle are κ_0, κ_1, $E_{min,l}$, a_ϕ, b_ϕ, $E_{max,l}$, and L_{lv}. The values for κ_0 and κ_1 in (6.4) of the elastance function are chosen in order to obtain systolic and diastolic time periods in close agreement with standard human pressure and flow profiles found in Noordergraaf (1978) and Guyton (1981). The minimal elastance $E_{min,l}$ provides ventricular filling, and the maximal elastance $E_{max,l}$, constants a_ϕ and b_ϕ, and inductance L_{lv} are chosen in order to obtain proper pressure and flow curves. The left ventricular elastance function (6.2) is shown in Figure 6.4.

6.3.4 Computed Results

In this section we will present some of the computed pressure and flow curves of the cardiovascular model. The purpose is to show that the computed pressure and flow profiles are in agreement with corresponding data found in the literature.

Table 6.1. *Average pressure and volume distribution among the vascular sections in the cardiovascular model. The first column specifies each section by its corresponding pressure, as defined in Figures 6.2 and 6.3. The fraction between unstressed and stressed volume is denoted by n. The approximate physiological location is shown in the last column. The total volume is 5139.8 ml.*

Section	Pressure [mmHg]	Volume [ml]	n	Location
				Systemic
p_{a1}	91.6	276	2.9	Large arteries
p_{a2}	84	507	2.7	Small arteries
p_{a3}	67.9	524	3.3	Arterioles
p_{v1}	7.3	693	6.2	Venules
p_{v2}	5.3	2328	5.0	Small veins
				Pulmonary
p_{p1}	18.7	91.4	1.2	Arteries
p_{p2}	16.6	55	1.2	Arteries
p_{p3}	11.8	74	2.6	Capillaries
p_{l1}	8.3	105.4	2.5	Veins
p_{l2}	6.0	105	2.5	Veins
				Heart
p_{ra}	2.9	78	0.6	Right atrium
p_{la}	4.9	95	0.5	Left atrium
p_{rv}	18.8	99	0.1	Right ventricle
p_{lv}	50.8	109	0.1	Left ventricle

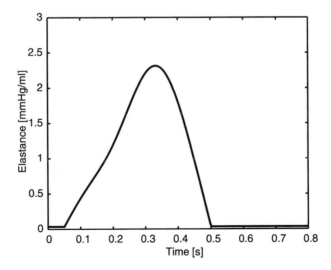

Figure 6.4. *The left ventricular time-varying elastance function $E_{lv}(t)$ given by (6.2). The heart period t_h is equal to 0.80 s.*

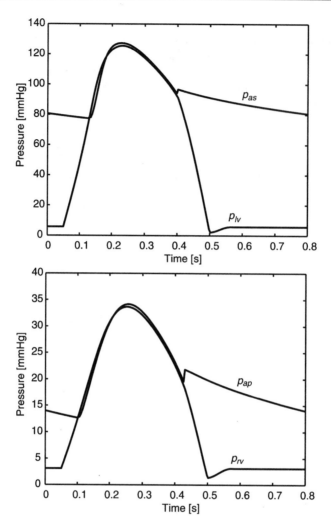

Figure 6.5. *The top panel shows the left ventricular pressure p_{lv} and the root aortic systemic pressure p_{as}. The bottom panel shows the right ventricular pressure p_{rv} and the root pulmonary pressure p_{ap}.*

The computed left ventricular pressure p_{lv} (bottom curve) and root aortic pressure p_{as} (top curve) during ejection are given in the top panel of Figure 6.5. This figure also shows the right ventricular pressure p_{rv} (bottom panel, bottom curve) and pulmonary trunk pressure p_{ap} (bottom panel, top curve) in the pulmonary circuit. Both pressures elegantly approximate representative curve shapes for normal human ventricles as found in Noordergraaf (1978) and Guyton (1981). Representative outflow curves from the right and left ventricles are given in Figure 6.6. The root aortic systemic and pulmonary pressures are 127/78 mmHg and 34/13 mmHg, respectively. The corresponding changes in the left and

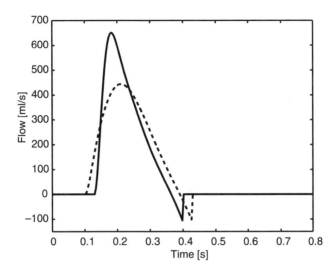

Figure 6.6. *Left ventricular outflow Q_{lv} (solid line) and right ventricular outflow Q_{rv} (dashed line). The negative outflows are accomplished by allowing volume to flow into the ventricles before the valves close.*

right ventricular volumes are shown in Figure 6.7. As a consequence of the lumped approach, the pressures in each section, p_{a1}, p_{a2}, p_{a3}, p_{v1}, p_{v2}, p_{p1}, p_{p2}, p_{p3}, p_{l1}, p_{l2}, p_{la}, p_{ra}, are not real physiological pressures but mathematical abstractions.

Classic pressure-volume curves are shown in Figure 6.7. The stroke volume V_s is 72.7 ml and the heart rate H is 1.25 Hz, resulting in a cardiac output CO equal to 90.9 ml/s. The left ventricular ejection fraction EF (EF is defined as SV/V_{ed}, i.e., stroke volume divided by end-diastolic volume) is 0.57. The results resemble data for a normal person found in Guyton (1981) and Despopoulos and Silbernagl (1991). The computed results are summarized in Table 6.2.

Hypertension and weak hearts can be studied using this model. A simulated state representing hypertension may be obtained by doubling the resistances in the systemic arteries (i.e., Ra_1–Ra_3). A weak heart may be simulated by setting the maximum elastance of the left ventricle equal to $E_{lv}/2$. Another example of a pathological condition is impaired filling of the right heart, which can be simulated by changing the filling resistance R_{ra} of the right heart. Obviously, one important feature is missing in these studies: the response from the nervous system. This response will be the topic of Chapter 7.

6.4 The Cardiovascular Model in Equations

The cardiovascular model comprises a system of differential equations. In this section we list all the equations involved. The parameter values are listed in section 6.5. The symbols and labels agree with those printed on Figures 6.2 and 6.3.

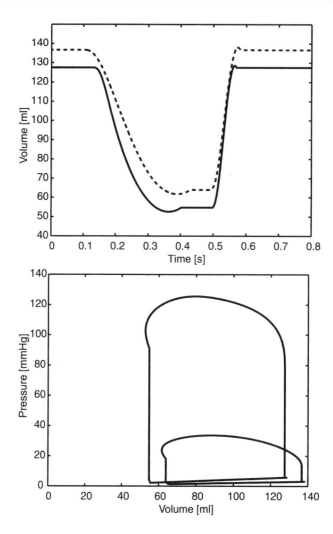

Figure 6.7. *Left ventricular volume V_{lv} (solid line) and right ventricular volume V_{rv} (dashed lines) during a cardiac cycle are shown in the top panel. The bottom panel shows pressure-volume work loops for the left (big loop) and right (small loop) ventricles. The areas encompassed by the work loops denote left and right ventricular work, respectively.*

Table 6.2. *Summary of the cardiovascular performance given by maximum systolic over end-diastolic pressures, end-diastolic volume V_{ed}, stroke volume V_s, and ejection fraction EF.*

Ventricle	Pressure (mmHg)	V_{ed} (ml)	V_s (ml)	EF
Left	127/78	127.6	72.7	0.57
Right	34/13	136.7	72.7	0.53

Left Heart Chamber

$$\frac{dQ_{la}}{dt} = \frac{1}{L_{la}}(p_{la} - p_{lv}) - \frac{R_{la}}{L_{la}}Q_{la} \qquad \text{when the mitral valve is open, } (P_{la} > P_{cv})$$

$$Q_{la} = 0 \qquad \text{when the mitral valve is closed, } (P_{la} < P_{cv})$$

$$\frac{dV_{la}}{dt} = Q_{l2} - Q_{la},$$

$$p_{la} = E_{la}(V_{la} - V_{un,la}),$$

$$\frac{dQ_{lv}}{dt} = \frac{1}{L_{lv}}(p_{lv} - p_{as}) \qquad \text{when the aortic valve is open,}^*$$

$$Q_{lv} = 0 \qquad \text{when the aortic valve is closed,}^*$$

$$\frac{dV_{lv}}{dt} = Q_{la} - Q_{lv},$$

where p_{as} is the root aortic pressure and is given by

$$p_{as} = R_{0s}Q_{lv} + p_{a1}.$$

We allow an amount of volume $V_{lv,b}$ to flow back into the left ventricle before complete closure of the aortic valve. The maximum volume we allow to flow back is denoted by $\tilde{V}_{lv,b}$. When $V_{lv,b}$ exceeds the volume $\tilde{V}_{lv,b}$, the valve closes. The volume $V_{lv,b}$ is given by

$$V_{lv,b} = -\int_{t^*}^{t} Q_{lv}dt, \quad t > t^*,$$

where t^* is the time when $p_{lv} - p_{as}$ becomes zero through positive values, to flow back into the left ventricle before complete closure of the aortic valve.

The relation between the left ventricular cavity pressure p_{lv} and the ventricular volume V_{lv} is described by

$$p_{lv} = E_{lv}(t)(V_{lv} - V_{l,lv}), \tag{6.8}$$

where $V_{d,lv}$ is the left ventricular volume at zero pressure. The elastance function $E_{lv}(t)$ in (6.8) is given by

$$E_{lv}(t) = E_{min,lv}(1 - \phi(t)) + E_{max,lv}\phi(t),$$

where

$$\phi(t) = \begin{cases} a_\phi \sin\left(\frac{\pi t}{t_{ce}}\right) - b_\phi \sin\left(\frac{2\pi t}{t_{ce}}\right) & \text{for } 0 \leq t < t_{ce}, \\ 0 & \text{for } t_{ce} \leq t \leq t_h. \end{cases} \tag{6.9}$$

---

*Conditions for opening and closing of the aortic valve are explained on page 149.

The parameters $E_{min,lv}$ and $E_{max,lv}$ are the minimal left diastolic and the maximal left systolic elastance, respectively; t_h is the heart period; and t_{ce} is the time of onset of constant elastance.

Right Heart Chamber

$$\frac{dQ_{ra}}{dt} = \frac{1}{L_{ra}}(p_{ra} - p_{rv}) - \frac{R_{ra}}{L_{ra}}Q_{ra} \quad \text{when the tricuspid valve is open, } (p_{ra} > p_{rv})$$

$$Q_{ra} = 0 \qquad\qquad\qquad\qquad \text{when the tricuspid valve is closed, } (p_{ra} < p_{rv})$$

$$\frac{dV_{ra}}{dt} = Q_{v2} - Q_{ra},$$

$$p_{ra} = E_{ra}(V_{ra} - V_{un,ra}),$$

$$\frac{dQ_{rv}}{dt} = \frac{1}{L_{rv}}(p_{rv} - p_{ap}) \qquad \text{when the pulmonary valve is open,}^*$$

$$Q_{rv} = 0 \qquad\qquad\qquad\qquad \text{when the pulmonary valve is closed,}^*$$

$$\frac{dV_{rv}}{dt} = Q_{ra} - Q_{rv},$$

where p_{ap} is the root aortic pressure and is given by

$$p_{ap} = R_{0p}Q_{rv} + p_{p1}.$$

We allow an amount of volume $V_{rv,b}$ to flow back into the right ventricle before complete closure of the pulmonary valve. The maximum volume we allow to flow back is denoted by $\tilde{V}_{rv,b}$. When $V_{rv,b}$ exceeds the volume $\tilde{V}_{rv,b}$, the valve closes. In mathematical terms this can be described as follows: Let $t = t^*$ when $p_{rv} - p_{ao}$ changes from being positive to zero; then the pulmonary valve closes, i.e., when

$$-\int_{t^*}^{t} Q_{rv}dt > \tilde{V}_{rv}, \quad t > t^*.$$

The relation between the right ventricular cavity pressure p_{rv} and the ventricular volume V_{rv} is described by

$$p_{rv} = E_{rv}(t)(V_{rv} - V_{d,rv}), \tag{6.10}$$

where $V_{d,rv}$ is the right ventricular volume at zero pressure. The elastance function $E_{rv}(t)$ in (6.10) is given by

$$E_{rv}(t) = E_{min,rv}(1 - \phi(t)) + E_{max,rv}\phi(t),$$

---

*Conditions for opening and closing of the pulmonary valve are described below.

where the function ϕ is defined by (6.9). The parameters $E_{min,rv}$ and $E_{max,rv}$ are the minimal right diastolic and the maximal right systolic elastance, respectively; t_h is the heart period; and t_{ce} is the time of onset of constant elastance.

Systemic Arterial System

First arterial section:

$$\frac{dQ_{a1}}{dt} = \frac{1}{L_{a1}}(p_{a1} - p_{a2}) - \frac{R_{a1}}{L_{a1}}Q_{a1},$$

$$\frac{dV_{a1}}{dt} = Q_{lv} - Q_{a1},$$

$$p_{a1} = \frac{1}{C_{a1}}(V_{a1} - V_{un,a1}).$$

Second arterial section:

$$\frac{dV_{a2}}{dt} = Q_{a1} - Q_{a2},$$

$$p_{a2} = \frac{1}{C_{a2}}(V_{a2} - V_{un,a2}).$$

Third arterial section:

$$\frac{dV_{a3}}{dt} = Q_{a2} - Q_{a3},$$

$$p_{a3} = \frac{1}{C_{a3}}(V_{a3} - V_{un,a3}).$$

Systemic Veins

First venous section:

$$\frac{dV_{v1}}{dt} = Q_{a3} - Q_{v1},$$

$$p_{v1} = \frac{1}{C_{v1}}(V_{v1} - V_{un,v1}).$$

Second venous section:

$$\frac{dQ_{v2}}{dt} = \frac{1}{L_{v2}}(p_{v2} - p_{ra}) - \frac{R_{v2}}{L_{v2}}Q_{v2},$$

$$\frac{dV_{v2}}{dt} = Q_{v1} - Q_{v2},$$

$$p_{v2} = \frac{1}{C_{v2}}(V_{v2} - V_{un,v2}).$$

Pulmonary Arterial System

First arterial section:

$$\frac{dQ_{p1}}{dt} = \frac{1}{L_{p1}}(p_{p1} - p_{p2}) - \frac{R_{p1}}{L_{p1}}Q_{p1},$$

$$\frac{dV_{p1}}{dt} = Q_{rv} - Q_{p1},$$

$$p_{p1} = \frac{1}{C_{p1}}(V_{p1} - V_{un,p1}).$$

Second arterial section:

$$\frac{dV_{p2}}{dt} = Q_{p1} - Q_{p2},$$

$$p_{p2} = \frac{1}{C_{p2}}(V_{p2} - V_{un,p2}).$$

Third arterial section:

$$\frac{dV_{p3}}{dt} = Q_{p2} - Q_{p3},$$

$$p_{p3} = \frac{1}{C_{p3}}(V_{p3} - V_{un,p3}).$$

Pulmonary Veins

First venous section:

$$\frac{dV_{l1}}{dt} = Q_{p3} - Q_{l1},$$

$$p_{l1} = \frac{1}{C_{l1}}(V_{l1} - V_{un,l1}).$$

Second venous section:

$$\frac{dQ_{l2}}{dt} = \frac{1}{L_{l2}}(p_{l2} - p_{la}) - \frac{R_{l2}}{L_{l2}}Q_{l2},$$

$$\frac{dV_{l2}}{dt} = Q_{l1} - Q_{l2},$$

$$p_{l2} = \frac{1}{C_{l2}}(V_{l2} - V_{un,l2}).$$

6.5 Parameter Values

Table 6.3. *Systemic parameter values of resistances, compliances, inductances, and unstressed volumes of the cardiovascular model of Figure* 6.2.

Parameter	Value	Units
R_{0s}	0.0334	mmHg $\cdot$ s/ml
R_{a1}	0.0824	mmHg $\cdot$ s/ml
R_{a2}	0.178	mmHg $\cdot$ s/ml
R_{a3}	0.667	mmHg $\cdot$ s/ml
R_{v1}	0.0223	mmHg $\cdot$ s/ml
R_{v2}	0.0267	mmHg $\cdot$ s/ml
C_{a1}	0.777	ml/mmHg
C_{a2}	1.64	ml/mmHg
C_{a3}	1.81	ml/mmHg
C_{v1}	13.24	ml/mmHg
C_{v2}	73.88	ml/mmHg
L_{a1}	0.00005	mmHg $\cdot$ s^2/ml
L_{v2}	0.00005	mmHg $\cdot$ s^2/ml
$V_{un,a1}$	205	ml
$V_{un,a2}$	370	ml
$V_{un,a3}$	401	ml
$V_{un,v1}$	596	ml
$V_{un,v2}$	1938	ml

Table 6.4. *The parameter values of the left heart chamber of Figure* 6.3. *The parameter* $\tilde{V}_{lv,b}$ *denotes the maximum volume that we allow to flow back into the left ventricle before complete closure of the valve.*

Parameter	Value	Units
H	1.25	Hz
$E_{max,lv}$	2.49	mmHg/ml
$E_{min,lv}$	0.049	mmHg/ml
E_{la}	0.075	mmHg/ml
$V_{d,lv}$	10	ml
$V_{d,la}$	30	ml
$\tilde{V}_{lv,b}$	2	ml
R_{la}	0.000089	mmHg $\cdot$ s/ml
L_{lv}	0.000416	mmHg $\cdot$ s^2/ml
L_{la}	0.00005	mmHg $\cdot$ s^2/ml

Table 6.5. *The parameter values of the right heart chamber of Figure 6.3. The parameter $\tilde{V}_{rv,b}$ denotes the maximum volume that we allow to flow back into the right ventricle before complete closure of the valve.*

Parameter	Value	Units
$E_{max,rv}$	0.523	mmHg/ml
$E_{min,rv}$	0.0243	mmHg/ml
E_{ra}	0.06	mmHg/ml
$V_{d,rv}$	10	ml
$V_{d,ra}$	30	ml
$\tilde{V}_{rv,b}$	2	ml
R_{ra}	0.0000594	mmHg $\cdot$ s/ml
L_{rv}	0.000206	mmHg $\cdot$ s^2/ml
L_{ra}	0.00005	mmHg $\cdot$ s^2/ml

Table 6.6. *Pulmonary parameter values of resistances, compliances, inductances, and unstressed volumes for the cardiovascular model of Figure 6.2.*

Parameter	Value	Units
R_{0p}	0.0251	mmHg $\cdot$ s/ml
R_{p1}	0.0227	mmHg $\cdot$ s/ml
R_{p2}	0.0530	mmHg $\cdot$ s/ml
R_{p3}	0.0379	mmHg $\cdot$ s/ml
R_{l1}	0.0252	mmHg $\cdot$ s/ml
R_{l2}	0.0126	mmHg $\cdot$ s/ml
C_{p1}	2.222	ml/mmHg
C_{p2}	1.481	ml/mmHg
C_{p3}	1.778	ml/mmHg
C_{l1}	6.666	ml/mmHg
C_{l2}	5	ml/mmHg
L_{p1}	0.00005	mmHg $\cdot$ s^2/ml
L_{l2}	0.00005	mmHg $\cdot$ s^2/ml
$V_{un,p1}$	50	ml
$V_{un,p2}$	30	ml
$V_{un,p3}$	53	ml
$V_{un,l1}$	75	ml
$V_{un,l2}$	75	ml

Table 6.7. *Summary of the parameters describing the function ϕ given in* (6.3) *and the division between the active and passive phases of the heart period in* (6.4).

Parameter	Value
a_ϕ	0.9
b_ϕ	0.25
κ_0	0.29
κ_1	0.2

Acknowledgments

Both authors were supported by High Performance Parallel Computing, Software Engineering Applications Eureka Project 1063 (SIMA—SIMulation in Anesthesia). In addition, M. Danielsen was supported by the Danish Academy for Technical Sciences, the Danish Heart Foundation (99-1-2-14-22675), and by Trinity College, Connecticut.

Chapter 7

A Baroreceptor Model

M. Danielsen and J.T. Ottesen

The primary goal of this chapter is to extend the cardiovascular model established in Chapter 6 with a baroreceptor model controlling the states (pressures) of the cardiovascular model. Baroreceptor regulation operates as a short-term control mechanism, which plays a leading role in combating, e.g., quick and moderate changes in blood volume during acute hemorrhages. It also has an important role during heart pacing, in which the heart rate varies.

Our aim in developing a model of baroreflex regulation is twofold: to obtain more insight into the underlying physiological mechanisms and to develop a model that can be inserted into an anesthesia simulator. Developing one model that fulfills both these goals may not be appropriate, since a model that tries to describe the physiology accurately may be too complex to be used in a simulator that has to run in real time. Consequently, this chapter includes two mathematical models: The first model includes only the basic effects of baroreflex regulation modeled as a function of changes in pressure, while the second model includes the effects of baroreflex regulation as a function of nervous activity (sympathetic and parasympathetic activity). In this model, nervous activity is then modeled as a function of observed changes in pressure. However, it is our intention to build two models that share part of the description. Both models will be built using experimental data and physiological arguments, and both models will be coupled to the circulatory system model described in Chapter 6. Finally, both models will include regulation of heart rate, cardiac contractility, and vessel tension.

The human circulation model is a lumped description of the cardiovascular system. Traditionally, and in order to obtain a balance in the complexity among submodels, this model should be connected to a simple baroreceptor model. The simplicity of the first baroreceptor model, presented in section 7.6, makes it possible to obtain such a balance. However, the complex second model contains a number of fundamental baroreceptor properties that are also useful in a lumped cardiovascular description.

We start this chapter with a basic introduction to the physiology of baroreceptors and to the overall performance of arterial blood pressure control provided by the baroreceptor mechanism. Section 7.1 offers an overview of the various control mechanisms of the human circulatory system; a brief introduction to the baroreceptor mechanism is given in section 7.2. The individual components of the mechanism are quantified in sections 7.3 and 7.4. Further physiological and historical details can be found in Ganong (1975), Noordergraaf (1978), Tortora and Anagnostakos (1990), Guyton (1991), and Acierno (1994).

Both baroreceptor models offer regulation of heart rate, cardiac contractility, peripheral arterial resistance, venous compliance, and venous unstressed volume by integrating various experimental observations obtained from the literature. In addition, the models include different time delays for each of the regulations. Section 7.6 presents the paradigm for the two baroreceptor models. The first model of the baroreceptor mechanism is described in section 7.6.1 and the second baroreceptor model is characterized in section 7.13.

Baroreceptor models are usually divided into a number of submodels. One submodel may describe the actual regulations of, e.g., the heart rate, whereas another submodel may provide details about how the baroreceptor firing rates depend on alterations in pressure. In previous models of the baroreceptor mechanisms the latter type of submodels tended to rely on ordinary differential equations, which embody some, and in a few cases, all of the nonlinear phenomena exhibited by the baroreceptor firing rates (Leaning et al., 1983b; Leaning et al., 1983a; Tham, 1988; Ursino, 1997; Ursino, 2000). However, two recent models by Taher et al. (1988) and Ottesen (1997b), termed the unified models, embody all the known nonlinear phenomena. In all of the above models the firing rate of the baroreceptors is not only sensitive to the average pressure but also to the change in the pressure. The approach taken in our first model is much simpler. The neural activities are related to the average arterial pressure by a sigmoidal function. In contrast, our second model adopts the unified model of the baroreceptor firing rates as a submodel.

We call this second model the *unified baroreceptor model*. The explicit coupling between the first model of the baroreceptor mechanism and the human circulation model established in Chapter 6 is described in section 7.7. The unified baroreceptor model and its regulation of human circulation are described as a whole in section 7.13. The strength of each regulation and thus our choices of parameter values in both models were mainly guided by results of the so-called open loop experiments. These key elements are experimentally observed sigmoidal curves that relate the regulations directly to the average arterial pressure. The adopted open loop experiments and time delays can be found in section 7.8. In sections 7.10 to 7.12 we will compare the published results from a number of experiments found in the literature to the corresponding computed results of the first baroreceptor model. We will show in sections 7.10 to 7.12 that the computed results during a 10% acute hemorrhage and heart pacing are in favorable agreement with corresponding experiments. In addition, the computed results display varying arterial pressure levels, depending on where sinus pulsation is superimposed on the arterial pressure. Also, we show that the control of systemic veins plays a leading role during an acute hemorrhage, whereas the remaining controls appear to have a minor function. The computed response of a 10% acute hemorrhage, using the unified baroreceptor model, can be found in section 7.14. Further details about both models are available in Danielsen (1998).

7.1 Control Mechanisms in the Human Circulatory System

Control of the human cardiovascular system involves a manifold of mechanisms whose explicit functions are not fully understood. However, we can roughly divide this complex environment into two types of control: the long- and short-term control of human circulation. *Long-term control* operates mainly via renal and humoral activities. The kidneys increase the output of water and salt in response to an enhanced arterial pressure. This action decreases blood volume and thus cardiac output. The net effect is a decline in arterial pressure. A drop in the arterial pressure promotes secretion of renin from the kidneys. Renin promotes the formation of the hormone angiotensin II, which enhances vessel constriction and thus increases arterial pressure. *Short-term regulation* is mainly mediated by the CNS and involves baroreceptors, mechanoreceptors, and chemoreceptors (Guyton, 1981; Guyton, 1991). The overall goal of neural control is to redistribute blood flow to the different areas of the body by innervating the heart and the vessels. The nervous activity from the CNS modifies the heart rate, the cardiac contractility, and the state of vessel constrictions. Chemoreceptors are sensitive to chemicals in blood and react to alterations in the concentration of oxygen, carbon dioxide, or hydrogen ions. A drop in arterial pressure may decrease the concentration of oxygen. The chemoreceptors respond by increasing cardiac strength and vessel constriction. Baroreceptors are stretch receptors that are sensitive to pressure alterations. The most important receptors are located in high pressure regions such as the carotid sinus and the aortic arch. Mechanoreceptors (or low pressure receptors) are located in the low pressure areas such as the atria and the pulmonary veins. Mechanoreceptors are also stretch receptors and provide arterial pressure control by combating alterations in venous volume. Baroreceptors are the best known and most easily accessible receptors; consequently they have been investigated extensively. Mechanoreceptors are less studied, and quantitative experimental data of these receptors are very sparse. The phenomenon of *autoregulation* is a local control mechanism independent of the CNS. Local tissues can control blood flow in response to moderate changes in cardiac output and arterial pressure via dilation or contraction of vessels (Guyton, 1991; Tortora and Anagnostakos, 1990). This may be due to a contractile response by the smooth muscles surrounding the vessels when blood vessels are stretched (Ganong, 1975).

7.2 Baroreceptor Mechanism

The baroreceptor mechanism demands the lion's share of our interest since this mechanism is believed to play the largest role in short-term pressure control. The baroreceptor mechanism provides rapid negative feedback control of arterial blood pressure. An instantaneous drop in arterial pressure is sensed by the baroreceptors, starting a chain of events leading to an increase in heart rate and cardiac contractility. This drop also stimulates the contraction of the vessels. These responses tend to alter the arterial pressure toward its previous value.

The baroreceptor mechanism has no long-term regulatory functions. An instantaneous step increase in the carotid sinus pressure is followed by enhanced firing activity in the baroreceptor nerves themselves. This firing activity declines significantly for the first few seconds and then decays more slowly. The decay continues and the time it takes the firing rate n to reach the prestimulation value can be 1 to 3 days (Guyton, 1991; Taher et al., 1988).

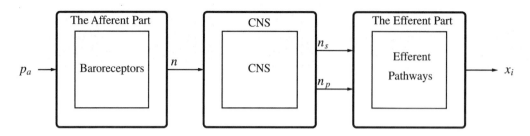

Figure 7.1. *The complete baroreceptor mechanism divided into three components: an afferent component, the central nervous system (CNS), and an efferent component. Alterations in the arterial pressure p_a generate the firing rates n in the afferent part. From the CNS the sympathetic n_s and the parasympathetic n_p nervous activities are transmitted via the efferent pathways to modify the heart and the vasculature x_i.*

The baroreceptor mechanism may be divided into three components, as shown in Figure 7.1. The first component is the *afferent part*, which contains the receptors. The firing rates n of the receptors are generated by alterations in the arterial pressure p_a. The second component is the CNS, which generates two nervous activities, the sympathetic n_s and the parasympathetic n_p. The third component, called the efferent part, consists of pathways to the individual organs in the cardiovascular system. From the CNS the two nervous activities are transmitted by the efferent pathways to alter heart rate, cardiac contractility, and vessel constriction. The state of vessel constriction is denoted x_i, where i designates the particular organ. Sections 7.3 and 7.4 quantify the three components in Figure 7.1 more carefully.

7.3 Afferent Part

As mentioned in the previous section, the baroreceptors are stretch receptors located in the vessel walls. The most accessible of these receptors are located in the carotid sinus and in the aortic arch. The carotid sinus baroreceptors are located in a distinctive part of the two common carotid sinus arteries, as shown in Figure 7.2. The aortic arch baroreceptors are located in the walls of the aortic arch. The carotid sinus receptors are the most studied, whereas the aortic arch baroreceptors have received less attention. But the aortic arch and the carotid sinus receptors are believed to be functionally equal, except that the aortic arch receptors operate at a higher pressure (Ganong, 1975). We restrict our studies to the carotid sinus baroreceptors.

Baroreceptors are nerve endings that respond to deformations in vessel walls (Ganong, 1975; Brown, 1980; Guyton, 1991). Nerve activity arises from two components: a pressure-mechanical component and a mechanical-electrical one (Brown, 1980). Baroreceptors react to deformations in the vessel walls by a pressure-mechanical mechanism. It is not known exactly how this mechanism is mediated. One suggestion is that the pressure alterations may cause changes in the cross-sectional areas of the vessels and thus deformations. The second component generates the nerve activity of the receptors via a mechanical-electrical

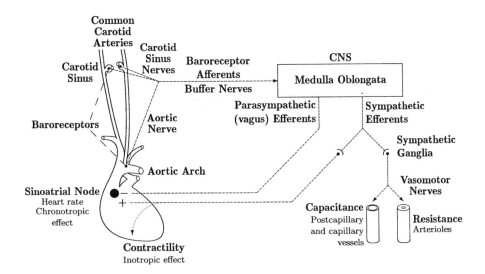

Figure 7.2. *The complete baroreceptor mechanism. The baroreceptors are located in the vessel walls of the carotid sinus and the aortic arch. Alterations in the arterial pressure generate wall deformation, which initiates the firing rates n from the receptors. The firing rates are pipelined by buffer nerves to the CNS. The CNS generates the sympathetic and parasympathetic nerve activities. The sympathetic activity plays a vital role by innervating most of the effector organs in the cardiovascular system. The parasympathetic activity modifies mainly the heart rate. Adapted from J. Ottesen (1997b), Modelling of the baroreflex-feedback mechanism with time-delay, J. Math. Biol.* **36**:1–63. *Used by permission from Springer-Verlag GmbH & Co. KG.*

mechanism within the receptors. We call this nerve activity from the carotid sinus receptors the *firing rates*, which are denoted by n. Signals from the carotid sinus and the aortic arch are transmitted from the receptors via the glossopharyngeal nerve and the vagus nerve, respectively. The two nerves are joined in the so-called buffer nerves, which direct the impulses to the CNS.

The firing rates n of the carotid sinus receptors vary with varying arterial pressure. This was first documented by Bronk and Stella (1932; 1935). Since then various experiments have revealed that the firing rates n exhibit a number of nonlinear phenomena associated with alterations in the carotid sinus pressure p_{cs}. These nonlinear phenomena include the following:

- *Threshold and saturation.* The firing rates n exhibit a threshold N below which they can be forced. In addition, the firing rates n increase with increasing carotid sinus pressure and display both low and high saturation.

- *Asymmetric response.* The firing rates n display a sigmoidal response as a function of carotid sinus pressure (Landgren, 1952). The response exhibits hysteresis or an asymmetric response.

- *Step response*. A step change in the carotid sinus pressure results in a step change in the firing rate followed by a resetting phenomenon (Brown, 1980).

- *Adaptation*. The firing rates n increase when carotid sinus pressure rises. Because the pressure is maintained at this higher level, the firing rates return to the threshold value N. This is called adaptation or resetting.

A more detailed discussion of nonlinear phenomena can be found (Taher et al., 1988; Ottesen, 1997b).

7.3.1 Models of the Firing Rates

Several efforts have been made to formulate the relation between firing rate n and carotid sinus pressure. Robinson and Sleight (1980) modeled the baroreceptor response to a step change in pressure as a simple function of time:

$$n = c_1 e^{-\frac{t}{\tau_1}} + c_2 e^{-\frac{t}{\tau_2}} + c_3,$$

where c_1, c_2, and c_3 are weighting parameters and τ_1 and τ_2 are time constants describing the resetting.

The first mathematical description of the relation between pressure and firing rates, $n - N$, was advanced by Warner (1958). Here N denotes the threshold value of the firing rates. Warner (1958) proposed a differential approach given by

$$n = k_1(p - p_0) + k_2 \frac{dp}{dt} + N, \tag{7.1}$$

where p is the pressure, k_1 and k_2 constant parameters, and p_0 the threshold pressure (i.e., the carotid sinus pressure that generates the threshold value N).

Low and high saturation, the asymmetric response, and the step response are not embodied in the model. In Warner (1958) (7.1) was extended to include the asymmetric response.

Some other models, all based on ordinary differential equations, include those of Secher and Young (1973), Spickler and Kezdi (1967), Franz (1969), and Srinivasan and Nudelman (1972). These models embody some, and in a few cases all, of the nonlinear phenomena (Taher et al., 1988; Ottesen, 1997b). However, the models consisting of differential equations make consistent use of set points, which are not physiologically based (Cecchini, Melbin, and Noordergraaf, 1981). Moreover, these models incorporate each of the nonlinearities separately, implying that they arise independently. In 1982 Cecchini, Melbin, and Noordergraaf (1981) demonstrated that this is not the case.

7.3.2 The Unified Models

Along another avenue of research we find two recent models that embody all the nonlinear phenomena listed in section 7.3 bred by adaptation, saturation, and threshold. Taher et al. (1988) proposed the first unified model:

$$\Delta n = \Delta p \left(k_1 e^{-\frac{t}{\tau_1}} + k_2 e^{-\frac{t}{\tau_2}} + k_3 e^{-\frac{t}{\tau_3}} \right) \sin^{\frac{1}{2}} \left(\frac{n}{M} \pi \right), \tag{7.2}$$

where Δn is the change in the firing rate to a step change Δp in pressure. The parameters k_1, k_2, and k_3 are weighting factors; M is the high saturation level of the firing rates; and τ_1, τ_2, and τ_3 are time constants that characterize the decay rate of Δn in response to Δp. Computation of a continuous firing rate n, using (7.2), requires a tremendous amount of bookkeeping. The second unified model, proposed by Ottesen (1997b), is a refined version of the model by Taher et al. Ottesen suggested that the firing rates n should be described by

$$n = N + \int_{-\infty}^{t} \dot{p}_{cs} \left(k_1 e^{-\frac{t-s}{\tau_1}} + k_2 e^{-\frac{t-s}{\tau_2}} + k_3 e^{-\frac{t-s}{\tau_3}} \right) \left[\frac{n(M-n)}{(M/2)^2} \right] ds, \qquad (7.3)$$

where p_{cs} is the carotid sinus pressure and $\dot{p}_{cs}$ the time derivative of p_{cs}. This is equivalent to a system of three nonlinear coupled differential equations given by

$$\begin{aligned}
\dot{\Delta n}_1 &= k_1 \dot{p}_{cs} \frac{n(M-n)}{(M/2)^2} - \frac{1}{\tau_1} \Delta n_1, \\
\dot{\Delta n}_2 &= k_2 \dot{p}_{cs} \frac{n(M-n)}{(M/2)^2} - \frac{1}{\tau_2} \Delta n_2, \\
\dot{\Delta n}_3 &= k_3 \dot{p}_{cs} \frac{n(M-n)}{(M/2)^2} - \frac{1}{\tau_3} \Delta n_3,
\end{aligned} \qquad (7.4)$$

where $n = \Delta n_1 + \Delta n_2 + \Delta n_3 - N$. The parameters k_1, k_2, and k_3 are weighting factors; τ_1, τ_2, and τ_3 are time constants describing the resetting phenomenon; M is the high saturation level of the firing rates; and N is the threshold value of the firing rates. The model contains eight parameters, k_1, k_2, k_3, τ_1, τ_2, τ_3, M, and N, found from data in the literature and through curve-fitting procedures. Typical values of the parameters are found in Table 7.1.

Table 7.1. *Typical parameter values for the unified model* (7.4).

Type	Value	Units
τ_1	0.5	s
τ_2	5.0	s
τ_3	500	s
k_1	0.5	Hz/mmHg
k_2	0.5	Hz/mmHg
k_3	1.0	Hz/mmHg
N	30	Hz
M	120	Hz

The values of the weighting factors and the time constants vary significantly in the literature (Taher et al., 1988). The time constant τ_3 may be assigned values in minutes, hours, or days. In addition, Ottesen (1997b) uses different sets of parameter values to simulate the nonlinear phenomena. The parameters may vary due to the physiological processes involved in each of the nonlinear phenomena.

According to (7.4), the firing rates n vary linearly with the instantaneous time derivative of the carotid sinus pressure p_{cs}. The history of the carotid sinus pressure is included by integration of $\dot{p}_{cs}$ via the three exponentials. The physiological interpretation of the three components Δn_1, Δn_2, and Δn_3 in (7.4) is that they are sensitive to different types of changes in the carotid sinus pressure. The component Δn_1 with the smallest time constant is sensitive to quick changes in $\dot{p}_{cs}$. The component Δn_3 with the greatest time constant τ_3 is sensitive to slower phenomena. It depends not only on the instantaneous value of $\dot{p}_{cs}$ but also on its integrated value. The component Δn_3 closely follows the pattern of behavior of the carotid sinus pressure p_{cs}. Intermediate responses are captured by Δn_2. Ottesen (1997b) hypothesizes that the three components represent different baroreceptor types A, B, and C, characterized by different transmission speeds. The parameters k_1, k_2, and k_3 are the corresponding weighting factors, which may be influenced by nervous control or other external conditions. The number of weighting factors should not exceed the number of known receptors. Three components Δn_1, Δn_2, and Δn_3 are included, three being the smallest number that can capture the nonlinear phenomena (Ottesen, 1997b). The model covers the range of nonlinear phenomena listed in section 7.3. Examples of modeling results are shown in Figures 7.3 and 7.4. Figure 7.3 shows the computed firing rate response to a step increase and decrease in pressure. The computed results agree well with data reported by Brown (1980). The only discrepancy is the absence of the postexcitatory depression gap. This gap is a short time period in which the firing rate is zero before the last ascending

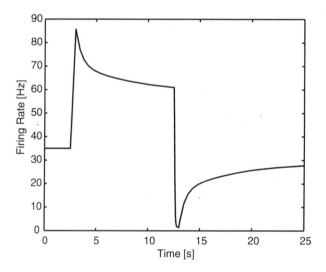

Figure 7.3. *The firing rate response n to a step increase in the pressure computed from (7.4). The pressure increases from 170 mmHg to 178 mmHg at time 2.5 s. At time 12.5 s the pressure is forced back to 170 mmHg. The computed response agrees well with the data found in Brown (1980). The only difference is the absence of the postexcitatory depression gap. Adapted from J. Ottesen (1997b), Nonlinearity of baroreceptor nerves,* Surveys Math. Ind. 7:187–201. *Used by permission from Springer-Verlag GmbH & Co. KG.*

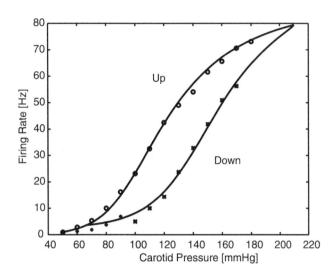

Figure 7.4. *Computed asymmetric response (solid line) based on (7.4) superimposed on experimental data found in the literature. Open circles denote measurements obtained during increase of pressure. x's indicate pressure lowered from 170 mmHg (n = 56 Hz) to 100 mmHg (n = 5 Hz). Plus signs denote pressure lowered from 100 mmHg (n = 23 Hz) to 50 mmHg (n = 1 Hz). Data adapted from Cecchini et al. (1982). The pressure increases by 1.37 mmHg/s and decreases by −1.37 mmHg/s. The parameter M is 103. Adapted from J. Ottesen (1997b), Nonlinearity of baroreceptor nerves,* Surveys Math. Ind. 7:187–201. *Used by permission from Springer-Verlag GmbH & Co. KG.*

part of the response curve in Figure 7.3. This discrepancy may be related to the model's assumption that the parameters k_1, k_2, and k_3 are independent of the Na^+ concentration. Figure 7.4 shows the asymmetric response. The asymmetric response arises from the second term on the right-hand side of (7.4), the resetting term. A more careful discussion of this model can be found in Ottesen (1997b).

7.4 CNS and the Efferent Part

The CNS and the various pathways to each of the effector organs are shown in Figures 7.1 and 7.2. The signal from the receptors arrives at the CNS via the buffer nerves. The information is then processed in the medulla oblongata of the CNS. Subsequently, the cardioinhibitory center and the vasomotor center of the medulla oblongata generate sympathetic n_s and parasympathetic n_p nerve activities, respectively. An enhanced firing rate n excites the cardioinhibitory center and inhibits stimulation of the vasomotor center. The net effect is an enhanced parasympathetic activity and a diminished sympathetic activity. The efferent pathways transmit the two nervous signals to the various parts of the cardiovascular system. (The sympathetic nerve fibers innervate most of the cardiovascular system, whereas the parasympathetic nerve fibers are restricted to the heart.) In summary, we have the following.

Enhanced sympathetic activity

- stimulates the heart rate and improves cardiac contractility,

- stimulates vessel constriction in the arteries, arterioles, and veins (Ganong, 1975; Guyton, 1991).

    Enhanced parasympathetic activity

- decreases heart rate,

- has little effect on cardiac contractility and vessel constriction (Guyton, 1991).

The entire operation of the afferent, the central, and the efferent neural systems may be illustrated using an example. An infusion of volume into the human circulatory system increases the arterial pressure, which in turn stimulates the carotid sinus baroreceptors. This stronger signal is then transmitted to the medulla oblongata via the buffer nerves. In the medulla oblongata the firing rates excite the cardioinhibitory center and inhibit stimulation of the vasomotor center. The result is a counteraction that decreases heart rate, decreases cardiac contractility, and stimulates arterial and venous vessel dilation. The latter, for example, enables blood to be stored in a venous reservoir. The net effect is a decrease in arterial pressure. This regulation is delayed by various mechanical and biochemical processes within the events listed above. Section 7.5.1 provides a more detailed discussion of this time delay.

7.5 Open Loop Descriptions of the Baroreceptor Mechanism

One part of our baroreceptor model is an empirically based function: the sigmoidal relationship σ^b between the effector responses (e.g., in the heart rate and in the peripheral resistance) and the average carotid sinus pressure $\bar{p}_{cs}$. These sigmoidal functions σ^b are typically obtained in vagotomized animals.[1] The carotid sinus pressure is increased in steps with the remaining hemodynamic variables free to move accordingly. The effector responses are measured 2 to 3 min after the step change in the carotid sinus pressure. The pressure is varied until the desired pressure range is covered. Subsequently, a sigmoidal curve is drawn showing the effector responses as functions of the average pressure. These responses are called *open loop responses*. In general, the absolute values of the open loop responses are found in animal experiments. These values may differ greatly between species. To eliminate these differences we report the relative change from the baseline value. As an example, Figure 7.5 shows the sigmoidal response in the maximum of the elastance function of the left and right ventricles. For simplicity, we assume that the low and high saturation levels are located symmetrically around the baseline value. Table 7.2 summarizes the adopted relative changes that we assume are representative for the intact human.

---

[1] In vagotomized animals the vagus is cut and the reflexes from the aortic arch receptors are eliminated.

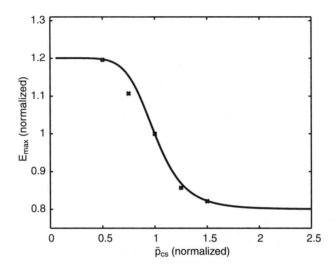

Figure 7.5. *Normalized values of the computed open loop responses of the maximum elastances $E_{max,lv}$ and $E_{max,rv}$ of the left and right ventricles as a function of the normalized average carotid sinus pressure $\bar{p}_{cs}$ superimposed on experimental data from Suga, Sagawa, and Kostiuk (1976) of the left ventricle (x). The pressures are normalized with respect to the computed steady state value (92 mmHg) and experimental data to the reported baseline values.*

Table 7.2. *Adopted low and high saturation levels for the control of heart rate H, cardiac contractility E_{max}, peripheral resistance R_{ps}, venous unstressed volume V_{un}, and venous compliance C_v.*

Parameter	Low value	High value
H [Hz]	0.25	1.75
E_{max} [mmHg/ml]	0.8	1.20
R_{ps} [mmHg·s/ml]	0.60	1.40
V_{un} [ml]	0.79	1.21
C_v [ml/mmHg]	0.90	1.10

7.5.1 Estimation of the Distributed Time Delay

As mentioned earlier, time delays occur in nervous control due to various biochemical and mechanical processes within the baroreceptor mechanism. The time delay may be viewed as the time from the perturbation of the cardiovascular system to the time when the related actions of the baroreceptor mechanism are completed. The control of the heart is a quick process, on the order of a few seconds, while regulation of the veins can take approximately 60 s. Arterial vessel constriction is intermediate between the two, taking around 10 to 15 s

(Donald and Edis, 1970; Rothe, 1983; Shoukas and Sagawa, 1973; Guyton, 1991). In addition, sympathetic activity exhibits a time delay relative to parasympathetic activity. We ignore this detail since our strategy is to obtain a simple model. Moreover, we are interested in steady states rather than the dynamics during transitions. Ottesen (1997a) offers a more detailed discussion of this topic.

7.6   The First Baroreceptor Model

In this section we present the first baroreceptor model. The structure of this model is inspired by the three building blocks of the baroreceptor mechanism shown in Figure 7.1. The approach using the unified model (7.4) follows the same structure and is presented as a whole in Section 7.13. The first baroreceptor model lumps the afferent part and the CNS into one model such that the sympathetic and parasympathetic activities are related directly to the arterial pressure. This relationship comes out of steady state experimental results, as described in section 7.6.1. The formulation of the efferent response is given in section 7.6.2. This model and the human circulation model established in Chapter 6 are connected in section 7.7.

7.6.1   Modelling the Sympathetic and Parasympathetic Activities Using a Steady State Description

We adopt a very simple approach to modeling the afferent part and the CNS. We assume that the sympathetic and parasympathetic activities are described by a sigmoidal function of the averaged carotid pressure, as reported by Korner (1971). Thus we describe the sympathetic activity n_s and the parasympathetic activity n_p by

$$n_s(\bar{p}_{cs}) = \frac{1}{1 + \left(\dfrac{\bar{p}_{cs}}{\mu}\right)^{\nu}}, \tag{7.5}$$

$$n_p(\bar{p}_{cs}) = \frac{1}{1 + \left(\dfrac{\bar{p}_{cs}}{\mu}\right)^{-\nu}}, \tag{7.6}$$

where $\bar{p}_{cs}$ is the average carotid sinus pressure, defined as the average value over one cardiac cycle. The constant μ is the average steady state arterial pressure at the baroreceptors (i.e., the adapted pressure of the baroreceptors). The parameter ν characterizes the steepness of the curves. Figure 7.6 shows n_s and n_p as functions of the carotid sinus pressure as predicted by the model.

Obviously, the model ignores all pulsatile information provided by the carotid sinus pressure that does not change the average carotid sinus pressure. In essence, the model is a very simple description of its real counterpart. In particular, the model (7.5)–(7.6) predicts that a given nervous activity follows from one and only one carotid sinus pressure, ignoring the dynamics that led to this pressure. The model (7.5)–(7.6) is justified by the low

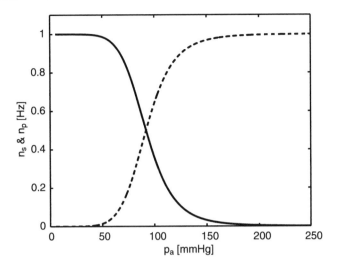

Figure 7.6. *The sympathetic (solid line) and the parasympathetic (dashed line) activity as predicted by the model (7.5)–(7.6).*

complexity of the experimental data used to establish the efferent part of this baroreceptor model. The model of the efferent part is based on experiments that relate changes in effector organs to the average carotid sinus pressure and not to the complex structure of nervous activities. In addition, we are not interested in studying the dynamic behavior during transitions, but only the behavior in steady states.

7.6.2 Formulation of the Efferent Responses

In both the first model and the unified model, the efferent responses consist of a static component and a dynamic component. The static component consists of the steady state responses σ^b. The dynamic component consists of a first order ordinary differential equation that describes the temporal dynamics. The efferent responses can be described by

$$\frac{dx_i(t)}{dt} = \frac{1}{\tau_i}(-x_i(t) + \sigma_i^b(\bar{p}_{cs})), \quad i \in E = \{H, E_{max}, R_{ps}, V_{un}, C_v\}. \qquad (7.7)$$

The index i denotes the particular efferent organ taken from the set E. The time constant τ_i characterizes the transition time for the efferent response i to take full effect. The time delay is called a distributed time delay and is distinct from a pure latency.

A sigmoidal response curve describes the response of the sympathetic n_s and parasympathetic n_p activities to changes in the average carotid sinus pressure $\bar{p}_{cs}$. Consequently, we assume that the function $\sigma_i^b(\bar{p}_{cs})$ can be expressed as a linear combination of n_s and n_p:

$$\sigma_i^b(\bar{p}_{cs}) = \alpha_i n_s(\bar{p}_{cs}) - \beta_i n_p(\bar{p}_{cs}) + \gamma_i, \quad i \in E, \qquad (7.8)$$

where α_i and β_i denote the strength of sympathetic and parasympathetic activities, respectively, on x_i. The constant γ_i is equal to x_i during complete denervation (i.e., $n_s = n_p = 0$). The parameter β_i is only nonzero when $i = H$. Combining (7.7) and (7.8), we obtain

$$\frac{dx_i(t)}{dt} = \frac{1}{\tau_i}(-x_i(t) + \alpha_i n_s(\bar{p}_{cs}) - \beta_i n_p(\bar{p}_{cs}) + \gamma_i), \quad i \in E. \tag{7.9}$$

By (7.9) we have completed the baroreceptor model. Obviously, alternative descriptions of the sympathetic n_s and parasympathetic n_p activities could be used in the model. This possibility follows directly from the model structure consisting of three components, as shown in Figure 7.1.

The model (7.9) contains a number of parameters: α_i, β_i, γ_i, and τ_i. Their values were chosen based on experimental data available in the literature. Unfortunately, firm experimental data do not exist allowing estimation of all of these parameters, and the amount of data is small. Thus determining parameter values is a balance between general experimental data available in the literature and data fitting to special topical cases such as acute hemorrhage.

7.7   The Baroreceptor Model and the Cardiovascular System

To allow the model of the baroreceptor mechanism to control a circulatory system we couple the model to the cardiovascular model established in Chapter 6. The coupling between the two models forms a closed negative feedback mechanism, as shown in Figure 7.7. The carotid sinus pressure in the baroreceptor model is set to be equal to the arterial pressure, p_{a1}, in the first section of the cardiovascular model of Figure 6.2. The individual components in this model are quantified in sections 7.7.1 and 7.7.2.

7.7.1   Control of the Two Ventricles

The two ventricles are modified by the inotropic and chronotropic effects. These effects are incorporated into the model, thereby modifying the heart's condition. The *inotropic effect* or the regulation of contractility is given by

$$\frac{dx_{E_{max,lv}}(t)}{dt} = \frac{1}{\tau_{E_{max,lv}}}(-x_{E_{max,lv}}(t) + \alpha_{E_{max,lv}} n_s(\bar{p}_{a1}) + \gamma_{E_{max,lv}}), \tag{7.10}$$

$$\frac{dx_{E_{max,rv}}(t)}{dt} = \frac{1}{\tau_{E_{max,rv}}}(-x_{E_{max,rv}}(t) + \alpha_{E_{max,rv}} n_s(\bar{p}_{a1}) + \gamma_{E_{max,rv}}), \tag{7.11}$$

where $E_{max,lv}$ and $E_{max,rv}$ are used in the elastance function (6.2) defined in section 6.3.1. By (7.10) and (7.11) we neglect the possible parasympathetic influence on cardiac contractility and follow the simplest approach.

The *chronotropic effect* or heart rate regulation is contained in

$$\frac{dx_H(t)}{dt} = \frac{1}{\tau_H}(-x_H(t) + \alpha_H n_s(\bar{p}_{a1}) - \beta_H n_p(\bar{p}_{a1}) + \gamma_H). \tag{7.12}$$

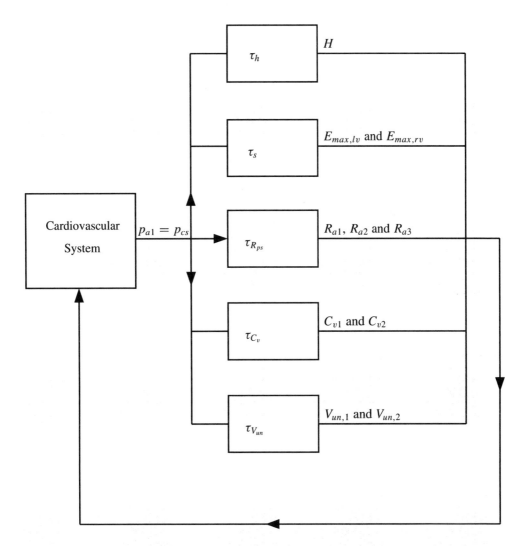

Figure 7.7. *The baroreceptor model coupled to the human circulation model established in Chapter 6. The arterial pressure p_{a1} enters the baroreceptor model and equals the carotid sinus pressure. The baroreceptor model offers control of the heart rate H and the cardiac contractility represented by the maximum of the elastance functions, $E_{max,lv}$ and $E_{max,rv}$, respectively. In addition, it modifies the three arterial resistances R_{a1}, R_{a2}, and R_{a3}; the venous compliances C_{v1} and C_{v2}; and the two venous unstressed volumes $V_{un,2}$ and $V_{un,1}$. The parameters τ_h, τ_s, $\tau_{R_{ps}}$, τ_{C_v}, and $\tau_{V_{un}}$ characterize the distributed time delay for the individual controls.*

7.7.2   Control of the Vasculature

The vascular efferent components involve the arterial resistances, the venous unstressed volumes, and the venous compliances of the systemic circulation. The control of the *peripheral resistance* is given by

$$\frac{dx_{R_{a1}}(t)}{dt} = \frac{1}{\tau_{R_{a1}}}(-x_{R_{a1}}(t) + \alpha_{R_{a1}}n_s(\bar{p}_{a1}) + \gamma_{R_{a1}}), \tag{7.13}$$

$$\frac{dx_{R_{a2}}(t)}{dt} = \frac{1}{\tau_{R_{a2}}}(-x_{R_{a2}}(t) + \alpha_{R_{a2}}n_s(\bar{p}_{a1}) + \gamma_{R_{a2}}), \tag{7.14}$$

$$\frac{dx_{R_{a3}}(t)}{dt} = \frac{1}{\tau_{R_{a3}}}(-x_{R_{a3}}(t) + \alpha_{R_{a3}}n_s(\bar{p}_{a1}) + \gamma_{R_{a3}}). \tag{7.15}$$

Thus we control the three resistances R_{a1}, R_{a2}, and R_{a3} of the systemic arterial system shown in Figure 6.2 of Chapter 6.

The control of the *unstressed volume* involves both sections in the systemic venous system and is directed by

$$\frac{dx_{V_{un1}}(t)}{dt} = \frac{1}{\tau_{V_{un1}}}(-x_{V_{un1}}(t) + \alpha_{V_{un1}}n_s(\bar{p}_{a1}) + \gamma_{V_{un1}}), \tag{7.16}$$

$$\frac{dx_{V_{un2}}(t)}{dt} = \frac{1}{\tau_{V_{un2}}}(-x_{V_{un2}}(t) + \alpha_{V_{un2}}n_s(\bar{p}_{a1}) + \gamma_{V_{un2}}), \tag{7.17}$$

where $V_{un,1}$ and $V_{un,2}$ are the unstressed volumes in the two venous sections of the systemic circulation shown in Figure 6.2.

The control of the *venous compliance* is defined by

$$\frac{dx_{C_{v1}}(t)}{dt} = \frac{1}{\tau_{C_{v1}}}(-x_{C_{v1}}(t) + \alpha_{C_{v1}}n_s(\bar{p}_{a1}) + \gamma_{C_{v1}}), \tag{7.18}$$

$$\frac{dx_{C_{v2}}(t)}{dt} = \frac{1}{\tau_{C_{v2}}}(-x_{C_{v2}}(t) + \alpha_{C_{v2}}n_s(\bar{p}_{a1}) + \gamma_{C_{v2}}), \tag{7.19}$$

where the compliances C_{v1} and C_{v2} are defined in the human circulation model of Figure 6.2. We allow the unstressed volumes and the compliances in both venous sections to be altered since the experimental data and the lumped approach do not allow individual measures of each of these volumes and compliances. The experiments do not distinguish between alterations in the arterial unstressed volume $\Delta V_{un,a}$ and the venous unstressed volume $\Delta V_{un,v}$. Since most of the circulatory volume resides in the veins, we assume that the entire change in unstressed volume is mediated by the veins.

7.8   Determination of Parameter Values

The values of the parameters α_i, β_i, γ_i, and τ_i for an intact human result from a compromise between the experimental open loop responses reported in section 7.5 and the experimental data from an acute hemorrhage by Hosomi and Sagawa (1979) (see section 7.10). The latter experiments are used because the open loop responses are not definitive, since they are obtained in vagotomized animals and thus the experiments are not representative of an intact subject. The values of the parameters α_i, β_i, and γ_i are first determined from the low and high saturation levels listed in Table 7.2. Later these or some of these values are modified in order to minimize the difference between the model-computed and experimental results during an acute hemorrhage.

7.9 Results

In this section we show the results obtained with our models. The computed open loop responses for the model, shown in Figures 7.8 to 7.10, are obtained by varying the carotid sinus pressure p_{cs} (sufficiently slowly) and observing the concomitant efferent responses. Figure 7.8 shows the open loop response in cardiac contractility $E_{max,lv}$ and $E_{max,rv}$ and heart rate H superimposed on the data by Suga, Sagawa, and Shoukas (1973) and Korner (1974); Bolter and Ledsome (1976); and Greene (1986), respectively. The computed heart rate response for the intact human is chosen to follow the experiments by Korner (1974) in the central range. The experimental data by Bolter and Ledsome (1976) and Greene (1986) are obtained with the contribution of the aortic arch eliminated. Thus they are not fully representative of an intact human.

The open loop responses for the three arterial resistances R_{a1}, R_{a2}, and R_{a3} are shown in Figure 7.9 superimposed on the experimental data by Greene (1986), Cox and Bagshaw (1975), and Shoukas and Brunner (1980). Cox and Bagshaw (1975) discriminate between the resistance in the femoral and celiac parts of the systemic arterial system. The resistance varies more in the outer part than in the more internal parts of the cardiovascular system. The lumped approach adapted here does not allow such a division.

The venous unstressed volume and the venous compliance exhibit the open loop responses shown in Figure 7.10. The top panel shows the open loop response of the venous compliance superimposed on the experimental data of Shoukas and Brunner (1980). The bottom panel shows the absolute venous unstressed volume superimposed on the absolute changes in the systemic unstressed volume (Shoukas and Sagawa, 1973; Shoukas and Brunner, 1980).

Figure 7.11 shows a first order validation of the baroreceptor model, a comparison between computed and experimental open loop responses of the carotid sinus pressure to the arterial pressure. The response from the baroreceptor model appears to be more powerful than that shown in the experimental data. It should be kept in mind that the experimental data are measured after the aortic arch receptors are eliminated. Consequently, this comparison is an imperfect validation of the model.

In sections 7.10 to 7.12 we compare animal experiments found in the literature with the corresponding computed results from the first model. This evaluation involves three different experiments: acute hemorrhage, heart pacing, and responses during different pulsatile carotid sinus pressures.

The regulated elements in the baroreceptor model are not equally important. To give the reader a better understanding of these differences, section 7.10.1 provides a sensitivity analysis studying the impact that the controls of H, R_{ps}, S, C_v, and V_{un} have on cardiovascular performance during an acute hemorrhage.

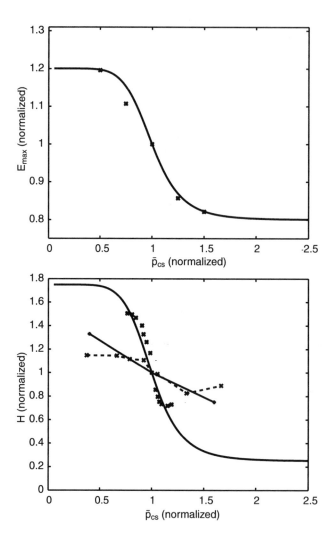

Figure 7.8. *The top panel shows the computed open loop responses of the maximum elastances $E_{max,lv}$ and $E_{max,rv}$ as functions of the normalized average carotid sinus pressure $\bar{p}_{cs}$ superimposed on the experimental data from Suga, Sagawa, and Kostiuk (1976) (x). The bottom panel displays the computed open loop responses of the heart rate H as a function of the normalized average carotid sinus pressure $\bar{p}_{cs}$ superimposed on experimental data from Korner (1974) (x), Bolter and Ledsome (1976) (dashed, x), and Greene (1986) (solid line, +). The pressures are normalized with respect to the computed steady state value (92 mmHg) and experimental data to the reported baseline values.*

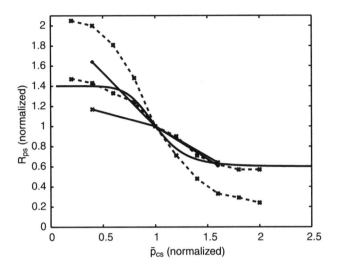

Figure 7.9. *Computed open loop response for the three arterial resistances R_{a1}, R_{a2}, and R_{a3} superimposed on the experimental data from Greene (1986) (solid line, x), Cox and Bagshaw (1975) (dashed lines, x), and Shoukas and Brunner (1980) (solid line, +). The computed pressures are normalized with respect to the steady state value (92 mmHg) and the experimental data to the reported baseline values.*

7.10   Acute Hemorrhage

An acute hemorrhage from the femoral arteries is followed by reduced cardiac filling, arterial pressure, cardiac output, and stroke volume, and increased heart rate and vessel constriction. These general effects were found in experiments by Hosomi and Sagawa (1979) in which 10% of the total circulatory volume was removed via the femoral arteries in 30 s. These experiments are simulated by a constant leak in the third section of the systemic arterial system. Figure 7.12 displays the computed and experimental results for an intact human. The figure shows the relative changes in average arterial pressure $\bar{p}_a$, cardiac output CO, heart rate H, and peripheral resistance R_{ps}. The agreement between computed and experimental results is striking. The response during complete denervation is shown in Figure 7.13. Again, the computed results agree well with the experiments. The computed results are obtained using only the average arterial pressure p_{a1}. This is consistent with the results reported by Kumada et al. (1970). They concluded that the pulsatile component of the carotid sinus pressure has little effect compared with that of the average carotid sinus pressure during a 20% acute hemorrhage.

7.10.1   Sensitivity Results during a Hemorrhage

Control of the venous unstressed volume has a profound impact on the overall performance of the circulatory system, as is clear from Figure 7.14. Figure 7.14 shows the computed

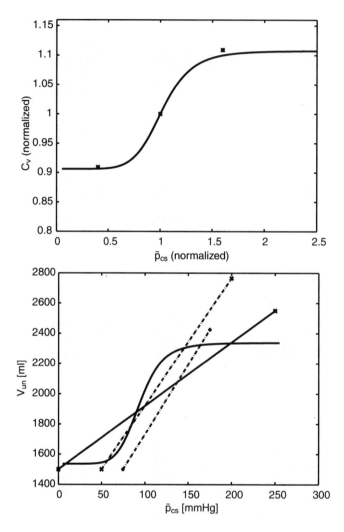

Figure 7.10. *The top panel shows the computed open loop response for the venous compliances C_{v1} and C_{v2} superimposed on data from Shoukas and Brunner (1980) (marked by x). The graphs are plotted as functions of the normalized average carotid pressure. The bottom panel shows the computed open loop response for the venous unstressed volume $V_{un,2}$ (solid line) and the total change in computed venous unstressed volume as functions of the nonnormalized average carotid pressure from 0 to 250 mmHg (solid line, x). The bottom panel also shows the corresponding experimental changes in the total venous unstressed volume obtained from Shoukas and Sagawa (1973) (dashed line, +) and Shoukas and Brunner (1980) (dashed line, x).*

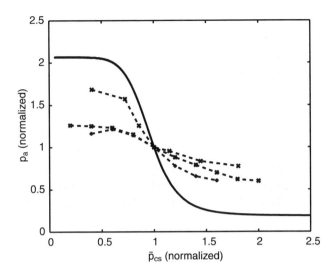

Figure 7.11. *Computed open loop responses for the arterial pressure p_{a1} super-imposed on experimental data by Cox and Bagshaw (1975) (dashed line, *), Bolter and Ledsome, (1976) (dashed line, x), and Shoukas and Brunner (1980) (dashed line, +). The computed pressures are normalized with respect to the steady state value (92 mmHg) and experimental data to the reported baseline values.*

responses during acute hemorrhage when control of the venous unstressed volume is weakened (in these experiments the values of α_i, β_i, and γ_i are reduced to 21%, 10%, or 0% of their original values). Discrepancies between experimental results and their computed counterparts arise as the strength of this control is reduced, while leaving the remaining regulatory strengths unaltered. The relative changes in arterial pressure p_a, cardiac output CO, peripheral resistance R_{ps}, and heart rate H increase. Heart rate H, however, stays within the same range as the experimental values. However, the response in H differs from the experiments when control of the unstressed volume is absent. These results are consistent with the computed results generated by Ursino, Artioli, and Gallerani (1994).

In contrast to unstressed venous volume, heart rate H, peripheral resistance R_{ps}, venous compliance, C_v, and cardiac contractility E_{max} play no major roles during an acute hemorrhage, as is evident from Figures 7.15 and 7.16. Figure 7.15 displays computed results simulated under various strengths of heart rate control. The results show that heart rate has only a weak influence after an acute hemorrhage. Figure 7.16 shows the computed results when the control of cardiac contractility E_{max}, peripheral resistance R_{ps}, and venous compliance C_v are eliminated. In particular, ignoring the control of peripheral resistance generated an improved cardiac output but no significant changes in average arterial pressure $\bar{p}_a$ or heart rate H compared with an intact human. Raising the cardiac contractility E_{max} or the heart rate H has only a weak impact on the circulatory response. This result makes sense because cardiac performance cannot be improved much due to the poor ventricular filling during an acute hemorrhage. A consequence may be that performance deteriorates

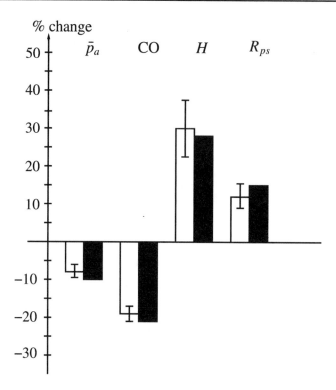

Figure 7.12. *Computed (black bars) and experimental (white bars) relative changes in average arterial pressure $\bar{p}_a$, cardiac output CO, heart rate H, and peripheral resistance R_{ps} during an acute 10% hemorrhage from the femoral arteries. The experimental results are adopted from Hosomi and Sagawa (1979) for an intact animal. The total circulatory volume is reduced by 10% in 30 s, and the experimental data are obtained 1 to 2 min after the hemorrhage is terminated. The computed results are obtained in the new steady state reached after 1 min.*

further by pumping more blood out of the heart. The minor contribution from the change in venous compliance may be related to minor absolute volume changes associated with this reflex.

The computed arterial pressure p_{a1} during hemorrhage, under conditions of different impairments of the efferent regulatory components, is shown in Figures 7.17 and 7.18. Essentially, the two figures summarize the results stated above. The top panel of Figure 7.17 shows the pivotal role played by the baroreceptor mechanism during an acute hemorrhage by displaying the computed results with and without an intact baroreceptor mechanism. The bottom panel shows the arterial pressure p_{a1}, when control of cardiac contractility is absent, and the computed result with an intact baroreceptor mechanism. No clear difference exists between these two results. This similarity demonstrates the weak role played by the control of cardiac contractility, $E_{max,lv}$ and $E_{max,rv}$, after an acute hemorrhage. The impact

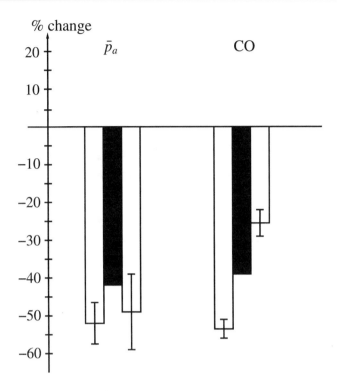

Figure 7.13. *Computed (black bars) and experimental (white bars) relative changes in average arterial pressure $\bar{p}_a$ and cardiac output CO during an acute 10% hemorrhage from the femoral arteries with complete denervation (i.e., no active carotid sinus and aortic arch receptors). The experimental results depend on the order of carotid sinus and aortic arch denervation. The aortic arch is denervated first in the left column. In the right column the carotid sinus is denervated first. The experiments are adopted from Hosomi and Sagawa (1979). The total circulatory volume is reduced by 10% in 30 s, and the experimental data are obtained 1 to 2 min after the hemorrhage are terminated. The computed results are obtained in the new steady state reached after 1 min.*

of the venous compliance can be neglected, as is evident from the top and bottom panels of Figure 7.18.

7.11 Heart Pacing

Stroke volume V_s drops as heart rate H increases. In addition, as heart rate increases cardiac output CO first increases, reaches a maximum, and finally declines (Kumada, Azuma, and Matsuda, 1967; Melbin et al., 1982). Figure 7.19 shows that computed stroke volume V_s and cardiac output CO during heart pacing are in qualitative agreement with these experimental

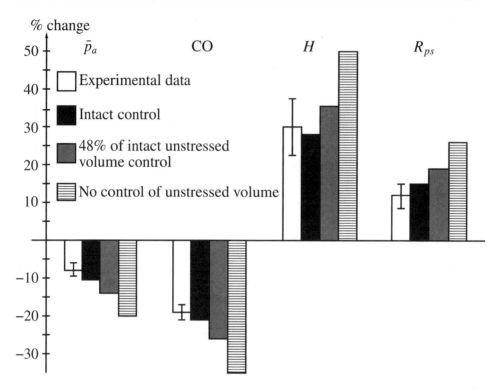

Figure 7.14. *Computed and experimental (white bars) relative changes in average arterial pressure $\bar{p}_a$, cardiac output CO, heart rate H, and peripheral resistance R_{ps} during an acute 10% hemorrhage from the femoral arteries with different control strengths of the unstressed volume. The experiments are adopted from Hosomi and Sagawa (1979). The total circulatory volume is reduced by 10% in 30 s, and the experimental data are obtained 1 to 2 min after the hemorrhage is terminated. The computed results are in the new steady state after 1 min.*

observations. Heart rate H is taken as the independent variable, whereas the remaining effector organs are influenced by the baroreceptor mechanism.

The computed results show that stroke volume consistently falls as heart rate increases. The computed result is in qualitative agreement with experiments reported by Kumada, Azuma, and Matsuda (1967) and Melbin et al. (1982). The computed result exhibits a reduced end-diastolic volume during the rising heart rate, which contributes to the decline in stroke volume and the behavior of cardiac output during atrial pacing (Kumada, Azuma, and Matsuda, 1967; Melbin et al., 1982). The increase in the number of strokes cannot compensate for the drop in stroke volume when heart rate exceeds approximately 2.3 Hz. Consequently, cardiac output and arterial pressure drop significantly, as shown in the bottom panel of Figure 7.19 and in the top panel of Figure 7.20, respectively. The bottom panel of Figure 7.20 shows that peripheral resistance increases when heart rate exceeds 2.3 Hz, an effect that may decrease stroke volume even further. The results show that the

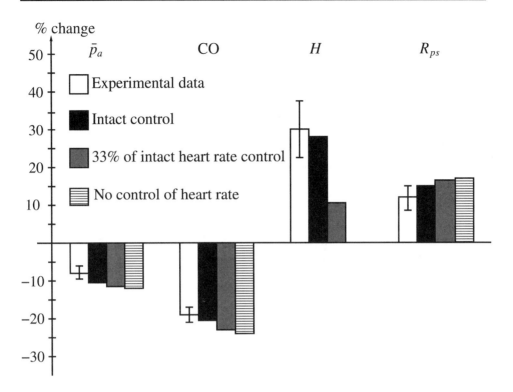

Figure 7.15. *Computed and experimental (white bars) results when the regulation of the heart rate is varied in strength. The figure shows the relative changes in average arterial pressure $\bar{p}_a$, cardiac output CO, heart rate H, and peripheral resistance R_{ps} during an acute 10% hemorrhage from the femoral arteries with different control strengths on the heart rate. The experiments are adopted from Hosomi and Sagawa (1979). The total circulatory volume is reduced by 10% in 30 s, and the experimental data are obtained 1 to 2 min after the hemorrhage is terminated. The computed results are obtained in the new steady state reached after 1 min.*

baroreceptor mechanism cannot effectively combat the drop in cardiac output when heart rate is sufficiently high. A sudden and sustained increase in peripheral resistance lowers stroke volume and cardiac output. Figure 7.19 shows the computed results when peripheral resistance is increased to $1.7R_{ps}$. This result agrees favorably with the experiments by Kumada, Azuma, and Matsuda (1967).

Experiments have shown that dV_s/dH as a function of heart rate H remains unaltered in a number of different alertness states. These include upright, awake, recumbent, and under the influence of various anesthetics. The curve is altered when, e.g., peripheral resistance is increased suddenly. Figure 7.21 displays the computed results superimposed on the data given by Melbin et al. (1982). The figure shows close agreement between the computed and experimental results when heart rate runs from 1.33 to 2.2 Hz. Discrepancies arise when heart rate increases above this range. The computed dV_s/dH declines significantly when

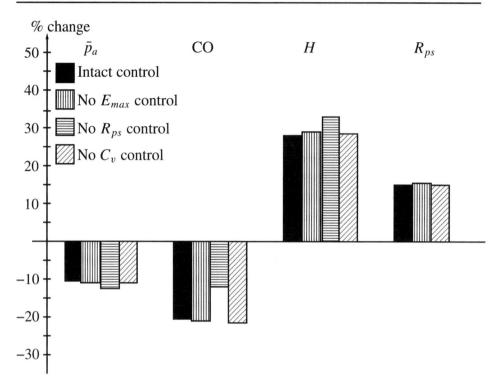

Figure 7.16. *Computed results when the control of cardiac contractility E_{max}, peripheral resistance R_{ps}, or venous compliance C_v is removed. The figure shows the relative changes in average arterial pressure $\bar{p}_a$, cardiac output CO, heart rate H, and peripheral resistance R_{ps} during an acute 10% hemorrhage from the femoral arteries. The computed results are obtained in the new steady state reached after 1 min.*

the heart rate exceeds 2.2 Hz. This decrease corresponds to an accelerated drop in stroke volume with heart rate. To understand the reason behind the discrepancies it may be useful to study some of the previous results and the heart rate model. In the previous computed results cardiac output displayed a drop when the heart rate exceeded 2.3 Hz. This fall started at a lower heart rate than in the animal experiments. The reasons for these discrepancies may be related to the simple model of ventricular performance. In the ventricular elastance function (6.2), the division between the active phase and the passive phase is specified by (6.4), which is a linear function of the heart period t_h. When the heart rate increases, diastole shortens much more than systole, which implies that ventricular filling is impaired (Tortora and Anagnostakos, 1990). This effect is included in the model. However, one reason for the discrepancy may be that diastole shortens too quickly, as a function of heart rate, when (6.4) is used. In addition, autoregulation and control from the chemoreceptors have not been included in the model. These two phenomena may play a role at this high value of the heart rate.

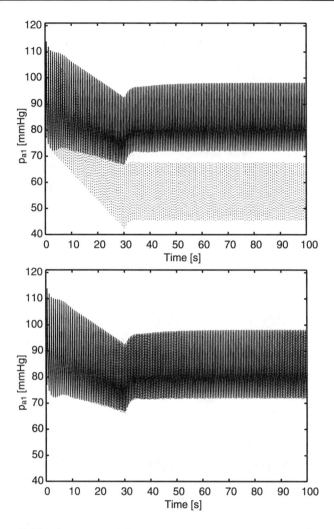

Figure 7.17. *The top panel shows the computed arterial pressure p_{a1} during a 10% acute hemorrhage with the entire baroreceptor mechanism active (solid line) and during complete denervation (dotted line). The bottom panel shows the computed arterial pressure p_{a1} with the entire baroreceptor mechanism active (solid line) and with no control of the cardiac contractility.*

7.12   Responses to Pulsatile Carotid Sinus Pressure Using the First Model

Exposure to different pulsatile carotid sinus pressures is a critical test of this model, since it is built exclusively on steady state results and is used in a pulsatile environment. Schmidt, Kumada, and Sagawa (1972) studied experimentally the response of average arterial pressure, peripheral resistance, and cardiac output to pulsations in the carotid sinus pressures in

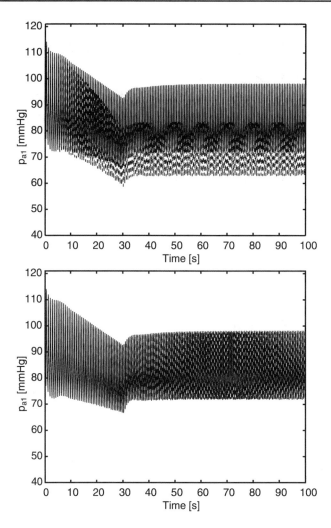

Figure 7.18. *The top panel show the computed arterial pressure p_{a1} during a 10% acute hemorrhage with the entire baroreceptor mechanism active (solid line) and with no control of the veins (dotted line) (i.e., no control of compliance, C_{v1} and C_{v2}, and unstressed volume, V_{un1} and V_{un2}). The bottom panel shows computed arterial pressure p_{a1} with the entire baroreceptor mechanism active (solid line) and with no control of the venous compliance, C_{v1} and C_{v2} (dotted line).*

vagotomized animals. In these experiments the carotid sinus pressure p_{cs} is taken as the independent variable (i.e., open loop) and the hemodynamic variables are altered accordingly. The carotid sinus pressure equals a static pressure $p_{cs,0}$ plus a pulsatile term $A \sin(2\pi f t)$, where the frequency is $f = 2$ Hz. Thus,

$$p_{cs} = p_{cs,0} + A \sin(2\pi f t), \tag{7.20}$$

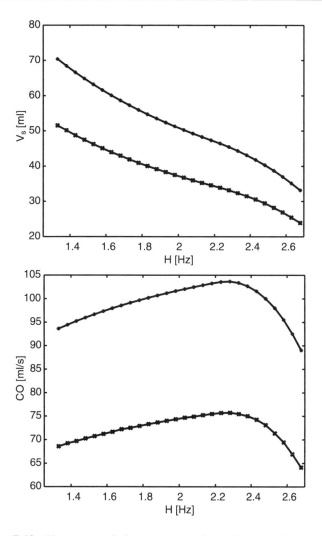

Figure 7.19. *The top panel shows computed steady state values of stroke volume V_s with the peripheral resistance R_{ps} (+) and with a higher peripheral resistance $1.7R_{ps}$ (x) during heart pacing. The bottom panel shows the corresponding computed cardiac output with the peripheral resistance R_{ps} (+) and with a higher peripheral resistance $1.7R_{ps}$ (x) during heart pacing.*

where A is the amplitude. In order to study the effects of pulsation, we replace the average pressure $\bar{p}_{cs}$ in the model of the sympathetic and parasympathetic activities, (7.5) and (7.6), with the instantaneous pressure p_{cs} in (7.20). We mimic the vagotomized condition by taking $n_p = 0$ and reduce the strength of heart rate control such that the low saturation level is 0.5 and the high level is 1.5 (see Table 7.2).

The computed results are shown in the top panel of Figure 7.22. This figure shows the average arterial pressure p_{a1} as a function of the static pressure $p_{cs,0}$ when carotid si-

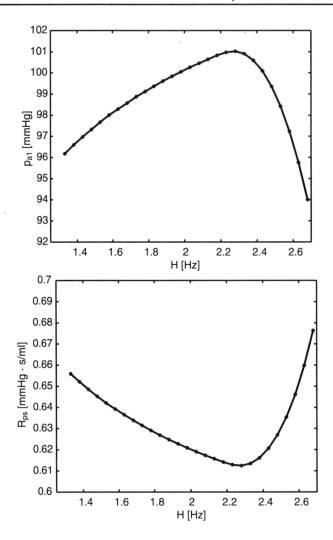

Figure 7.20. *The top panel shows computed steady state values of the average of arterial pressure p_{a1} during heart pacing. The bottom panel shows the corresponding computed steady state values for the peripheral resistance.*

nus pressure is given by (7.20) with $A = 0$ superimposed on the predicted response when $A = 25$ mmHg. In these computations the static pressure $p_{cs,0}$ has been varied from 30 to 250 mmHg in steps of 15 mmHg. The displayed results are obtained in the steady state. According to the computations, shown in the top panel of Figure 7.22, the effect of pulsation in carotid sinus pressure is a drop in arterial pressure when the static pressure $p_{cs,0}$ is below 90 mmHg. In contrast, arterial pressure is unaltered when $p_{cs,0}$ is equal to 90 mmHg and when arterial pressure is in the high and low saturation regions. When the static pressure $p_{cs,0}$ exceeds 90 mmHg, the effect of pulsation in carotid sinus pressure is a rise in the arterial pressure. The same behavior pattern is predicted in peripheral resistance and in cardiac

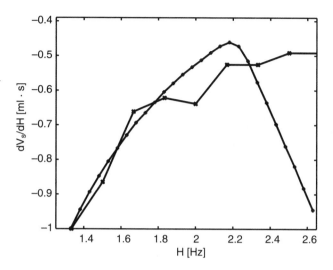

Figure 7.21. *Computed dV_s/dH (+) as a function of heart rate H superimposed on data by Melbin et al. (1982) (x) in the heart rate range from 1.33 Hz to 2.6 Hz.*

output. The bottom panel of Figure 7.22 shows the experimental results by Schmidt, Kumada, and Sagawa (1972). These results also show a drop in average arterial pressure when the carotid sinus (7.20) becomes pulsatile. The effect of pulsatility is a marked drop in arterial pressure when $p_{cs,0}$ falls below 150 mmHg, but arterial pressure is practically unaltered when $p_{cs,0}$ is equal to 150 mmHg. The behavior above 150 mmHg and below 75 mmHg is not available. Thus the computed sigmoidal behavior pattern cannot be directly compared with the experiments. However, the drop in arterial pressure is clearly more pronounced in the experiments than in the computations. The experiments also display a fall in peripheral resistance and in cardiac output. Again, the fall is more accentuated in the experiments than in the corresponding computations. One probable reason for the differences between the experiments and the computations may be the simple description (7.5) of the relation between sympathetic activity and the average arterial pressure. Explicit knowledge of the processes in the CNS is not available yet, but knowledge does exist regarding the relation between firing rates n of carotid sinus receptors and the carotid sinus pressure p_{cs}. This relation is not a building block in the model but is lumped into the formulation (7.5). Experimentally, the firing rates are affected by carotid sinus pressure (7.20) in a characteristic fashion, as shown by Chapleau and Abboud (1987). These experiments show that the firing rates n of baroreceptors increase sigmoidally with the static carotid sinus pressure when the pressure is varied from 40 to 200 mmHg. When a pulsatile pressure is superimposed on the static carotid sinus pressure, the curve can become practically linear. Thus the effect of pulsatility in the experiments is that firing rates rise when the average pressure is below approximately 100 mmHg and drop when the average pressure exceeds approximately 100 mmHg. The firing rates are almost unaltered for high pressures and at 100 mmHg. In essence, the computed pressure in the top panel of Figure 7.22 is consistent with the experiments by Chapleau and Abboud (1987).

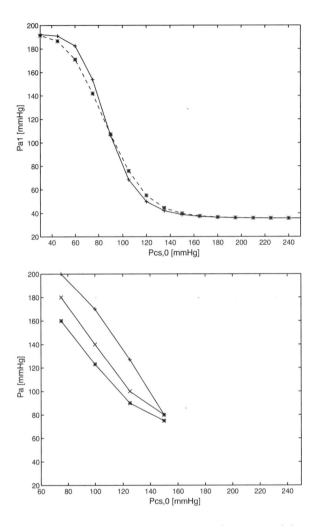

Figure 7.22. *The top panel shows the computed average of the arterial pressure p_{a1} as a function of the static pressure $p_{cs,0}$ when the carotid sinus pressure is given by (7.20) with $A = 0$ (solid line, +) after vagotomy. This is superimposed on the predicted pressure when $A = 25$ mmHg (dashed line, $*$). The results are obtained in the steady state. The bottom panel displays the corresponding experimentally obtained arterial pressure p_a, when $A = 0$ (+, the upper curve) in (7.20), as a function of $p_{cs,0}$. This is superimposed on the obtained arterial pressure when $A = 12.5$ (x, the middle curve) and when $A = 25$ ($*$, the lower curve). The experiments are adopted from Schmidt, Kumada, and Sagawa (1972).*

The behavior of the computed results when the pressure is given by (7.20) can be explained from the sigmoidal shape of the curve relating sympathetic activity n_s to arterial pressure (7.5). The symmetry around the central point of the sigmoidal curve generates no changes in the average pressure. As the static term $p_{cs,0}$ decreases the top portion of the pulsatile term is reduced relative to the lower part, due to the bending of the sigmoidal curves. This implies a lower average value of n_s. In contrast, the lower portion of the pulsatile term is reduced when the static term $p_{cs,0}$ is higher than the baseline value.

7.13 Unified Baroreceptor Model

The unified baroreceptor model distinguishes itself from the first model by modeling the afferent part and the CNS. These two models are established in sections 7.13.1 and 7.13.2. The model of the efferent responses is the same as the formulation in the first model, with minor differences, as explained in detail in section 7.13.3. The parameter values of the unified model are discussed in section 7.13.4.

7.13.1 Model of the Afferent Part

The model of the afferent part assumes that the firing rates n of the carotid sinus receptors can be described by the model (7.4), repeated here for convenience:

$$\dot{\Delta n_1} = k_1 \dot{p}_{cs} \frac{n(M-n)}{(M/2)^2} - \frac{1}{\tau_1}\Delta n_1, \tag{7.21}$$
$$\dot{\Delta n_2} = k_2 \dot{p}_{cs} \frac{n(M-n)}{(M/2)^2} - \frac{1}{\tau_2}\Delta n_2,$$
$$\dot{\Delta n_3} = k_3 \dot{p}_{cs} \frac{n(M-n)}{(M/2)^2} - \frac{1}{\tau_3}\Delta n_3,$$

where

$$n = \Delta n_1 + \Delta n_2 + \Delta n_3 - N.$$

The parameters k_1, k_2, and k_3 are weighting factors; τ_1, τ_2, and τ_3 are time constants describing the resetting phenomenon; M denotes the saturation level of the firing rates; and N is the threshold value of the firing rate n. The model contains eight parameters k_1, k_2, k_3, $\tau_1, \tau_2, \tau_3, M$, and N, found from data in the literature and through curve-fitting procedures. Further details of the unified model (7.22) are offered in section 7.3.2.

7.13.2 Generation of the Sympathetic and Parasympathetic Activities

Information processing in the CNS is complex, and the explicit interaction between firing rate n, sympathetic activity n_s, and parasympathetic activity n_p is not available yet. Accordingly, and out of a desire for simplicity, we describe n_s as decreasing for increasing n by

$$n_s(\bar{n}) = \frac{1}{1 + \left(\dfrac{\bar{n}}{\mu_n}\right)^{\nu_n}}, \tag{7.22}$$

where $\bar{n}$ is the average of the firing rates and the constant μ_n is the average of the firing rates generated by the steady state carotid sinus pressure (i.e., generated by the adapted pressure at the carotid sinus receptors). The parameter ν_n characterizes the steepness of the curve. The parasympathetic activity is given by

$$n_p(\bar{n}) = 1 - n_s \tag{7.23}$$

$$= \frac{1}{1 + \left(\dfrac{\bar{n}}{\mu_n}\right)^{-\nu_n}}. \tag{7.24}$$

Recent experiments by Wang, Brandle, and Zucker (1993) show that the sympathetic activity n_s may decrease exponentially with the average $\bar{n}$ of the firing rates. The difference between (7.22) and Wang, Brandle, and Zucker (1993) is that (7.22) predicts n_s to be higher for small values of $\bar{n}$. These experiments may only provide guidance since they relate n_s to $\bar{n}$ and not to the real dynamics.

7.13.3 Efferent Responses

The mathematical formulation of the efferent responses is quantified by (7.10) to (7.19). However, when the unified baroreceptor is used, $n_s(\bar{p}_{cs})$ and $n_p(\bar{p}_{cs})$ should be replaced with $n_s(\bar{n})$ and $n_p(\bar{n})$, respectively. Thus the efferent responses can be described by

$$\frac{dx_i(t)}{dt} = \frac{1}{\tau_i}(-x_i(t) + \alpha_i n_s(\bar{n}) - \beta_i n_p(\bar{n}) + \gamma_i), \quad i \in E = \{H, E_{max}, R_{ps}, V_{un}, C_v\}. \tag{7.25}$$

The parameters are explained in section 7.6.2.

7.13.4 Parameter Values

The unified baroreceptor model contains ten parameters, k_1, k_2, k_3, τ_1, τ_2, τ_3, N, M, μ_n, and ν_n, which are not included in the first model. These parameters are determined as follows: The parameter value of μ_n is equal to the average firing rate generated by the steady state arterial pressure (the adapted pressure of baroreceptors) and ν_n is determined as in section 7.13.2. The values of the time constants τ_1, τ_2, τ_3 are taken from Ottesen (1997b) and are given in Table 7.1. The values of the weighting factors k_1, k_2, k_3 are determined from a comparison between the computed and experimental results during an acute hemorrhage, taking into account the physiological significance of the parameters. During the 30-s hemorrhage we assume that the slow component has the dominant role ($k_3 = 1.5$ Hz/mmHg), the intermediate component plays a minor role ($k_2 = 0.5$ Hz/mmHg), and the quick component is of no importance ($k_1 = 0$ Hz/mmHg).

7.14 Acute Hemorrhage Using the Unified Baroreceptor Model

In this section we compare the results of a simulated acute hemorrhage with the corresponding experiments by Hosomi and Sagawa (1979). In these experiments 10% of the total

circulatory blood volume is removed via the femoral arteries in 30 s. The experiments are
simulated as explained in section 7.10.

The first direct comparison is given in Figure 7.23. It shows the computed and
experimental relative changes in average arterial pressure $\bar{p}_a$, cardiac output CO, heart rate
H, and peripheral resistance R_{ps}. The results display a striking agreement between the
computed and experimental results.

Note that these results are very similar to the ones obtained with the first baroreceptor
model presented in section 7.10; compare Figures 7.12 and 7.23. Closer scrutiny of the
results reveals minor differences; the main difference is that the cardiac output CO and the
heart rate H obtained with the unified model are slightly lower than (less than one percent
of) the values obtained with the first model. From these results we can conclude that, even

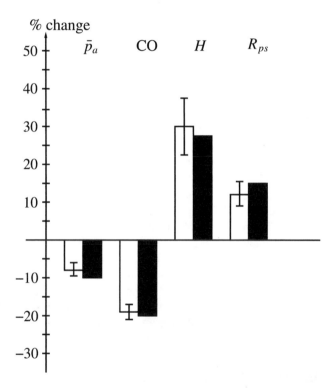

Figure 7.23. *Computed (black bars) and experimental (white bars) relative
changes in average arterial pressure $\bar{p}_a$, cardiac output CO, heart rate H, and peripheral
resistance R_{ps} during an acute 10% hemorrhage from the femoral arteries. The experimental results are adopted from Hosomi and Sagawa (1979) for an intact animal. The total
circulatory volume is reduced by 10% in 30 s, and the experimental data are obtained 1 to
2 min after the hemorrhage is terminated. The computed results are obtained in the new
steady state reached after 1 min. Note that this figure is very similar to Figure 7.12. However, in this figure the heart rate H and cardiac output CO are approximately 1% smaller
than the corresponding values in Figure 7.12.*

though the unified model includes more detailed aspects of the physiology, the first model is equally good at describing the net effect of the regulation responding to a 10% hemorrhage. However, the dynamics described by the two models are different; for example, the two curves for the arterial pressure P_a as functions of time are different, even though their end-points coincide, i.e., at the end of the hemorrhage and one minute thereafter when the system reaches steady state. One reason for this deviation is the nonlinear description of the function of the baroreceptor nerves themselves in the unified model. Here the firing rates exhibit a hysteresis, as shown in Figure 7.4, among other phenomena described in section 7.3.

Adaptation is built into the model, as discussed in section 7.3.2. The adaptation after the hemorrhage is shown in Figures 7.24 and 7.25. Figure 7.24 shows the instantaneous changes in the heart rate H and the resistance R_{a3} during and after the hemorrhage. Figure 7.25 shows the concomitant change in the average arterial pressure p_{a1} and the average firing rates n. The adaptation directs the firing rates n, the heart rate H, and the resistance R_{a3} toward their approximate prehemorrhage values. In contrast, the arterial pressure will reach a lower value, close to the pressure level obtained, if control from the baroreceptor mechanism is absent during the hemorrhage. This is shown in Table 7.3. We stress that this adaptation results exclusively from the baroreceptors, so the computations exclude many other contributing mechanisms.

As seen in Figures 7.24 and 7.25, the arterial pressure p_{a1} decays faster than the decline in the heart rate and the peripheral resistance and the rise in the firing rates. The reason for this difference is that the total effect on the pressure consists of a nonlinear interaction between the controls; the total effect is not simply the sum of the individual effects. Figures 7.24 and 7.25 show that a knot appears in the curves of the heart rate, the arterial pressure, and the firing rates in the early stages of the hemorrhage. The knot is only weakly displayed in the curve of the peripheral resistance because of the higher value of the distributed time delay. The knot is also observed in the first model, but is less pronounced. In contrast, the knot is not observed if the bleeding is carried out in veins instead of in the third section in the cardiovascular model. When the controls from the baroreceptors are absent, the computed firing rates n, and the pressure, exhibit a small change in the decay rate in the early stage of the hemorrhage. We conclude therefore that the appearance of the knot depends on where the hemorrhage is carried out and evolves out of the interaction between the cardiovascular system and the baroreceptor mechanism.

7.15 Summary and Discussion

This chapter has presented two models of the human baroreceptor mechanism, based on experimental data and physiological arguments, and described the explicit coupling of the two models to the human circulation model in Chapter 6. Both models include control of heart rate, cardiac contractility, peripheral resistance, venous compliance, and venous unstressed volume. The first model of the baroreceptor mechanism exhibits the principal features of short-term pressure control during an acute hemorrhage, alterations in peripheral resistance, and changes in the heart rate in the range below 2.2 Hz. Due to its simplicity, the model is also computationally fast. The unified baroreceptor model displays a striking agreement between the computed and experimental results during a 10% hemorrhage. One major

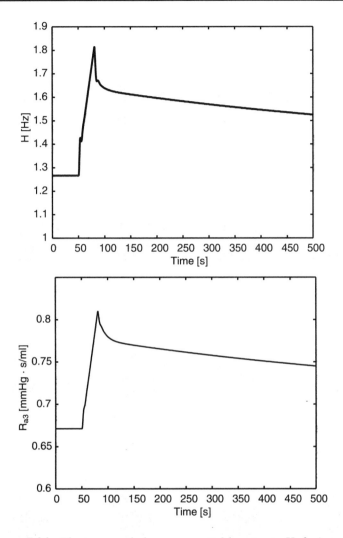

Figure 7.24. *The top panel shows computed heart rate H during and after the 10% acute hemorrhage, which starts at 50 s and ends at 80 s. The bottom panel shows the concomitant changes in the resistance R_{a3}.*

Table 7.3. *The values of the firing rates n, the heart rate H, the peripheral resistance R_{a3}, and the average arterial pressure $\bar{p}_{a1}$ before the hemorrhage and after the hemorrhage when the adaptation is completed.*

Parameter	Value before	Value after	Units
n	49.1	47	Hz
H	1.27	1.37	Hz
R_{a3}	0.67	0.70	mmHg· s/ml
$\bar{p}_{a1}$	94	63	mmHg

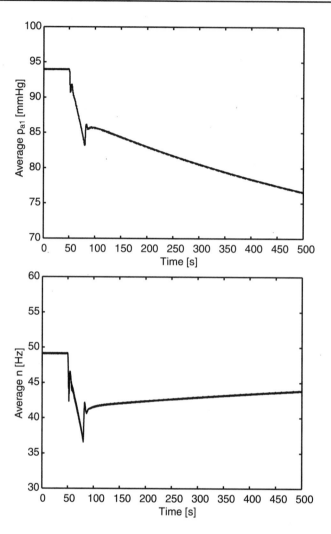

Figure 7.25. *The top panel shows the computed average of the arterial pressure p_{a1} during and after the 10% acute hemorrhage, which starts at 50 s and ends at 80 s. The bottom panel shows the concomitant changes in the average of the firing rates n.*

difference between the first model and the unified model is that the unified model exhibits adaptation, e.g., after a hemorrhage. A knot appears clearly in the graphs of arterial pressure and of efferent controls with the lowest transient time when the hemorrhage is carried out via the third section of the cardiovascular model. The knot disappears, however, when the hemorrhage is carried out via the veins. The appearance of the knot may depend on where the hemorrhage is carried out and evolves from the interaction between the cardiovascular system and the baroreceptor mechanism. Future animal experiments may reveal whether the knot is a real phenomenon.

The model paradigm divides the baroreceptor mechanism into three distinct physiological components: the afferent part, the CNS, and the efferent part. The first model lumps the first two components and relates the sympathetic activity n_s and the parasympathetic activity n_p directly to the average arterial pressure by two sigmoidal functions. The unified model adopts a more physiological approach and describes each component separately. In both models the efferent part is divided into static and dynamic components. The dynamic component consists of a first order differential equation with a time constant characterizing the transition time required for the efferent controls to take full effect. Both models allow discrimination between sympathetic activity n_s and parasympathetic activity n_p in the static component, such that the heart rate is computed from a linear combination of n_s and n_p, while the remaining controls are linear in n_s. One could argue that the validation of the model should be expanded. Indeed, a number of other experiments exist that could validate the model. Examples of these experiments include hemorrhages with weaker strength, tilting of patients, and pulsation of the carotid sinus pressure with different frequencies and different responses in atrial pressures. Also, computed responses to heart failures and changes in the vascular parameters could have been considered. Finally, it would be interesting to show the impact of the parameters on the sigmoidal curves.

The control from *mechanoreceptors* (or low pressure receptors) has been ignored. This control can combat alterations in the venous volume before activation of the baroreceptors (Guyton, 1991). The model by Ursino (2000) includes control from baroreceptors and mechanoreceptors. His computed results suggest that control mediated by the mechanoreceptors plays a major role in the first stage of an acute hemorrhage, while the drop in the circulatory volume is less than 150 to 200 ml. In this range he observed no evident fall in arterial pressure. When the drop in circulatory volume exceeds this level, the dominant control comes from the baroreceptor mechanism. Our model reacts to changes in circulatory volume only if they affect the arterial pressure. Thus control mediated by the mechanoreceptors constitutes a possible extension of the model and it may be relevant in applications such as an anesthetic simulator in which a hemorrhage is a topical case. Unfortunately, published data relevant to this control are very sparse. The literature offers only qualitative descriptions and describes no explicit dependence of the mechanoreceptors on the pressure changes.

The effect of *autoregulation* is omitted in the model. In fact, the model assumes the response in the efferent organs to be controlled entirely by the neural signals of n_s and n_p. However, during an acute hemorrhage, the vessel may dilate, by local controls, in order to provide blood flow to tissue with a high demand for oxygen, such as the brain. Control from the *chemoreceptors* is also ignored since they are not believed to play a major role in short-term pressure control, compared with the baroreceptor mechanism's role. These aspects may be a part of future model improvements.

The estimation of parameter values needs to be critically addressed in the unified baroreceptor model. The unified model contains eight parameters, k_1, k_2, k_3, τ_1, τ_2, τ_3, M, and N. Except for the values of the weighting parameters k_1, k_2, and k_3, these parameter values are adopted from Ottesen (1997b). He obtained the values from data in the literature and through curve-fitting procedures. We altered the values of the weighting factors k_1, k_2, k_3 in order to obtain a better agreement between the computed and experimental acute hemorrhage results. For the acute hemorrhage we assumed that the slow component has the dominant role ($k_3 = 1.5$ Hz/mmHg), the intermediate component plays a minor role

($k_2 = 0.5$ Hz/mmHg), and the quick component is of no importance ($k_1 = 0$ Hz/mmHg). The computed results are sensitive to alterations in the values of k_2 and k_3. Adaptation after hemorrhage is accelerated when k_2 and k_3 have higher values. If the acceleration is too strong, the computed results will not be in agreement with the corresponding experimental results. In addition, the parameter values of the unified model appear to depend on the particular physiological processes involved in each experiment, as also pointed out by Ottesen (1997b). This dependence is not yet available.

How simple can a baroreceptor model be and still include the main behaviors of its real counterpart? As discussed previously in this chapter, the first baroreceptor model contains an afferent part that lacks the sophistication used in many previous models. Explicit knowledge of the CNS part of the baroreceptor mechanism is not yet available. Thus tentative descriptions cannot be replaced with descriptive models before further experimental studies are available. However, we do have knowledge about the relation between the firing rates of the carotid sinus receptors and the arterial pressure. We have not fully used this knowledge, but merely lumped it into the description of the sympathetic and parasympathetic activities. The sensitivity to the change in pressure $\dot{p}_{cs}$ is not described in the first model. This may be one reason for the results we observed during pulsation in the carotid sinus pressure. According to the results of Chapleau and Abboud (1987), the effect of pulsation can change the relation between baroreceptor activity and arterial pressure from being sigmoidal to being practically linear. Studies of the effects of pulsation require a more advanced description of the afferent part, including sensitivity to change in the carotid sinus pressure $\dot{p}_{cs}$. The unified baroreceptor model, which embodies all the nonlinear phenomena of the carotid sinus baroreceptors, can be used in such a study. Alternatively, less comprehensive models can be used, as in the recent proposal by Ursino (2000), which is based on a high-pass filter approach. In essence, including sensitivity of the firing rates to $\dot{p}_{cs}$ in the first model will broaden the range of the description of the baroreceptor mechanism. Gratifying as a comprehensive mathematical description of the afferent part may seem, we still have to defeat at least one other obstacle, namely, ignorance about the CNS. In general, the description of the CNS shows up as a weak link in the baroreceptor models due to the lack of available data. If in the future more experimental knowledge is provided about the relation between firing rates and the sympathetic and parasympathetic activities, the information provided by, e.g., the unified model can be better utilized.

Acknowledgments

Both authors were supported by High Performance Parallel Computing, Software Engineering Applications Eureka Project 1063 (SIMA—SIMulation in Anesthesia). In addition, M. Danielsen was supported by the Danish Academy for Technical Sciences, the Danish Heart Foundation (99-1-2-14-22675), and Trinity College, Connecticut.

Chapter 8

Respiration

T.G. Christensen and C. Dræby[1]

8.1 Introduction

The models of the previous chapters have been concerned with the circulation and its regulation. In this chapter we complete this system of physiological models with models of the transport of blood gases and respiration. For this system, it seems natural to divide the modeling effort into two parts: creating a model of the lung and creating a model of the blood gas transport. Below we identify relevant literature and give our rationale for our choice of modeling elements.

A survey of the literature reveals several models of different aspects of the respiratory system. Some of these models describe the total system, while others provide detailed insight into the function of specific subsystems. Our review will focus on how the models treat some crucial aspects of the physiology: the gas flow in and out of the lung and the mixing of gas inside the lung; the gas exchange between the lung and blood; the blood transport system and the modeling of gas dissociation in the blood and the tissue; and the modeling of the relation between gas dissociation and pH value in the blood.

In models of both the blood transport system and the lung a special kind of differential equation model, the compartment model (Chiari et al., 1995; Olofsen, 1994; Rideout, 1991; Jacquez, 1985), is applicable. A compartment model uses a compartment to represent a region holding an amount of matter that is instantaneously mixed inside the compartment. The differential equation describing the progression of matter in the compartment follows from the principle of mass conservation. The rate of change of matter in the compartment equals the total flow into the compartment minus the total flow out of the compartment. Compartment models are widely used in the modeling of biology and medicine (Jacquez, 1985) and can describe both physical flows like the gas flow over the lung membrane and more abstract flows such as a substance changing into another by chemical reaction.

---

[1] Edited by V. Andreasen and J. Larsen from Christiansen and Dræby (1996).

8.1.1 Lung Modeling

The aim with the lung model derived in this chapter is to be able to replicate the connection between the atmosphere and alveoli in such a way that, e.g., simulations of defective lungs are possible. During surgery, the aspect of ventilation that can be measured is the gas composition of the expired air. During artificial respiration, a respiratory mask measures the pressure and tidal volume. These quantities are usually plotted dynamically in a pressure-volume loop. Disorders of the lung can be observed by disturbances in these outputs.

It is our goal to construct a model that can reproduce pressure-volume diagrams and keep track of the partial pressures of gases in the expired air and in the alveoli. Even though alveolar pressures are not measurable during an operation, their values will affect the gas status of the blood and will cause changes in the tensions of the gas in the arterial blood. Hence an indirect goal of the model is that it keep track of alveolar partial pressure.

The relevant defects to simulate are lungs with abnormal compliance, lungs with increased resistance to airflow, and lungs with reduced permeability of the membrane dividing the pulmonary capillaries and the alveoli.

These cases will be simulated for both an artificially ventilated patient and a naturally breathing one. Under various clinical circumstances, the air inspired by the patient is mixed with anesthetic agents. Thus in the model the composition of inspired air will be a changeable input parameter.

Recent models of the respiratory system are found in the work of Olofsen (1994) and Chiari et al. (1995). Older models are found in the work of Fincham and Tehrani (1983), Hoppenstaedt and Peskin (1992), Longobardo, Cherniack, and Fishman (1966), and Saunders, Bali, and Carson (1980). A common feature among these models is that the control system has a prominent place in each of them. Even though we do not treat the control system in any detail here, we find these models relevant to our work because they are the most general models of the respiratory system we have found. All the models use compartments to represent areas of the body and have mass balance equations for each compartment.

The first model, by Olofsen (1994), focuses on the special situations in which ventilation is absent (apnea). The primary results are the blood oxygen saturation curve and the CO_2 pressure curve in blood as a function of time. The curves show that the oxygen supply to the tissues is acceptable for up to 8 min without ventilation if the lung is initially filled with pure oxygen.

The second model, by Chiari et al. (1995), is presented as a general model of the respiratory system. It is claimed to compare well to experimental data when simulations are done with an increased CO_2 and decreased O_2 level in inspired air.

Both models use a single compartment for the lung, describing both the gas-filled alveoli and the blood-filled capillaries. Thus instant equilibrium between the gas phase and the blood phase is assumed. Olofsen represents the body by a single compartment, while Chiari et al. use a brain compartment and a compartment representing the rest of the body. The brain compartment is used for regulation of the system, reflecting the fact that the central chemoreceptor for carbon dioxide is situated in the brain.

The transport of oxygen and carbon dioxide between compartments is determined by dissociation of the gases in the blood. The theoretical foundations for the exact nature of this dissociation differ significantly between the two models; see section 8.3.4. Olofsen uses a detailed dissociation model by Siggaard-Andersen. Chiari et al. use a piecewise linear curve for the dissociation of O_2 and a model of their own for the dissociation of CO_2. It is not clear how Olofsen incorporates the acid/base balance of the blood into the dissociation curves. Chiari et al. neglect it for oxygen, but their model of carbon dioxide dissociation does include the effect.

The simplest models of the lung are the models used in the compartment models of the respiratory system. Here the lung is modeled as a single compartment with a constant volume, representing the mean lung volume. The compartment represents both the alveoli and the blood in the capillaries. There is instantaneous equilibrium between the gas and blood phases, and the flow in and out of the compartment describes gas carried by blood and gas carried by air. The airflows in and out of the lung are modeled as separate nonpulsatile flows (Chiari et al., 1995; Olofsen, 1994).

All models more complex than a single compartment partition the lung into several sections and describe a pulsative gas flow in and out of these sections. Such models, modeling the actual "bellows" of the lung, are called *models of the mechanics of the lung*. Examples of such models are Rideout (1991, Chap. 5); Golden, Clark, and Stevens (1973); and Jackson and Milhorn (1973). Furthermore, we have examined an unpublished model developed by Galster, who kindly sent us this model in the form of a computer program source code listing (Galster, 1995). All models of the mechanics of the lung are expressed in terms of an equivalent electrical network. They differ, however, in how they partition the lung and in whether they use constant or variable parameters for compliances and airflow resistances.

While the single compartment models of the lung consider a uniform gas mix to exist in the compartment, and thus easily calculate the partial pressures of gases in the air mix, models of the lung mechanics are normally not concerned with the mix of gases. The sole exception is the model in Rideout (1991) in which the total pressures in various lung sections of the network model are used to drive the flows in a compartment model. The compartment model also keeps track of the gas mix in the lung sections.

Models of gas exchange fall into two categories. One type of model describes the gas exchange based on the ratio between ventilation and perfusion (Evans, Wagner, and West, 1974; Hoppensteadt and Peskin, 1992; Poon and Wiberg, 1981; Riley and Cournand, 1949; West, 1974). The ventilation-perfusion ratio is much used in the literature of physiology, and a main concern of all these models is the situation in which the ratio is not uniform throughout the lung.

Other models have a more direct approach. Here the focus is on the composition of gas on each side of the lung membrane and the flux through the membrane (Granger et al., 1987; Piiper and Scheid, 1981).

In this chapter we will describe a model that includes descriptions of the lung and the blood transport system. Since the lung model must allow simulations with different compliances and airflow resistances and produce pressure-volume diagrams, a single constant flow ventilated compartment is not sufficient. The approach of describing airflows between separate sections of the lung seems more appropriate. This approach must be combined with a model of the mix of gases in the lung sections because we need both the partial pressures, for deciding the transport over the lung membrane, and the composition of the expired air, for output. This is similar to the approach of Rideout (1991), but the fundamental assumptions of his model are not clear. Therefore, we will use the idea of a pressure model and a gas composition model and develop a model of the gas mix from the overall flows.

Our lung membrane model will not be based on the ventilation-perfusion ratio, even though that is a commonly used parameter in the literature; instead we will model the membrane flux explicitly. If the ventilation-perfusion ratio is needed later, it can be calculated from the gas flow in the lung and the blood flow. Furthermore, the explicit approach gives the possibility of modeling impairment of transport due to membrane limitations as well as inhomogeneous ventilation-perfusion ratios.

8.1.2 Blood Gas Transport

A model of the blood transport system must keep track of the different quantities describing the status of CO_2 and O_2 content in blood at various places in the body. The most important quantities are those observed during surgery: the tension of oxygen and carbon dioxide, the saturation of hemoglobin with oxygen, and the pH of blood plasma. During surgery, these quantities are normally measured in both arterial and venous blood.

When changes or abnormalities in the monitored data are observed, they are normally ascribed to disturbances in ventilation, metabolism, or the blood circuit, or to an abnormal concentration of blood components that interact with the respiratory gases.

Changes in the blood concentration of O_2 and CO_2 are complex because of the chemical interactions between respiratory gases, hemoglobin, and hydrogen ions. It is our goal that the model output a pH based on the concentration of carbon dioxide. If the gas dissociation model includes variables representing the important gas-carrying blood components, e.g., hemoglobin, better simulations will be allowed of a patient with abnormal levels of these.

In a clinical setting, two possible countermeasures to disturbances in blood chemistry are changing the composition of inspired air and artificially ventilating the patient. It is our goal to develop a model that can reproduce the changes in the measured blood data that are effected by these countermeasures.

Some respiratory system models aim at describing pharmacokinetics along with the blood transport system (Andreasen et al., 1994; Lerou et al., 1991; Hull, 1979; Bischoff and Dedrick, 1968).

These transport models are compartment models like the respiratory models, but generally have more compartments. The structure of the models is the same as that of the respiratory models, except that the movement of matter by the blood is governed by different dissociation curves.

A completely different approach to modeling the kinetics of matter in the blood transport system is found in the one-, two-, or three-compartment models used to describe pharmacokinetics (Gibaldi and Perrier, 1982; Hull, 1979; Jacquez, 1985). These models claim no direct connection to the physiology but are fitted to produce good approximations of the time course of the concentration of an anesthetic agent in the blood. The basic dissociation model is known in physics as Henry's law. It states that if no chemical reaction takes place between solute and solvent, then the solubility is constant at low concentrations of the solute (Atkins, 1990). Even in situations in which chemical reaction takes place, Henry's law is normally regarded as valid for the part of the solute that has not reacted chemically.

The level of detail varies greatly among models of oxygen and carbon dioxide dissociation. Examples of oxygen dissociation models are a three-piece linear curve (Longobardo, Cherniack, and Fishman, 1966), a ten-piece linear curve (Rideout, 1991), a quotient between two fourth degree polynomials (Kelman, 1966), and a rather complicated expression by Siggaard-Andersen (Siggaard-Andersen et al., 1988; Siggaard-Andersen et al., 1984; Siggaard-Andersen and Siggaard-Andersen, 1995). Siggaard-Andersen models the dissociation of carbon dioxide as well. The model by Chiari et al. (1995) is based on considerations of how the dissociation of carbon dioxide interacts with the acid/base balance of the blood. The dissociation models by Siggaard-Andersen are unique in their level of detail and in the number of effects they include. They are the only models we have encountered that include the mutual effects of oxygen and carbon dioxide on the dissociation of each other.

The model of blood acid/base balance and pH by Chiari et al. (1994) is a set of chemical equations describing reactions between components of blood. Other treatments of the acid/base balance of blood are found in Siggaard-Andersen and Gøthgen (1989), Siggaard-Andersen (1974), and Singer and Hastings (1948). However, these are not really models, but rather theoretical considerations that could be used in modeling.

We use a compartment model to describe the dynamics of the blood transport system. The movement of matter in and out of such compartments involves both diffusion and bulk flow, and thus we will need to determine both tensions and concentrations. Therefore, we need models of dissociation of the respiratory gases. Though we have not previously stated our goals for the level of detail such dissociation functions must have, one requirement is that the model return a pH value, given the content of gases in the blood. We therefore find it reasonable to demand that our dissociation functions include the effects of pH, since the literature agrees that pH affects the dissociation of both carbon dioxide and oxygen significantly. Without a proper link between the amount of dissociated oxygen and carbon dioxide and the acid/base balance, we find it unlikely that we will be able to construct an acceptable model of pH.

Finally, we will need a model of metabolism that can maintain an appropriate metabolic rate under normal circumstances, produce anaerobic metabolism in situations with insufficient oxygen, and increase the metabolic rate as a response to external input, e.g., from the temperature model.

8.2   Modeling the Lung

This section presents a model of the airway dynamics. As opposed to a model that is averaged across the respiratory cycle (cf. Keener and Sneyd, 1998), this model reflects instantaneous states of respiration. One of the key parameters in anesthesia and intensive care management is the instantaneous value of CO_2 in the expired air. To be able to estimate these parameters, we have developed a pulsatile respiratory model.

The task of the lung model is to connect the atmosphere (or the respirator mask) and the alveoli with a model of airflow. The model must divide the lung into various parts and calculate the flow of oxygen, carbon dioxide, and possibly anesthetic gases between these parts. The output of the lung model is the pressure at the mouth and the tidal volume, which are in data that can be measured using a spirometer; the partial pressures of the expired air, usually measured with a gas analyzer; and the partial pressures in the alveoli. The partial pressures of the alveoli are not directly measurable but must be calculated in order to model the exchange of gases through the lung membrane. Membrane transport will be part of both the lung model and the transport model, as it is via membrane transport that the two models interact.

When modeling airway dynamics, the respiratory model is split into two parts: the pressure model and the gas model. The former model describes the total pressure in parts of the lung, while the latter model keeps track of the gas composition in the different parts of the lungs. Hence the outputs from the pressure models are used to create pressure-volume diagrams, and the gas compositions together with the total pressure allow one to compute the required partial pressures in various parts of the lung. A basic assumption of both parts of the lung model is that the gases of the model are considered to be ideal, obeying the ideal gas law

$$pV = n\mathcal{R}T, \tag{8.1}$$

where p is pressure, V is volume, n is the number of moles, $\mathcal{R}$ is the gas constant, and T is temperature on the Kelvin scale. Real gases are better described by the van der Waals equation (Atkins, 1990, Chap. 1). However, at room temperature and atmospheric pressure the deviation from the ideal gas law is insignificant for the gases included in the current model.

In both models the net transport of gases through the pulmonary membrane is assumed to be zero, so the exchange of gas with the blood does not cause any differences in the total pressure. Although this might not be true when the metabolism is anaerobic, it is a valid approximation, since the carbon dioxide diffusion into the lung is small compared to the alveolar volume.

8.2.1 Pressure Model

In the model of flow in the airways we neglect turbulent effects and assume Poiseuille flow throughout the system. This implies proportionality between flow and pressure differences. Consequently, we can use an analog electrical network model of the airflow in the airways, as explained below.

The relation between a pressure difference ΔP and laminar bulk flow Q of a fluid through a tube is described by Poiseuille's formula (Atkins, 1990, p. 741)

$$Q = \frac{1}{R}\Delta P, \tag{8.2}$$

where the proportionality factor $R = 8\mu l/(\pi r^4)$ expresses a resistance to the airflow. R is determined by the viscosity of the gas μ and the length l and radius r of the tube. The radii of the airways change during the breathing cycle due to the elasticity of the pipes. Consequently, so does the airway resistance, but this variation is neglected in the model.

During natural inspiration, airflow into the lung is caused by the pressure difference that arises from an enlargement of the lungs. When the lung volume is decreased at expiration, air is pressed out of the lungs. This effect is described by the ideal gas law, which states that the product of pressure and volume pV is constant if the same number of molecules n

are present and the temperature is constant. The latter assumption is not necessarily valid, but we use this approximation throughout.

In Figure 8.1 the relationship between lung volume and the pressure difference between alveoli and the interpleural fluid is shown. The pressure difference is often called the transmural pressure Δp_t. The compliance of the lung C_l is found as the slope of the curve, which is almost constant for lung volumes during the normal ventilation cycle. Thus C_l measures the ability of the lung to enlarge when the pressure of the interpleural fluid decreases. Consequently, the relationship between the transmural pressure and the volume of the lung can be written as

$$V = C_l \Delta p_t + V_0, \tag{8.3}$$

where V_0 is the volume of the unstretched lung, found at the end of an unforced expiration, when the pressure in the lungs equals the pressure of the interpleural fluid.

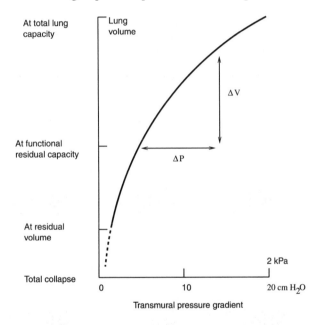

Figure 8.1. *Plot of the lung volume as a function of the pressure difference between the interpleural pressure and the lung pressure. From* Nunn's Applied Respiratory Physiology *by J.F. Nunn. Reprinted by permission of Elsevier Science Ltd.*

The relationships in (8.2) and (8.3) suggest that an equivalent electrical network is an appropriate framework for modeling pressure in the lungs. Airflows are equivalent to currents, pressures are equivalent to potentials, flow resistances are resistances R, compliances are capacitances (C), and volumes are equivalent to charges q. Inertances, or resistances to accelerations of the flow, are neglected.

The instantaneous relationship between current I and voltage U in the electrical network is given by Ohm's law $U_R = RI$, equivalent to (8.2). There exists an analogous law for a linear capacitance $CU_C = q$, to which is added the term V_0 because the zero of the transmural pressure in order to prevent a collapse of the lung differs from the atmospheric pressure. The law thus becomes equivalent to (8.3). In the electrical model, capacitors are compartments storing matter and currents are flows. Like airflow in the lung, flows in the networks change direction when the pressure differences change sign.

The electrical network modeling pressure and airflows in the lung is used in several models in the literature. In this section two examples of such models of the mechanical lung are discussed. We chose these two because they represent two different principles of dividing the lung. The model of Rideout (1991) focuses on the mix of inspired gas with the air in the dead space before the exchange of gases with blood in the alveoli. This model also describes serially connected parts of the lung representing the conducting airways. In contrast, the second model (Galster, 1995) focuses on the nonuniform ventilation of alveoli throughout the lung and therefore models a lung divided into several parts containing alveoli.

The model by Rideout (1991) is shown in Figure 8.2 and consists of four serially connected sections of the airways: the larynx, the trachea, the bronchi, and the alveoli. Between each section is a resistance. The driving force is implemented as an oscillating generator that expands the last three chambers, as only these are expanded with the thorax. This network allows modeling of restricted airflow by changing one or several resistances in the network.

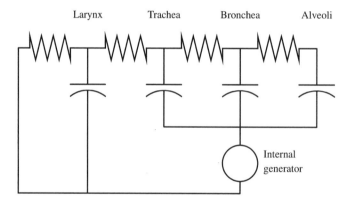

Figure 8.2. *Electrical circuit representing Rideout's lung model. In this model only internal generation, i.e., natural breathing, is included.*

The model shown in Figure 8.3 is extracted from Galster (1995). The model splits the lung into 10–50 parallel sections and describes each with an analog electrical network of a resistance in series with a capacitor. With this network Galster simulates a breathing cycle that consists of an inspiration period, an optional pause, and an expiration period. The simulation can use either of two modes: one in which the inspiration is driven by an external source and one in which the natural inspiration works with a constant inspired and expired volume at each breath. With this model nonuniform ventilation of the alveoli can be modeled.

To be sufficient for use in a medical simulator, our model must be able to change the properties of the upper airways and the alveoli for different scenarios. Consequently,

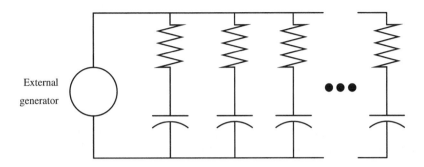

Figure 8.3. *Electrical circuit representing Galster's lung model. The model has both artificial and natural ventilation through special operation modes of the generator. This model allows for nonuniform ventilation of the alveoli.*

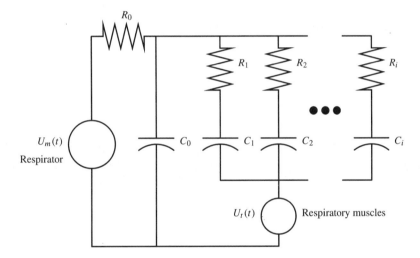

Figure 8.4. *Electrical circuit representing the lung model discussed in this chapter. This model includes two generators, one for the simulation of artificial ventilation and one for the simulation of natural ventilation. The model allows for nonuniform ventilation of the alveoli.*

we combine aspects of both of the models above. As in Galster's model, several alveolar branches are placed in parallel, each consisting of a resistance in series with a capacitor. This electrical network can simulate the phenomena that the alveoli differ in their compliances or that airflow is restricted in some parts of the lung. All these branches connect to a generator U_t (t for thorax), generating the transmural pressure induced by the respiratory muscles. External pressure is the driving force in artificial respiration and is represented by a generator U_m (m for mouth); see Figure 8.4. In parallel to the alveoli we have a capacitor

C_0 representing the anatomical dead space, i.e., the volume of air that does not reach the areas in the lungs where gas exchange with the blood occurs. In the physiology literature the anatomical dead space is considered to be constant, or at least not to vary during a breath, and is therefore modeled as the unstressed volume of the capacitor C_0, which has a low capacitance.

R_0 models the resistance of the upper airways. This is the largest resistance in the model, as the resistance of the airways peaks at the fourth generation of branching, according to West (1979). The upper airways are, in this context, the generations of branching in which exchange of gases with blood does not occur—generations 0 to 19. The alveoli represent generations 19 to 23. The pulmonary tree is depicted in Figure 8.5.

The internal generator U_t is assumed to be sinusoidal. It is possible to switch off this sine wave, thus simulating situations in which the natural respiratory drive is inactivated. The external pressure U_m must mimic the pressure made by the respirator, which is often found to be a serrated curve.

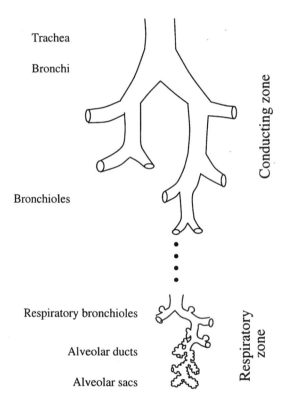

Figure 8.5. *Branching of the pulmonary tree. The figure shows some of the 20–25 generations of the pulmonary tree starting at the trachea and ending with the alveolar sacs. The scaling shown on the figure does not accurately depict the scales observed in the real lung, where the trachea has a diameter of approximately 2.5 cm and the alveolar sacs have a diameter of 0.2 mm.*

The model equations derived from the circuit in Figure 8.4 express the rate of change of the pressure p in various parts of the lung. The index of parameters and variables of the upper airways is 0, while the n alveolar parts are indexed with i, from 1 to the number of alveolar sections n_a:

$$\dot{p}_0 = \frac{1}{R_0 C_0} \left(U_m - p_0 - R_0 \sum_{i=1}^{n} C_i \dot{p}_i \right), \tag{8.4}$$

$$\dot{p}_i = \frac{1}{R_i C_i} (p_0 - p_i - U_t), \qquad i = 1, 2, \ldots, n_a. \tag{8.5}$$

Equation (8.4) expresses the change in the pressure of the upper airways $\dot{p}_0$ calculated from the pressure in this part of the lung p_0, the external pressure U_m, and the change in pressure in the alveolar sections $\dot{p}_i$. The n_a equations in (8.5) express the change in pressure in the alveolar sections using the internal pressure U_t and the pressure in the alveoli themselves p_i. The RC products of the equations have dimension time and yield a characteristic time that expresses how fast the pertinent part of the lung empties or fills with air.

In order to plot the required pressure-volume diagrams, we need to calculate the volumes of different parts of the lung. To do this calculation we use the relationship given in (8.3). Thus the alveolar volumes V_i can be found from the various alveolar pressures p_i:

$$V_i = C_i p_i + V_{0i},$$

where V_{0i} is the unstressed volume of the ith alveolus. The same relationship holds for the volume V_0. The sum of these volumes yields the total volume of the lungs and enables the plot of the pressure-volume loops.

8.2.2 Gas Model

On top of the pressure model, we need a model to calculate the partial pressures of the different lung sections, especially the alveolar partial pressures and the partial pressures of the expired air. We intend to model the partial pressures without keeping track of all the gases in the lung. For instance, nitrogen is of no particular interest either to the other parts of the model or as output during simulations. This approach implies that we neglect the change in the total pressure that might occur in the alveoli as a result of a nonzero net diffusion through the lung membrane.

The lung compartments represented by the capacitors in the pressure model, detailed in the last section, are the compartments of the gas model as well. We want to know the composition of the gas in each section. Thus the layout of the gas compartment model is determined by the layout of the network in the pressure model: The central compartment, represented by C_0, connects to a number of alveolar compartments and to the mouth or respirator mask; each alveolar compartment connects only to the central compartment and to the blood in the capillaries; see Figure 8.6.

In the gas model of the lung, flows between the compartments depend on total pressures and partial pressures of gases in the compartments. The transport of gas between

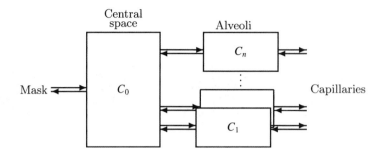

Figure 8.6. *The gas model. Arrows symbolize the gas exchange between the different compartments, i.e., between the central compartment and the n compartments representing the alveoli.*

lung sections includes two effects: airflow and diffusion. When a pressure difference exists throughout the lungs, airflow is normally the most significant. Diffusion might be important when bulk flow is absent, as in the case of no breathing (apnea). These effects will superimpose. In the next two sections we consider the two effects separately.

Laminar Flow

We wish to model how changes in the composition of air in different parts of the lungs depend on the total pressures, which are output from the pressure model, and on the composition of gas in these parts. To keep track of several gas types, we will use vector variables in which each coordinate refers to a specific gas. The first coordinate represents carbon dioxide and the second oxygen. Two necessary vector variables are the vector $\mathbf{x}$, which expresses the amount of each relevant gas, and the fraction vector $\mathbf{f}$, which expresses the fraction of the total amount of matter in moles (n) that each gas forms. Thus the relation between the two vectors is given by $\mathbf{x} = \mathbf{f}n$, and the rate of change of matter can be expressed as

$$\frac{d\mathbf{x}}{dt} = \frac{d\mathbf{f}}{dt}n + \mathbf{f}\frac{dn}{dt}. \tag{8.6}$$

For a model of all gases in the lungs the components in $\mathbf{f}$ always add to one. But as we do not explicitly keep track of substances like N_2 in our model, the sums of fractions may be less. The equation for each compartment must express the relationship between the total pressures and the gas compositions in that compartment and the surrounding compartments. In order to find such an expression, we will first examine how the composition of gases in a compartment is affected when this compartment is connected to another compartment j with a different composition of gases, illustrated in Figure 8.7.

A flow leaving a compartment will not change the mix of gas in the compartment, as the flow has the same mix, and therefore $\frac{d\mathbf{f}}{dt}$ in this case. For a flow into a compartment with a different mix, matters are a little more complicated. An inflow will result in an increase in the total number of molecules, $\frac{dn}{dt} > 0$. The composition of the inflowing molecules is the same as the composition of gas in the originating compartment. Thus the change in the

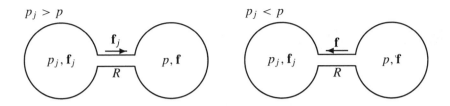

Figure 8.7. *Change of gas fraction in a compartment. The two situations, $p_j > p$ and $p_j < p$, are depicted with flow direction and composition of the flowing gas.*

amount of the gases will be

$$\frac{d\mathbf{x}}{dt} = \mathbf{f}_j \frac{dn}{dt}, \tag{8.7}$$

where $\mathbf{f}_j$ is the fraction vector of the gas in the inflowing air. The subscript j indicates that the vector refers to the gas fractions in the compartment from which the gas flows.

Laminar airflow between the compartments, discussed in section 8.2.1, is determined by the pressure difference $(p_j - p)$ and the airflow resistance R:

$$p_j - p = R \frac{dn}{dt}. \tag{8.8}$$

Inserting $\frac{d\mathbf{x}}{dt}$ and $\frac{dn}{dt}$ from (8.7) and (8.8) into (8.6) and isolating the term $\frac{d\mathbf{f}}{dt}$ yields

$$\frac{d\mathbf{f}}{dt} = \frac{(\mathbf{f}_j - \mathbf{f})(p_j - p)}{nR}. \tag{8.9}$$

To obtain a differential equation with only one unknown variable, $\mathbf{f}$, we express the number of molecules in moles, n, in (8.9) as the total pressure of the compartment, which is known as input from the pressure model. Gases in our model are still assumed to obey the ideal gas law. Assuming that the change in volume and pressures does not introduce a change in temperature, we can express the number of molecules as

$$n = \frac{pV}{\mathcal{R}T}, \tag{8.10}$$

where p is pressure, V is volume, n is the number of molecules, $\mathcal{R}$ is the gas constant, and T is temperature. A change in the size of a compartment will introduce a change in the pressures inside the compartment. We assume that the compliance of the lung is constant. When a gas is in an elastic compartment with compliance C and unstressed volume V_0, the relation between pressure and volume is $V = Cp + V_0$. Combining this expression with (8.10) yields the following expression for the amount of gas in a compartment:

$$n = \frac{Cp^2 + V_0 p}{\mathcal{R}T}. \tag{8.11}$$

Combining (8.11) and (8.9), we obtain an equation describing the change in the mixture when gas flows into a compartment:

$$\frac{d\mathbf{f}}{dt} = \frac{(p_j - p)(\mathbf{f}_j - \mathbf{f})\mathcal{R}T}{R(Cp^2 + V_0 p)}, \tag{8.12}$$

where $p \neq 0$ follows from the physical situation.

Combining the special cases above, the contribution to the change in the fraction vector of a compartment, when air flows between this and another compartment, can be expressed as

$$\frac{d\mathbf{f}}{dt} = \begin{cases} 0 & \text{when } p \geq p_j \text{ (outflow)}, \\ \dfrac{(p_j - p)(\mathbf{f}_j - \mathbf{f})\mathcal{R}T}{R(Cp^2 + V_0 p)} & \text{when } p < p_j \text{ (inflow)}. \end{cases} \tag{8.13}$$

To simplify our notation we introduce the function I_+, which is 0 for negative arguments, and the identity function otherwise:

$$I_+(x) = \begin{cases} 0 & \text{for } x < 0, \\ x & \text{otherwise}. \end{cases}$$

With this definition, (8.13) can be written

$$\frac{d\mathbf{f}}{dt} = \frac{I_+(p_j - p)(\mathbf{f}_j - \mathbf{f})\mathcal{R}T}{R(Cp^2 + V_0 p)}.$$

Diffusion

The diffusion of gases between two compartments will cause changes in the composition of gases in both compartments unless both compartments initially have the same distribution. We will in this section estimate the extent to which diffusion contributes to the change in gas content of the compartments. We first assume that the molecules of a gas behave independently of the other gases in the system, since diffusion is caused by the random thermal motions of the molecules. The diffusive flux expresses the net number of molecules of a gas passing through a unit area in a unit time. The flux is given by Ficks's first law of diffusion, which states that the flux is proportional to the concentration gradient of the gas (More, 1972).

The diffusion coefficients are specific to each gas. Hence the net diffusion of gases between two compartments is not necessarily zero. Different diffusion coefficients of ideal gases imply a net flow and hence induce a difference in the total pressure of the two compartments. The pressure difference will give rise to a laminar flow. However, the diffusion coefficient of all atmospheric gases is approximately 10^{-5} m^2/s (Alonso and Finn, 1979). In the case of anesthetic gas mix the diffusion coefficient might differ from this order of magnitude. But in situations in which a patient inhales an anesthetic gas mixture one does not find periods without respiration, and a laminar flow is dominant.

We therefore find it acceptable to assume that the gases have the same diffusion coefficient, which ensures that the net diffusion is close to zero. If the alveoli are filled with 100% oxygen, respiration may be absent for 5 to 6 min without the patient becoming undersupplied with oxygen (hypoxic) (Nunn, 1987). The question is whether part of the oxygen in the anatomical dead space is utilized, or only the supply in the alveoli is used. And we want to know how the effect of diffusion between the lung sections should be modeled in our lung configuration. We model the conducting zone in the pulmonary tree as a central compartment, which is connected to the alveoli. The central compartment covers generations 0 to 19 of the upper airways. The alveolar compartment covers generations 20 to 23. An average distance[2] between these compartments is assumed to be 0.1 m, and the total cross-sectional area is less than 0.01 m^2. With these parameter values it is obvious that the effect of diffusion between the compartments in our model is negligible, even in a long period without breathing. This is a consequence of the rough partition of the lung we have made; i.e., on this scale we can neglect diffusion. However, the ratio of area to length between the 19th and 20th generations is approximately 500. Hence the diffusion of a gas is about $0.005\Delta c/s$, and approximately a fourth of the amount of oxygen in the 19th generation will, during 300 s, diffuse into the alveoli from the upper compartment. Hence only a model with a more detailed description of the lung would benefit from a model of diffusion. In our model diffusion is included in the instantaneous mix of gas inside each compartment, and only the diffusion through the alveolar membrane is explicitly modeled.

Compartment Equations

Neglecting diffusion, we can summarize the compartmental equations as follows. The total change in the fraction vector of each compartment is the sum of the contributions of the laminar flow from all connections to other compartments.

The central compartment has connections to the respiratory mask reservoir and to each of the alveolar compartments. Hence considering the contribution of the flows to and from these compartments (cf. (8.13)), the mass balance for the central compartment yields

$$\frac{d\mathbf{f}_0}{dt} = \frac{\mathcal{R}T}{C_0 p_0^2 + V_{00}p_0}\left(\frac{I_+(U_m - p_0)(\mathbf{f}_e - \mathbf{f}_0)}{R_0} + \sum_{i=1}^{n}\frac{I_+(p_i - p_0)(\mathbf{f}_i - \mathbf{f}_0)}{R_i}\right), \quad (8.14)$$

where U_m and $\mathbf{f}_e$ are the pressure and the fraction vector of the atmosphere (or the respiratory mask in case of artificial ventilation). Each alveolar compartment connects to the central compartment and exchanges gas with the blood through diffusion over the membrane to the capillaries. We model membrane diffusion as $\frac{d\mathbf{x}}{dt} = \kappa(\mathbf{p}_{cp} - p\mathbf{f})$, where $\mathbf{p}_{cp}$ is the gas tension in the lung capillaries. Thus the exchange of gases depends on the pressure difference across the membrane and the elements of the matrix κ, which expresses the permeability of the membrane with respect to the particular gas. We do not model the effect of one gas on the permeablity of another. This effect is small and modeling it would require us to keep track of all gases, including nitrogen. Because of this κ is diagonal.

Under the assumption of no net transport of gas over the membrane, we obtain, by adding the contributions of (8.13) and the membrane diffusion, the following equation for

---

[2]All parameters used in this section are estimated on the basis of Table 8.1, below.

the gas fraction vector of the alveolar compartment i:

$$\frac{d\mathbf{f}_i}{dt} = \frac{\mathcal{R}T}{C_i p_i^2 + V_{0i} p_i} \frac{I_+(p_0 - U_t - p_i)(\mathbf{f}_0 - \mathbf{f}_i)}{R_i}$$

$$+ \frac{\mathcal{R}T}{C_i p_i^2 + V_{0i} p_i} \kappa(\mathbf{p}_{cp} - p_i \mathbf{f}_i), \tag{8.15}$$

where U_t is the pressure caused by the respiratory muscles. Equations (8.14) and (8.15) constitute the gas model. Together with the pressure model, given by (8.4) and (8.5), these equations constitute our model of the lung.

8.2.3 Parameters in the Lung Model

The physiological lung parameters are compliance C and the unstretched volume V_0 for each section of the lung; the resistance R, estimated from the area A; and the length l of each passage between the lung sections. Additional parameters are the external pressure source U_m representing the respirator and the internal pressure source U_t representing the respiratory muscles.

Parameters Concerning Pressure Sources

The graph of the external pressure source U_m representing the respirator is found in Bardoczky et al. (1996). Here are plotted typical graphs of respirator pressure as a function of time. We have implemented similar graphs in the model; see section 8.4. The pressure curves oscillate between atmospheric pressure and 3 kPa above atmospheric pressure in a continuous but nondifferentiable way. However, the precise pressure curve of the respirator depends on the respirator type and the conditions of the patient. When the present model is implemented with other models of the simulator, the lung model will receive the pressure as input from the respirator that is used in the operating theater during the simulations.

According to Rideout (1991), the internal pressure source U_t representing the respiratory muscle creates a pressure that is well approximated by a sine curve. The amplitude and the frequency are found so that a physiological tidal volume and ventilatory rate are obtained. The tidal volume at natural ventilation is 0.5 to 0.8 l, and the frequency is 10 to 15 breaths/min.

Parameters Concerning Laminar Flow

The $1 + n_a$ resistances, representing the airflow resistances between the lung sections, can be calculated from the dimensions of the tubes the air flows through or found in the literature.

The first resistance, R_0, represents the resistance of the conducting airways of branching generations 0 to 19. The remaining parallel resistances represent the resistances of the respiratory zone of the lung, branching generations 20 to 23. Since each generation of the airways consists of 2^g tubes, where g is the generation number, the total resistance of a generation can be calculated if one knows the resistance of each tube in the generation.

When the pressure model was described (see section 8.2.1), the resistance of a laminar airflow through a tube with length l and radius r was found to be

$$R = \frac{8\,l\,\eta}{\pi\,r^4}, \tag{8.16}$$

where η is the viscosity of the gas. From (8.16) the resistance in each generation of tubes in the branched airways is calculated as

$$R = \frac{8\,l\,\eta}{2^g\pi\,r^4}, \tag{8.17}$$

where 2^g is the number of tubes in the gth generation; see Table 8.1. Thus an estimate of the resistances in our model can be calculated as the sum of the serially connected resistances in the generations of the tubes. Table 8.1 also provides some literature values for resistances of each generation of branching (West, 1979).

Our lung compartments represent the upper airways (generations 0–19) and the alveoli (generations 20–23). We have used the resistance from the upper 12 generations (0–11) to represent the resistance from the atmosphere to the center of compartment 0. The resistances in generations 12 to 21 have been used to estimate the resistance between the center of the central compartment and the center of the alveoli. One can see that the calculated resistances in Table 8.1 do not match the values from West (1979) well in the first generations. Two possible reasons for this discrepancy are that the simplification of the airways to cylindrical tubes is too rough and that the measured values include some effects of turbulence that increase the resistance. Another reason might be that the resistance of the elastic tubes is not constant during the ventilatory cycle. None of the references discuss the variation in the size of the airways with the pressure.

In situations of artificial ventilation the resistance of the upper airways, R_0, must include the resistance of the endotracheal tube, since this is inserted through the mouth to the upper airways when the respiratory mask is connected (Bardoczky et al., 1996). We have used the physiological values found in the literature (West, 1979; Rideout, 1991) for the experiments with our model. We also performed some fitting to obtain reasonable pV diagrams; see section 8.4.1.

8.3 Models of the Blood Transport System

In summary, the blood transport model must keep track of the tension[3] and the concentration of carbon dioxide, oxygen, and various other substances transported by arterial and venous blood flow. We model the transport of all substances in one compartment model so the influence of the substances, especially the respiratory gases, on one another can be dynamically modeled. In this presentation we will concentrate on CO_2 and O_2.

The state variables of the model are therefore vector variables. One vector describes the content of matter in each compartment; each vector component represents one substance. The compartment model therefore consists of a set of differential equations, one for each

---

[3]The term tension is used for partial pressure in blood and tissue to distinguish between the partial pressure in a gas phase and the partial pressure of a dissolved gas.

Table 8.1. *The dimensions of the airways Grodins and Yamashiro* (1978) *and the calculated resistances to airflow in the generations (g) of branching compared to values of the resistances from West* (1979).

g	d [mm]	l [m]	R [kPa·s/l] Calculated	R [kPa·s/l] Literature
0	18.00	0.1200	0.00088	0.00680
1	12.20	0.0476	0.00083	0.00700
2	8.30	0.0190	0.00077	0.00740
3	5.60	0.0076	0.00075	0.00790
4	4.50	0.0127	0.00150	0.00830
5	3.50	0.0107	0.00172	0.00760
6	2.80	0.0090	0.00177	0.00610
7	2.30	0.0076	0.00164	0.00440
8	1.86	0.0064	0.00162	0.00330
9	1.54	0.0054	0.00145	0.00230
10	1.30	0.0046	0.00122	0.00130
11	1.09	0.0039	0.00104	0.00100
12	0.95	0.0033	0.00077	0.00050
13	0.82	0.0027	0.00056	0.00025
14	0.74	0.0023	0.00036	0.00020
15	0.66	0.0020	0.00025	0.00020
16	0.60	0.0017	0.00015	0.00015
17	0.54	0.0014	0.00010	
18	0.50	0.0012	0.00006	
19	0.47	0.0010	0.00003	
R_0 Generations 0–11			0.01521	0.06340
20	0.45	0.0008	0.00001	
21	0.43	0.0007	0.00001	
22	0.41	0.0006	0.00000	
23	0.41	0.0005	0.00000	
R_1 Generations 12–21			0.002	

compartment, describing changes in the vector components. These equations all take the same form: Rate of change equals total flow into the compartment minus total flow out of the compartment; i.e.,

$$\frac{dx}{dt} = \sum I - \sum O.$$

Inside a compartment instantaneous mixing is assumed. Until now we have not decided on the actual number and configuration of the compartments. Obviously there must be an arterial and a venous blood compartment in order to calculate the tensions of oxygen and carbon dioxide in the blood. These two compartments must be connected, so that the flows between compartments represent the pulmonary circuit and the systemic circuit.

Since the transport model eventually will include the distribution of drugs in the body, we have decided to use a model with a high level of detail, originally from Lerou et al.

(1991). This is almost the same model as the one in "Model 10" (Andreasen et al., 1994), an earlier work on pharmacokinetics of the SIMA simulator. The compartment configuration is shown in Figure 8.8. In this model the compartments, which represent organs in the systemic circuit, consist of a tissue part and a blood part. In addition, the model includes peripheral blood pools of different sizes. These cause a difference in circulation times

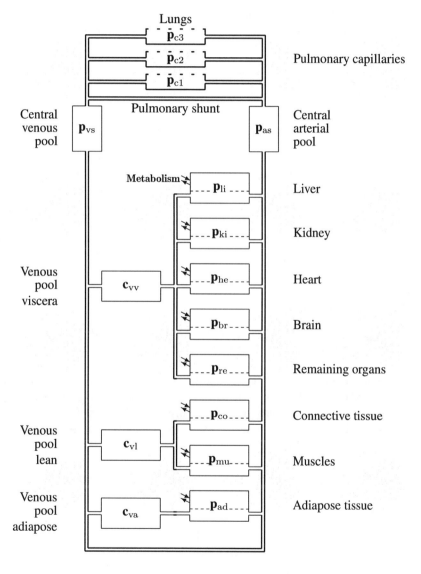

Figure 8.8. *Blood transport model. The model contains* $13 + n_a$ *compartments. The compartments include five blood pools, eight organ compartments, and* n_a *capillary compartments. p's indicate the tension (the partial pressure in blood and tissue), and c's indicate the gas concentrations.*

through the systemic circuit. In the pulmonary circuit we have extended the single capillary compartment of Lerou et al. (1991) with multiple pulmonary capillary compartments, one for each alveolar compartment of the lung model. The blood model connects to the lung model by each capillary compartment exchanging gas with one alveolar compartment and vice versa. By such a configuration inhomogeneous ventilation-perfusion ratios throughout the lung can easily be simulated. The blood flow through the two central blood pools is determined by cardiac output Q. The distribution of blood to the various organs in the circuit, described by z_i, determines the fraction of Q that each compartment receives. Both cardiac output Q and the fractions z_i can be extracted from the cardiovascular model, but here we will use constant mean values found in the literature, e.g., Lerou et al. (1991).

8.3.1   Mass Balance Equations

As seen in Figure 8.8, our model contains $13 + n_a$ compartments: five blood pools, eight organ compartments, and n_a capillary compartments. The different types of compartments have different in- and outflows. The venous and arterial blood pools have only blood flows. The pulmonary capillaries have blood flows and gas flows to and from the alveoli. And finally, the organs have blood flows and metabolic flows representing the consumption and production of matter.

The following sections describe each type of compartment. They also explain how the differential equation describing a compartment has either pressure or concentration as the state variable, depending on the nature of the compartment. The differential equation of each compartment is 2D, with state variable p or c vectors consisting of a CO_2 component and an O_2 component.

Organ Compartments

An organ compartment (Lerou et al., 1991; Olofsen, 1994; Rideout, 1991) consists of both a tissue and a blood part; see Figure 8.9. We use this dual compartment approach to model regions or parts of the body with an equilibrium distribution of blood-carried substances. The volumes of tissue and blood V_T and V_B in each compartment are assumed to be constant. The blood flow inside an organ branches into capillaries in which diffusion of gases between blood and tissue takes place. We assume instant equilibrium over the membrane between blood and tissue inside an organ. Thus the tension of each gas is uniform throughout both

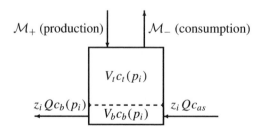

Figure 8.9. *An organ compartment, with expressions for the gas content in the blood and the tissue phase and the flows in and out.*

tissue and blood inside an organ. This equality of tensions, however, does not imply that the concentration is equal in tissue and blood, as the solubility may differ.

The dissociation of gases is described by dissociation curves, which constitute a vector function containing a specific function for each gas and each solvent. The dissociation function yields a concentration c as a function of the tension p of the substance and might also depend on other variables of the model. For example, it might depend on the tensions of other gases, pH, or temperature T. The dissociation function depends on the solvents, and thus we call the blood dissociation function c_b and the tissue dissociation function c_t. The tissue dissociation function c_t may differ from compartment to compartment, while the blood dissociation function for gases is common to all the compartments in the model.

The ability of the respiratory gases to combine with blood components makes the nature of the dissociation of these gases rather complex. We have chosen to use a model from the literature (Siggaard-Andersen et al., 1988) that includes much of this complexity. The functions in this model are rather long expressions and depend on several variable quantities. In the case of carbon dioxide the function is given only implicitly. We will present the dissociation models in a separate section; see section 8.3.3. In order to find the differential equations of the compartments, it is sufficient to know that the dissociation models exist and that the dissociation functions are continuous and monotonically increasing. However, it is worth pointing out that, because of the implicitness of the dissociation models, we have transformed the mass balance equations from ones describing the change of matter $\mathbf{x}$ in the compartment to an equation describing the change of tensions $\mathbf{p}$ in the compartment. By this method we avoid finding the inverse of the dissociation function, which is the method used in the model of the blood transport of respiratory gases in the literature. Many of the models, we have seen, use approximated curves to describe the gas dissociation. This makes easier the task of finding the tension of a respiratory gas of a certain concentration but is an inadequate model of gas dissociation for our purposes.

Each organ compartment is instantaneously mixed and thus in equilibrium. This means that the tensions in the blood and the tissue are always the same. The assumption of local equilibrium, however, does not imply equilibrium between different compartments of the model, since the transport limitation of blood flow will allow dynamic differences. Each organ compartment receives a fraction of the total systemic blood flow z_i, all of it flowing from the arterial compartment. Since the blood volume inside the organ is considered constant, the blood flows into and out of the compartment are both $z_i Q$, where Q is cardiac output.

In a compartment represented by a vector with a component for each type of matter (CO_2 and O_2) the differential equation of the organ compartment states that the change in an amount of matter $\mathbf{x}$ equals the matter flowing into the compartment minus the matter flowing out of the compartment:

$$\frac{d\mathbf{x}}{dt} = z_i Q(\mathbf{c}_{as} - \mathbf{c}_b(\mathbf{p})) + \mathcal{M}_+(\mathbf{p}) - \mathcal{M}_-(\mathbf{p}). \tag{8.18}$$

The variable $\mathbf{p}$ is the tension vector in the compartment. The inflows and outflows of gas with the blood flow in (8.18) are calculated from the concentration of gas in the blood and the total blood flow through the compartment $z_i Q$. Concentrations in blood are found as a function of the tension using the blood dissociation functions $\mathbf{c}_b(\mathbf{p})$. The subscript as indicates that this term in the equation is the gas concentration vector of the systemic arterial compartment

$c_{as} = c_b(p_{as})$. The two metabolic functions, $\mathcal{M}_+$ and $\mathcal{M}_-$, describe the tissue consumption and production of a gas and are detailed in the section on metabolism; see section 8.3.2.

The gases have uniform tension throughout the compartment, and thus the distribution of matter between different phases can be found via the dissociation functions. The equation for the total amount of matter in the compartment is

$$\mathbf{x} = V_t \mathbf{c}_t(\mathbf{p}) + V_b \mathbf{c}_b(\mathbf{p}),$$

where V_t and V_b are the tissue and blood volumes and $\mathbf{c}_t$ and $\mathbf{c}_b$ are the tissue and blood dissociation functions. By substituting this expression into the differential equation (8.18), we obtain

$$\frac{d}{dt}(V_t \mathbf{c}_t(\mathbf{p}) + V_b \mathbf{c}_b(\mathbf{p})) = z_i Q(\mathbf{c}_{as} - \mathbf{c}_b(\mathbf{p})) + \mathcal{M}_+(\mathbf{p}) - \mathcal{M}_-(\mathbf{p}).$$

By the chain rule we can thus express the change in terms of $\frac{d\mathbf{c}}{d\mathbf{p}}$ and $\frac{d\mathbf{p}}{dt}$:

$$V_t \frac{d\mathbf{c}_t}{d\mathbf{p}} \frac{d\mathbf{p}}{dt} + V_b \frac{d\mathbf{c}_b}{d\mathbf{p}} \frac{d\mathbf{p}}{dt} = z_i Q(\mathbf{c}_{as} - \mathbf{c}_b(\mathbf{p})) + \mathcal{M}_+(\mathbf{p}) - \mathcal{M}_-(\mathbf{p}).$$

Isolating $\frac{d\mathbf{p}}{dt}$, we finally obtain

$$\frac{d\mathbf{p}}{dt} = \left(V_t \frac{d\mathbf{c}_t}{d\mathbf{p}} + V_b \frac{d\mathbf{c}_b}{d\mathbf{p}}\right)^{-1} (z_i Q(\mathbf{c}_{as} - \mathbf{c}_b(\mathbf{p})) + \mathcal{M}_+(\mathbf{p}) - \mathcal{M}_-(\mathbf{p})). \tag{8.19}$$

In contrast to the models we have found in the literature, our compartment equation does not contain the inverse of the dissociation functions. This is because we use the tension $\mathbf{p}$ as a state variable and resolve the question of distribution of matter between the two phases by the Jacobian matrices $\frac{d\mathbf{c}_t}{d\mathbf{p}}$ and $\frac{d\mathbf{c}_b}{d\mathbf{p}}$. Under this assumption, we have arrived at a compartment equation that describes the content of gases in the compartment by the state variable $\mathbf{p}$ and the dissociation functions.

Pulmonary Capillary Compartment

The pulmonary capillary compartments are the site of the exchange between the transport model and the gas model of the lung. In the lung model oxygen and carbon dioxide are in gas phase and not dissolved in liquid. Thus the pulmonary capillaries have blood flows carrying dissolved gases, and they also have net diffusion of each gas through the membrane separating the capillaries and the alveoli; see Figure 8.10. The amounts of gas flowing out of the capillaries are determined by the tensions (partial pressures) in the blood and the permeability κ of the membrane with respect to the gases. The gas flows into the capillaries are determined by the partial pressures in the alveoli and thus depend on the output of the gas model and the permeabilities.

Normally, the gas flow through the membrane is sufficiently fast to ensure that equilibrium is reached, and thus the partial pressures in the alveoli and the capillaries are equal (Nunn, 1987; Widdicombe and Davies, 1991). This, however, is not the case if there are defects in the lungs or in the case of injected anesthetic agents, some of which do not penetrate the membrane. To enable our model to simulate these cases the alveoli and pulmonary capillaries are modeled by separate compartments. In that way instant equilibrium over

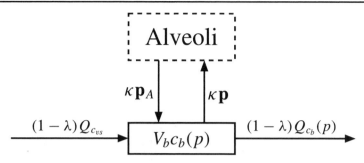

Figure 8.10. *Pulmonary capillary compartment with expressions for contents and flows. The blood flow through the compartment is $(1 - \lambda)Q$. The exchange between the capillary and the alveoli is goverened by the permeability κ as shown.*

the membrane is not assumed. In section 8.2.2 we saw the alveoli modeled as a number of compartments in parallel. The same number of capillary compartments are chosen and connected to one alveolar compartment each. Hence we can model different abilities of gas exchange, ventilation, or blood perfusion in different parts of the lung.

Based on the above considerations about in- and outflows to the pulmonary capillaries, we obtain the following mass balance equation determining the amount of substance:

$$\frac{d\mathbf{x}}{dt} = Q(1 - \lambda)(\mathbf{c}_{vs} - \mathbf{c}_b(\mathbf{p})) + \kappa(\mathbf{p}_A - \mathbf{p}), \qquad (8.20)$$

with the factor λ being the pulmonary shunt factor (the part of the blood shunted past the lung). The permeability κ is a diagonal matrix describing the diffusion rate of the gases through the membrane. $\mathbf{c}_{vs}$ is the concentration in the blood of the central venous compartment; $\mathbf{c}_{vs} = \mathbf{c}_b(\mathbf{p}_{vs})$.

Again we need the tensions $\mathbf{p}$, since the diffusion through the membrane is determined by these. By a calculation equivalent to the one leading to the equation for the organ compartment, (8.19), we transform the mass balance equation (8.20) so that $\mathbf{p}$ is the state variable:

$$\frac{d\mathbf{p}}{dt} = \left(V_b \frac{d\mathbf{c}_b}{d\mathbf{p}}\right)^{-1} (Q(1 - \lambda)(\mathbf{c}_{vs} - \mathbf{c}_b(\mathbf{p})) + \kappa(\mathbf{p}_A - \mathbf{p})). \qquad (8.21)$$

Blood Pools

A part of the blood is stored in the blood pools. There are two central pools, one on the venous side and one on the arterial side. Three peripheral blood pools simulate the different circulation times that exist in the body (Lerou et al., 1991).

All flows in and out of the blood pools are blood flows, except when injections are involved, making concentrations an excellent system variable. Thus the following equation governs the change in the concentration in a blood pool compartment:

$$\frac{d\mathbf{c}}{dt} = \frac{Q(\mathbf{c}_j - \mathbf{c})}{V_b}, \qquad (8.22)$$

where $Q\mathbf{c}_j$ is the amount of matter per unit time flowing into the compartment, since Q is the blood flow through the compartment and c_j is the concentration of the inflowing blood.

The outflow is thus calculated as $Q\mathbf{c}$, with $\mathbf{c}$ being the concentration of the compartment in focus and V_b being the volume of the compartment.

When blood is arriving from several compartments, the overall concentration of inflowing blood $\mathbf{c}_i$ is found as the weighted average of the concentrations. Hence the concentration of blood when two flows merge is

$$\mathbf{c}_i = \frac{Q_1\mathbf{c}_1 + Q_2\mathbf{c}_2}{Q_1 + Q_2},$$

where $\mathbf{c}_i$ is the resulting concentration, Q_1 and $\mathbf{c}_1$ the flow and concentration of one branch, and Q_2 and $\mathbf{c}_2$ the flow and concentration of the other. The flow after the merger is $Q_1 + Q_2$.

Since the tensions of the arterial and venous pools are explicitly required as output from the model, we have chosen to use tension as a state variable in these compartments, even though there are no flows that can be determined by knowledge of this value only. The reason is that we found it easier to apply the same transformation of state variable than to have to introduce the inverse of the blood dissociation function $\mathbf{c}_b$. Thus by an equivalent transformation to the one used to reach (8.19) we find the following equation for the tension change in the central pools:

$$\frac{d\mathbf{p}}{dt} = \left(V_b \frac{d\mathbf{c}}{d\mathbf{p}} \right)^{-1} Q(\mathbf{c}_x - \mathbf{c}_b(\mathbf{p})), \qquad (8.23)$$

where $\mathbf{c}_x$ is the concentration of the inflowing blood.

The Transport Model So Far

We have now stated four different equations for the different kinds of compartments. Each type of differential equation contains terms that have not been specified yet: the metabolism terms $\mathcal{M}_+$ and $\mathcal{M}_-$ and the dissociation functions $\mathbf{c}_b(\mathbf{p})$ and $\mathbf{c}_t(\mathbf{p})$. In the next sections we will present these submodels of the blood transport model. The dissociation curves depend on the pH, so a model of pH is presented as well. The blood volume parameters used in this work are given in Table 8.2.

Table 8.2. *Standard blood volume parameters.*

Function	*Organ*	*Fraction of blood flow*	*Tissue volume*	*Volume of blood in tissue*
Units			*[l]*	*[l]*
	Liver	0.283	2.973	1.106
Viscera	Kidney	0.222	0.270	0.051
	Heart	0.048	0.307	0.040
	Remaining	0.041	0.217	0.015
Brain		0.135	1.300	0.105
Muscles	Connective	0.101	8.182	0.653
	Muscles	0.107	26.773	0.700
Adiapose		0.062	14.786	0.562

8.3.2 Metabolism

Metabolism takes place everywhere in the body. In a model of the respiratory system metabolism is an important part; together with ventilation it constitutes the sinks and sources of the system. In our model we neglect metabolism in the blood and let only the organ tissue consume oxygen and produce carbon dioxide. The metabolic function in the organs is modeled by the terms $\mathcal{M}_+$ and $\mathcal{M}_-$ in the differential equation for the organ compartments. These are vector functions modeling metabolic production and consumption, respectively. We have chosen to model metabolism in these two different vector functions, because in this way we maintain the general principles of compartment models having only positive flows. The sign of each term in the differential equations determines whether it is an inflow or an outflow of the compartment.

Like the state vectors of the system, each coordinate in the vector function concerns metabolism of a particular substance. The first coordinate is CO_2, the second O_2.

Carbon dioxide is removed only by ventilation, so there is no consumption of CO_2 anywhere. Thus $\mathcal{M}_-$ consists of a zero as a first element. As long as oxygen is present in a compartment, there will be consumption. A simple view of the complicated chemical processes involved is that the metabolic consumption must tend to 0 as the concentration of O_2 tends to 0, and at high levels of oxygen some other factors will limit the metabolic rate. The O_2 metabolism is therefore modeled by a function of the Michaelis–Menten kinetic form (Bischoff and Dedrick, 1968):

$$\frac{\mathrm{d}x}{\mathrm{d}t} = M \frac{c}{\beta + c}, \tag{8.24}$$

where M is a constant representing the maximum metabolic rate and β is a parameter representing the oxygen concentration when the metabolic rate is half of the maximum value. It is easy to verify that the equation tends to the maximum metabolic rate M for c tending to infinity, and to 0 for c tending to 0.

$\mathcal{M}_-$ in organ compartments models only consumption of oxygen:

$$\mathcal{M}_- = \begin{pmatrix} 0 \\ M_{O_2} \frac{c_{O_2}}{\beta_{O_2} + c_{O_2}} \end{pmatrix}. \tag{8.25}$$

M_{O_2} differs from compartment to compartment. The sum of the maximum consumption rates over all organ compartments must under steady state conditions equal the rate at which oxygen enters the body by ventilation.

The production of CO_2 depends to a lesser extent on the substances present in the blood, since the tissue is metabolizing whether oxygen is present or not. Thus the metabolism may be either aerobic (with oxygen) or anaerobic (without oxygen), but in both cases the waste product is carbon dioxide. The total energy produced per amount of carbon dioxide produced is not the same for different types of metabolism. Thus the CO_2 production is different for different types of metabolism, provided that the body produces a constant amount of energy. We have not modeled this difference.

The metabolic rates depend on the work that the body performs. At rest and under normal conditions, the production of CO_2 equals the rate of CO_2 eliminated through the lungs. $\mathcal{M}_+$ is therefore modeled as a constant vector function specific to each organ compartment:

$$\mathcal{M}_+ = \begin{pmatrix} M_{CO_2} \\ 0 \end{pmatrix}. \tag{8.26}$$

By normal conditions we mean conditions in which there is no reason for the body to perform more work. In conditions of too-low temperature, muscles begin to shiver and thereby increase CO_2 production, and as long as oxygen is present O_2 consumption also increases. Thus the metabolic consumption and production, $\mathcal{M}_-$ and $\mathcal{M}_+$, need to be modeled as functions of temperature. It is only necessary to model a temperature effect on metabolism if the temperature is too low. Otherwise the normal metabolic rate is expected to be maintained.

The metabolic rate will also increase in case of the allergic reaction malignant hyperthermia, in which the metabolic rate increases in response to a hypersensitivity to an anesthetic. Malignant hyperthermia will be modeled as a specific scenario in which $\mathcal{M}_-$ is changed to a higher rate. This is an example of the dual nature of temperature and metabolism, because a clinical indication of malignant hyperthermia is an increasing body temperature resulting from the increased metabolism. During malignant hyperthermia, the ion balance of the cells and the intracellular fluid is disturbed. Thus the scenario requires that a model of the electrolyte and fluid balance be coupled to the metabolic model and the model of pH.

8.3.3   Gas Dissociation and pH Value

When writing the mass balance equations of the compartments in the blood transport models, we have introduced functions that convert the partial pressures $\mathbf{p}$ (also called tension) to concentrations $\mathbf{c}$. Furthermore, when we change variables in the mass balance equation, we assume the existence of the derivative $\frac{\partial \mathbf{c}}{\partial \mathbf{p}}$ of the dissociation functions.

The dissociation functions are specific with respect to the dissolved substances as well as to the solvents. Thus for any substance that we track we have to include a model of the dissociation of the substance in the different solvents in the model. For each solvent in the model (the solvents being blood and different kinds of tissue) a dissociation vector function must be specified in which each component is a dissociation function of one solute.

The dissociation of substances in tissue is considered in section 8.3.4. In this section we will describe the transport of substances by the blood and interactions of gases in the blood with each other and with pH. Section 8.3.4 presents the models of the gas dissociation functions and pH. The requirements for the simulator do not include any specification of gas dissociation, but the pH value of the blood is required as an output of the blood transport model.

The pH value is the negative logarithm of the concentration of hydrogen ions $[H^+]$. This is an important quantity measured during anesthesia, as it indicates the gas status of the blood. A measurement of pH gives information about the acid-base balance, since hydrogen ions are produced by acids, which dissociate. One acid in the blood is carbonic acid, formed by CO_2 dissolved in water, and thus there are close relations between the CO_2 level and pH. Furthermore, hydrogen, carbon dioxide, and oxygen all form reversible combinations with hemoglobin, and thus changes in the blood content of one of these substances will displace the equilibrium of the reactions with hemoglobin and change the concentration of all involved reactants.

A model of gas dissociation in the blood must for any two tensions of carbon dioxide and oxygen give the concentrations of the gases and the blood pH. The chemical interactions between these quantities imply that a change in one of the quantities causes at least three others to change as well. The only independent quantities of the five are the two gas tensions, or the oxygen tension and the pH.

The models (Siggaard-Andersen et al., 1988) were developed for analyzing blood data and are consequently based on measurements of pH, the tensions of gases, and other quantities that are not included in our work. Thus using these functions as submodels in our blood transport model requires a model of pH. The pH value depends on carbon dioxide concentration, and therefore the expression for the carbon dioxide concentration is only implicitly given when using the models of Siggaard-Andersen et al. (1988) combined with a model of pH.

The normal values for pH are 7.41 in arterial blood and 7.37 in venous blood (Nunn, 1987). Situations where $[H^+]$ is raised are called *acidosis*. If the concentration is below the normal level (corresponding to a high pH), one speaks about *alkalosis*. These changes in $[H^+]$ can be due to a respiratory defect in which the elimination of CO_2 in the lung is too slow or too fast. Alternatively, a change could be a metabolic acidosis or alkalosis, situations in which the deviation in pH is not primarily due to respiratory problems. Metabolic acidosis/alkalosis may be caused by electrolyte and fluid imbalance, which we do not include in our modeling. Our concern is therefore restricted to modeling the respiratory disturbances and the respiratory compensations to metabolic disturbances of the pH system, which will be described in the following as the respiratory buffer system.

The physiological range of pH in blood for a normal person is 7.0 to 7.8 (Widdicombe and Davies, 1991). Blood contains several buffer systems, which ensure that $[H^+]$ varies only slightly when the concentration of an acid in the blood changes. The buffer systems immediately neutralize excess H^+ by combining with it, and the excess hydrogen ions are then eliminated by the body via the kidneys. We will model only the respiratory buffer system and not include the metabolic excretion of hydrogen ions. Excretion is a slow process that we assume does not influence the pH level during anesthesia.

The most important blood buffers are bicarbonate (produced by carbonic acid) and proteins, which mean that the buffer systems of the blood are closely connected to O_2 and CO_2 transport in plasma and erythrocytes.

Carbon Dioxide Transport

CO_2 is produced in the tissues when metabolism takes place. As a result of the tension gradient, CO_2 diffuses from the tissue into the blood system. The diffusion process will tend toward an equilibrium, and thus the same tension is found in tissue and blood. Now the question is which concentration the equilibrium tension will imply. If no chemical reactions were taking place between carbon dioxide and blood components when carbon dioxide dissolved into the blood, the relation between concentration **c** and tension **p** could be expected to obey Henry's law:

$$\mathbf{p} = \alpha^{-1}\mathbf{c}, \tag{8.27}$$

which describes ideal solutions (Atkins, 1990, p. 163). The proportionality factor α is called the solubility.

When some carbon dioxide reacts with the solvents, more carbon dioxide can diffuse into the solvents until Henry's law (8.27) again is obeyed. Therefore, the total concentration of carbon dioxide, some of it bound to solvents, is increased. In the case of blood transport of CO_2, it is the total amount transported that is of interest since the reactions are reversible. When the net diffusion of the dissolved part goes in the opposite direction, the reactions will do so as well.

Dissolved CO_2 reacts in the following way:

$$CO_2 + H_2O \leftrightharpoons H_2CO_3 \leftrightharpoons HCO_3^- + H^+. \tag{8.28}$$

The first reaction from dissolved CO_2 to carbonic acid (H_2CO_3) is rather slow, but it is speeded up by the enzyme carbonic anhydrase. A large part of the carbonic acid is ionized into bicarbonate and hydrogen ions. The equilibrium of these reactions is given by the K_a value:

$$K_a = 10^{-6.1} = \frac{[H^+][HCO_3^-]}{[CO_2] + [H_2CO_3]}. \tag{8.29}$$

Since the amount of dissolved CO_2 is much larger than the amount of carbonic acid [H_2CO_3], the concentration of the latter can be ignored, and (8.29) is thus written

$$K_a = 10^{-6.1} = \frac{[H^+][HCO_3^-]}{[CO_2]}. \tag{8.30}$$

The condition (8.30) is important for carbon dioxide dissociation as well as for the pH value because it describes equilibrium of [H^+] and [CO_2]. Using Henry's law (8.27), the equilibrium condition in (8.30) can be written

$$K_a = 10^{-6.1} = \frac{[H^+][HCO_3^-]}{\alpha p_{CO_2}}, \tag{8.31}$$

since the dissolved CO_2 tension is proportional to the CO_2 tension, with a factor of proportionality α, which differs between plasma and erythrocytes.

Under normal physiological conditions, approximately 10% of the total CO_2 in venous blood is dissolved; 60% has reacted with water and is found as bicarbonate and hydrogen ions. Most of these ions are found in the erythrocytes because the catalyzing enzyme for the reaction is present there. The last 30% of CO_2 has reacted with hemoglobin to form carbamino compounds (Vander, Sherman, and Luciano, 1990) and hydrogen ions. Carbon dioxide can bind oxyhemoglobin as well as deoxyhemoglobin (hemoglobin bound or not bound by oxygen), but the latter is 3.5 times as effective (Nunn, 1987, p. 211). This difference causes the Haldane effect, which means that a fall in oxygen gives more deoxyhemoglobin and thus a rise in carbon dioxide carried as carbamino. Therefore, the ability of the blood to carry CO_2 is increased on the venous side.

When the $[H^+]$ level rises for some reason, the reactions are driven to the left side of the expression in (8.28) to obtain the equilibrium written in (8.29). Therefore, the level of dissolved CO_2 in blood rises and more CO_2 is eliminated in the lungs. Even though there is no elimination of $[H^+]$ in this process, it results in a reduced amount of carbamino and carbonic acid, which would have been donating $[H^+]$ ions if present.

The regulatory system of ventilation is also affected by the pH. By increasing ventilation in response to a decreased pH, indicating an increased CO_2 level in the blood, the respiratory system acts as a physiological buffer system by means of a faster elimination of CO_2 in the lungs.

Oxygen Carriage

Oxygen dissolves poorly in blood. Nearly all oxygen is carried by hemoglobin Hb, which is found in the erythrocytes. The total concentration of oxygen c_{O_2} carried by blood is found as

$$c_{O_2} = \alpha p_{O_2} + [HbOO], \tag{8.32}$$

where the first term expresses the concentration of dissolved oxygen, by use of Henry's law (8.27), and the last term expresses the concentration of oxygen bound to hemoglobin. Each hemoglobin molecule can carry four molecules of O_2. However, oxygen transport is not the only function of the hemoglobin. Deoxyhemoglobin Hb reacts with hydrogen ions, and therefore oxygen competes with hydrogen ions in binding hemoglobin. The competition is described by the equilibrium reaction (Vander, Sherman, and Luciano, 1990, p. 454)

$$HbOO + H^+ \leftrightarrows HbH + O_2.$$

This reaction causes the pH to influence the O_2 affinity. With an increased concentration of hydrogen ions, the equilibrium conditions drive the reaction to the right, and the ability of oxygen to bind hemoglobin decreases. This is called the Bohr shift and is a very useful effect for delivering oxygen to the tissues. A decrease in pH of only 0.2 units (a physiologically acceptable change) can increase the O_2 release by 25% at low oxygen tensions (Widdicombe and Davies, 1991, p. 65). Under conditions of acidosis, the CO_2 increase may be caused by an increase in metabolism, and the improved delivery of oxygen is therefore an advantageous physiological response.

8.3.4   Models of Gas Dissociation and pH Value

In the previous section we stated the basic reactions in blood involving carbon dioxide, oxygen, and hydrogen. We also described the interactions of these three substances in the blood. In this section we will present the models of pH and gas dissociation, which are all based on the reactions in blood and the equilibrium conditions of these reactions.

pH Model

In this section we present a model of pH in blood, implemented as one of the submodels of the respiratory model. We have based our pH model directly on a model developed by

Chiari et al. (1994) that relates $[H^+]$ and carbon dioxide concentration c_{CO_2}. (When we refer to the concentration of carbon dioxide c_{CO_2}, we mean all the CO_2 transported by blood, both as dissolved CO_2 and as bicarbonate ions, HCO_3^-.)

The pH model of Chiari et al. (1994) is based on four chemical reactions, all involving hydrogen. The chemical reactions are reversible processes such that the concentrations will adjust to a well-defined equilibrium. Thus a change in one reaction will disturb reactions that involve one or several of the same quantities. This is exactly what happens in the system described by the four reactions of the model. The reactions of Chiari's model are the following:

$$CO_2 + H_2O \leftrightarrows H^+ + HCO_3^-, \tag{8.33}$$

$$HPr \leftrightarrows H^+ + Pr^-, \tag{8.34}$$

$$NaOH + CO_2 \leftrightarrows Na^+ + HCO_3^-, \tag{8.35}$$

$$NaOH + HPr \leftrightarrows Na^+ + Pr^- + H_2O, \tag{8.36}$$

where Pr represents the proteinates.

The first two reactions seem obvious to use because they describe the blood buffers related to the transport of respiratory gases. Including the last two reactions yields a pH model in which the pH value depends on the total concentration of carbon dioxide carried by the blood.

Based on the above reactions, Chiari et al. (1994) obtain the following equations. Expressing equilibrium for the dissociation of CO_2 and deoxyhemoglobin, (8.33) and (8.34), gives

$$K_{a,CO} = \frac{[HCO_3^-][H^+]}{[CO_2]},$$

$$K_{a,Pr} = \frac{[Pr^-][H^+]}{[HPr]}.$$

Conservation of charge gives

$$[H^+] + [Na^+] = [HCO_3^-] + [Pr^-].$$

Finally, mass balance of the chemical substances yields

$$c_{CO_2} = [CO_2] + [HCO_3^-],$$

$$[NaOH]_0 = [Na^+],$$

$$[HPr]_0 = [Pr^-] + [HPr].$$

These six equations have six variables, with four initial concentrations and two equilibrium constants as parameters.

We have solved the six equations with respect to $[H^+]$ and obtained the following third degree polynomial equation, with combinations of the initial conditions and equilibrium constants as coefficients:

$$0 = [H^+]^3 + a_2[H^+]^2 + a_1[H^+] + a_0,$$

$$a_2 = K_{a,\text{Pr}} + [\text{NaOH}]_0 + K_{a,\text{co}},$$

$$a_1 = K_{a,\text{co}}([\text{NaOH}]_0 - c_{\text{CO}_2}) + K_{a,\text{Pr}}(K_{a,\text{co}} + [\text{NaOH}]_0 - [\text{HPr}]_0),$$

$$a_0 = K_{a,\text{co}}K_{a,\text{Pr}}([\text{NaOH}]_0 - [\text{HPr}]_0 - c_{\text{CO}_2}).$$

In general the equation has three solutions, each of which may be complex. However, it is possible to prove the existence of a unique positive real solution in an appropriate range of the carbon dioxide concentration (Christiansen and Dræby, 1996).

The Carbon Dioxide Dissociation Function

In this section a model of the carbon dioxide dissociation curve is given. The dissociation models for oxygen and carbon dioxide in blood are both from Siggaard-Andersen et al. (1988). The dissociation functions are rather complex, and the fact that the derivatives of the functions are used in the differential equation of the transport model also makes this part of the model quite complex. We have chosen these dissociation functions because it is possible to understand them physiologically and because they contain physiological parameters. These two characteristics enable simulation of a response of the transport of respiratory gases to changes in the blood components. According to Siggaard-Andersen et al. (1988), the total concentration of CO_2 in blood, c_{CO_2}, can be calculated as a weighted sum of the concentrations of CO_2 in the plasma and the erythrocytes. In the following equations, superscripts Ery and Pla refer to erythrocytes and plasma, respectively. Subscripts CO_2 and Hb refer to carbon dioxide and hemoglobin, respectively.

$$c_{\text{CO}_2}(\mathbf{p}, \text{pH}) = c_{\text{CO}_2}^{\text{Ery}}(\mathbf{p}, \text{pH})\frac{c_{\text{Hb}}}{c_{\text{Hb}}^{\text{Ery}}} + c_{\text{CO}_2}^{\text{Pla}}(\mathbf{p}, \text{pH})\left(1 - \frac{c_{\text{Hb}}}{c_{\text{Hb}}^{\text{Ery}}}\right). \tag{8.37}$$

The total concentration of CO_2 in the erythrocytes, $c_{\text{CO}_2}^{\text{Ery}}$, is a sum of the bicarbonate concentration and dissolved CO_2 in erythrocytes:

$$c_{\text{CO}_2}^{\text{Ery}} = [\text{HCO}_3^-] + \alpha_{\text{CO}_2}^{\text{Ery}} p_{\text{CO}_2}. \tag{8.38}$$

Using the equality $K_a/[H^+] = 10^{(\text{pH}-p_K)}$, the bicarbonate concentration $[\text{HCO}_3^-]$ is found from (8.31):

$$[\text{HCO}_3^-] = \frac{K_a}{[H^+]}\alpha_{\text{CO}_2}^{\text{Ery}} p_{\text{CO}_2}, \tag{8.39}$$

which implies that

$$[\text{HCO}_3^-] = \alpha_{\text{CO}_2}^{\text{Pla}} p_{\text{CO}_2} 10^{(\text{pH}-p_K)}. \tag{8.40}$$

Thus the total concentration is

$$c_{CO_2}^{Ery}(\mathbf{p}, pH) = \alpha_{CO_2}^{Pla} p_{CO_2} \left(1 + 10^{(pH^{Ery}(\mathbf{p},pH) - p_K^{Ery}(\mathbf{p},pH))}\right). \tag{8.41}$$

By equivalent calculations, the concentration in the plasma, $c_{CO_2}^{Pla}$, is given by

$$c_{CO_2}^{Pla}(\mathbf{p}, pH) = \alpha_{CO_2}^{Pla} p_{CO_2} \left(1 + 10^{(pH - p_K^{Pla}(\mathbf{p},pH))}\right). \tag{8.42}$$

The p_K and pH values of the erythrocytes and plasma are also given by Siggaard-Andersen et al. (1988) as

$$p_K^{Ery}(\mathbf{p}, pH) = 6.125 - \log_{10}\left(1 + 10^{pH^{Ery}(\mathbf{p},pH) - 7.84 - 0.06\, s_{O_2}(\mathbf{p},pH)}\right), \tag{8.43}$$

$$pH^{Ery}(\mathbf{p}, pH) = 7.19 + 0.77(pH - 7.4) + 0.035(1 - s_{O_2}(\mathbf{p}, pH)), \tag{8.44}$$

$$p_K^{Pla}(\mathbf{p}, pH) = 6.125 - \log_{10}(1 + 10^{pH - 8.7}). \tag{8.45}$$

The term s_{O_2} is the oxygen saturation of the hemoglobin, which will be detailed in the next section. The pH of the plasma is output from the pH model, which was detailed in the previous section. By these equations the CO_2 concentration as a function of the tension $c = \phi_b(\mathbf{p})$ is only implicitly given. The reason is that pH value varies with CO_2 concentration: $pH(c_{CO_2})$.

Model of Oxygen Dissociation in Blood

The total concentration of oxygen, c_{O_2}, is calculated as the sum of dissolved oxygen and oxyhemoglobin:

$$c_{O_2}(\mathbf{p}) = \alpha_{O_2} p_{O_2} + c_{Hb} s_{O_2}(\mathbf{p}), \tag{8.46}$$

where α is the solubility constant, p_{O_2} the O_2 tension, and c_{Hb} the hemoglobin concentration of the blood. $s_{O_2}(\mathbf{p})$ yields the fractional saturation of the hemoglobin with oxygen, which is defined as the oxyhemoglobin concentration divided by the total hemoglobin concentration: [HbOO]/([Hb] + [HbOO]). The model of saturation states the relation between the saturation s_{O_2}, the O_2 tension p_{O_2}, the CO_2 tension p_{CO_2}, and the pH.

In Figure 8.11 the saturation curves are shown at different pH values, illustrating the Bohr effect. When the pH or the carbon dioxide tension is raised, the saturation curve is shifted to the right because of a decrease in affinity of hemoglobin for oxygen. If the pH or p_{CO_2} is lowered, the shift is to the left.

If the ability of the hemoglobin molecule to bind O_2 were the same for each hemoglobin binding site, we would expect the saturation to be a logistic function of the oxygen tension. But in a plot of $\log(p_{O_2})$ versus $\log(s_{O_2}/(1 - s_{O_2}))$, measurements of oxygen saturation do not follow a linear curve, but show a symmetrical S shape, which reflects the different affinities of the hemoglobin for the first through the fourth O_2 molecules (Nunn, 1987). The curves measured under different circumstances can be characterized by the point of symmetry $(x_0, 1.875)$, the slope of the symmetry point n_0, and the distance $2h$ between the tangent slopes for $\log p_{O_2} \to \pm\infty$ (Siggaard-Andersen et al., 1984).

These parameters are in the model of Siggaard-Andersen et al. (1988) used to fit measured curves. The model of oxygen saturation of hemoglobin is given by the following

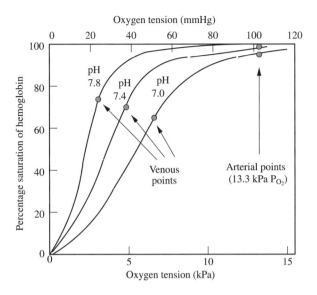

Figure 8.11. *Oxygen saturation curve. From* Nunn's Applied Respiratory Physiology *by J.F. Nunn. Reprinted by permission of Elsevier Science Ltd.*

equations, where the saturation is expressed as a function of the gas tension vector $\mathbf{p}$:

$$s_{O_2}(\mathbf{p}) = \frac{1}{1 + e^{-y(\mathbf{p})}},$$

$$y(\mathbf{p}) = 1.875 + x(\mathbf{p}) - x_0(\mathbf{p}) + h(\mathbf{p}) \tanh(0.5343(x(\mathbf{p}) - x_0(\mathbf{p}))),$$

$$h(\mathbf{p}) = (3.5 + a(\mathbf{p})),$$

$$x(\mathbf{p}) = \log(p_{O_2}/\text{kPa}),$$

$$x_0(\mathbf{p}) = 1.946 + a(\mathbf{p}) + 0.055(T/C - 37),$$

$$a(\mathbf{p}) = -0.72(\text{pH}(c_{CO_2}(\mathbf{p})) - 7.4) + 0.09 \log(p_{CO_2}/5.33 \text{ kPa})$$
$$+ (0.07 - 0.03 x_{Hbf})(c_{dpg}/\text{mol}/1 - 5)$$
$$- 0.368 x_{HbCO} - 0.174 x_{Hi} - 0.28 x_{Hbf},$$

where T is the temperature in degrees Celsius. The model parameters are the substance fraction of fetal hemoglobin x_{Hbf}, the substance fraction of hemoglobin x_{Hi}, the substance fraction of carboxyhemoglobin x_{HbCO}, the concentration of 2,3-diphosphoglycerate in the erythrocytes c_{dpg}, and the temperature T. According to Siggaard-Andersen et al. (1984), the equations represent a very good fit to the Servinghaus standard oxygen dissociation curve; however, the parameters may easily be fitted to other oxygen dissociation curves as well.

The equations do not give an explicit expression for the saturation, as the expression depends on pH and pH depends on carbon dioxide concentration $c_{CO_2}(\mathbf{p})$, and that in turn depends on the saturation.

Dissociation in Tissue

The models of dissociation in tissue are less complicated than the blood dissociation functions. In Chiari et al. (1995) the blood dissociation function is used for CO_2 in tissue and a simple linear solution is used for O_2.

Our blood dissociation function (Siggaard-Andersen et al., 1988) differentiates between the dissociation in plasma and in erythrocytes, and we use the plasma part for tissue dissociation, after direct advice from Siggaard-Andersen (private communication). For oxygen we use a linear solubility with the coefficient α_{O_2} found in (8.46). This reflects that no hemoglobin is present in the tissue, so the solution is simpler, but this method may fail to capture the effect of other oxygen-binding proteins. In the muscles an oxygen-binding protein called myoglobin is present, but we have not found any quantitative information about the binding of myoglobin with oxygen. Changes in the oxygen stores in the muscles happen slowly because the muscles are low perfused, i.e., the blood flow relative to the volume of the muscles is small. The effects of such change on the rest of the body are very small.

Parameters for Gas Dissociation and the pH Value

In Table 8.3 we have listed the important parameters of the gas dissociation functions and the pH-model.

Table 8.3. *Parameters of the pH Model.*

Symbol	Standard value		Description	Source
K_{aCO}	$10^{-6.1}$	mol/l	Equilibrium constant	
K_{aPr}	$10^{-7.3}$	mol/l	Equilibrium constant	
$[NaOH]_0$	46.2	mmol/l	Concentration of NaOH	(Chiari et al., 1994)
$[HPr]_0$	39.8	mmol/l	Concentration of proteinate	(Chiari et al., 1994)
$\alpha_{O_2}^{Ery}$	0.195	mmol/l/kPa	Solubility of O_2 in erythrocytes	(Siggaard-Andersen et al., 1988)
$\alpha_{CO_2}^{Pla}$	0.230	mmol/l/kPa	Solubility of CO_2 in plasma	(Siggaard-Andersen et al., 1988)
α_{O_2}	9.83	μmole/l/kPa	Solubility of CO_2	(Siggaard-Andersen et al., 1988)
c_{Hb}	9.30	mmol/l	Total concentration of hemoglobin	(Siggaard-Andersen et al., 1988)
x_{Hbf}	0		Fraction of Hb that is fetal hemoglobin	(Siggaard-Andersen et al., 1988)
x_{HbCO}	0.005		Fraction of Hb that is carboxyhemoglobin	(Siggaard-Andersen et al., 1988)
x_{Hi}	0.005		Fraction of Hb that is hemoglobin	(Siggaard-Andersen et al., 1988)
c_{DPG}	5.01	mmol/l	Concentration of 2,3-diphosphoglycerate	(Siggaard-Andersen et al., 1988)
c_{Hb}^{Ery}	21.0	mmol/l	Concentration of Hb in erythrocytes	(Siggaard-Andersen et al., 1988)
κ	∞		Lung membrane flux coefficient	(Siggaard-Andersen and Gøthgen, 1989)
β_{O_2}	0.01	mmol/l	Concentration where metabolism is half of max	—
λ	0.02		Pulmonary shunt	(Lerou et al., 1991)

8.3.5   Control of Respiration

The substances carried by the blood transport system are vital for the body. Lack of O_2 for even short periods can cause brain damage. Damaging effects are normally avoided, though, because the system is controlled to match the pulmonary and metabolic gas exchange rates. In this section we present a brief discussion of the control of the respiratory system, which is performed by the CNS.

Before we discuss the control system, we will draw some attention to the dynamics of the transport system as we have modeled it. The system already contains some mechanisms to improve utilization of oxygen and elimination of carbon dioxide in response to disturbances. When the metabolic rate increases, the diffusion of oxygen and carbon dioxide between blood and tissue increases because of an increased tension gradient. An additional improvement in oxygen delivery is caused by the Bohr shift. Increased metabolic production of CO_2 affects the pH of blood. At decreased pH the oxygen dissociation curve is shifted to the right. This impairs the oxygenation of blood in the lung, but the negative effect is small compared to the improved release of oxygen in tissue (Nunn, 1987). In the same way, the Haldane effect shifts the carbon dioxide curve with respect to the oxygen level in blood and yields a difference in quantity of carbon dioxide carried in oxygenated and reduced blood (at constant carbon dioxide tension). This results in an improvement in gas transport as well.

The principal function of the respiratory control system is to reduce the sensitivity of the transport system to external disturbances. Thus the signal from the CNS counteracts changes in the body.

The behavior of the respiratory control system can be effectively described by looking at a healthy person's response to exercise, though one must be aware of the fact that the principal controlling mechanism is different under different conditions (Nunn, 1987, p. 72).

When faced with an increased metabolic rate due to moderate exercise, the respiratory control system increases ventilation and blood flow. A sufficient oxygen supply is thus maintained and the system finds a new steady state. At heavy exercise the oxygen supply becomes too small and anaerobic production will occur. How long work can continue depends on the level of lactate in the arterial blood. Lactic acid is the principal product of anaerobic metabolism, but ionizes to lactate and hydrogen ions. If the lactate remains constant, a steady state can exist during exercise. After work has finished, there will be a period of recovery when the oxygen debt has to be repaid in order to oxidize the products of the anaerobic metabolism.

The Controlled Variables

The CNS affects the respiratory system by changing ventilation, blood flow, and cardiac output. However, other systems affect the heart through the CNS, and hence the effect on cardiac output is a combination of stimuli from the respiratory system, the baroreceptors (described in detail in Chapter 7), and other systems.

The Controlling Variables

The most important input for the control of ventilation is from the peripheral and central chemoreceptors. The peripheral receptors are called the carotid and aortic bodies. They are sensitive to changes in the arterial blood. Specifically, they are stimulated by an increase in hydrogen ion concentration, an increase in the tension of CO_2, or a decrease in the tension of O_2. The central chemoreceptors are located in the medulla oblangata. They are stimulated by changes in the hydrogen ion concentration.

As mentioned previously, the heart is controlled by variables from both the transport model and the cardiovascular model. The controlling variables from the transport model are arterial tensions of O_2 and CO_2.

8.4   Results

In this section we describe tests of the models of the respiratory system. The aim of this section is to give an overview of the possible simulations of the respiratory system that can be carried out with the models described in this chapter. The simulations we will present are thus not a complete description of the behavior of the models under all circumstances, but rather a highlight of some interesting behaviors. Systems of ordinary differential equations in the models are solved using the Runge–Kutta–Fehlberg method.

First the lung model is tested in order to produce characteristic pressure-volume diagrams, found in Bardoczky et al. (1996). In the simulation we use the parameters found in section 8.2.3. These diagrams are output from the pressure model, which is a submodel of the lung model. The other part of the lung model, the gas model, shows the partial pressures in the expired air and the alveoli under different circumstances.

The evaluation of the blood transport model is made first by testing the submodels of the blood transport system. Then the blood transport model is used for simulations of distribution of respiratory gases and anesthetic. We show the dynamics of the blood transport model, with and without metabolism. For respiratory gases we describe the influence of the Bohr–Haldane effect.

8.4.1   Lung Model

The lung model consists of the pressure model and the gas model, and we will test each submodel individually.

The Pressure Model

One of our goals was that our lung model can produce the pV diagrams normally encountered in the monitoring of anesthetized patients. The technique of measurement is described in Bardoczky et al. (1996), which also shows some characteristic diagrams.

To compare the output of the model with the diagrams of Bardoczky et al. (1996) we have used an approximation of a respirator pressure curve for input; see Figure 8.12. Using the pressure curve shown in Figure 8.12, $U_m(t)$, and the parameters from Table 8.4, we have been able to compute a pressure-volume diagram for a normal lung; see Figure 8.13.

The effect of a low compliance, simulated by halving C_1, can be seen in Figure 8.14. The characteristic effect is that the tidal volume is lower than in the normal lung and thus that the loop is more flat. Another significant feature is the "ducktail" at the end of the expiration. The pressure falls to zero, but since the lung contains a small amount of air relative to the resistance, the air leaves easily, and at the last of the expiration the pressure inside the lung is almost equal to the external pressure.

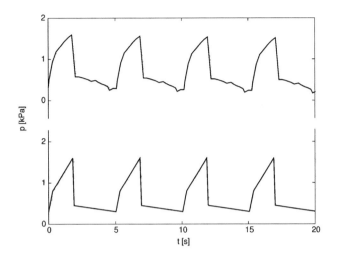

Figure 8.12. *The respirator pressure from Bardoczky et al. (1996) (upper curve)
and our approximation (lower curve).*

Table 8.4. *The parameters of the standard lung.*

	R [$l \cdot s/(kPa \cdot min)$]	C [$l \cdot s/kPa$]	V_0 [l]
Central compartment	1.00	0.02	0.1
Alveolar compartment	0.02	0.80	2.7

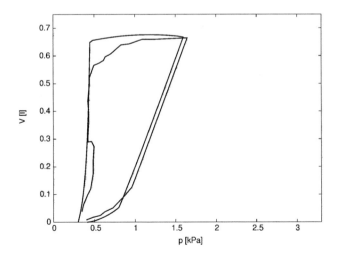

Figure 8.13. *The model-generated pV diagram for a normal lung and the diagram
from Bardoczky et al. (1996) (irregular curve).*

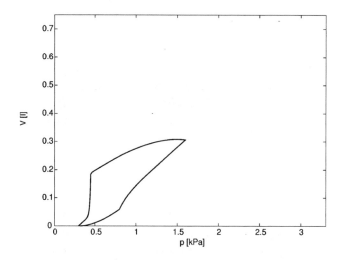

Figure 8.14. *pV diagram generated with low compliance $C_1 = C_{1,std}/2$.*

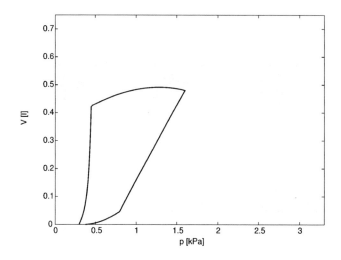

Figure 8.15. *pV diagram generated with high airway resistance $R_0 = 2R_{o,std}$.*

Increasing the resistance of the upper airways by doubling R_0 hampers the flow of air into and out of the lung; see Figure 8.15. Thus the tidal volume decreases, since less air enters during inspiration. Notice that the ducktail effect observed with low compliance does not occur.

The Gas Model

The interesting aspect of the gas model is its ability to renew the air in the alveoli. To show the dynamics of the ventilation we have set the gas flow over the lung membrane constant at

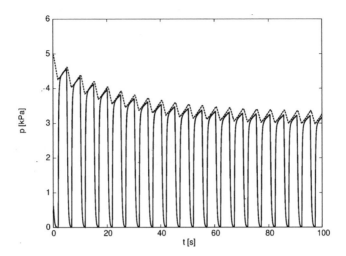

Figure 8.16. p_{CO_2} in a normal lung. Solid line: central compartment; dashed line: alveolar compartment.

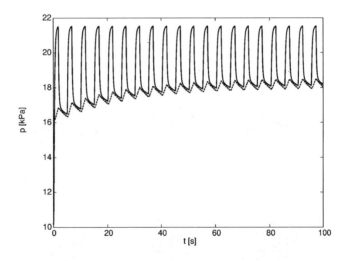

Figure 8.17. p_{O_2} in a normal lung. Solid line: central compartment; dashed line: alveolar compartment.

a level that matches normal metabolism (260 ml O_2 and 160 ml CO_2 per minute) and started the model with a high level of carbon dioxide and a low level of oxygen in the alveoli.

In Figures 8.16 and 8.17 can be seen the behavior of the partial pressures of CO_2 and O_2 in both the central and the alveolar compartments. The standard parameters from Table 8.4 and the pressure curve from Figure 8.12 are used as inputs. The central compartment shows large variation, since the partial pressures tend toward the partial pressures of the atmosphere during inspiration and the partial pressures of the alveoli during expiration. The variation in

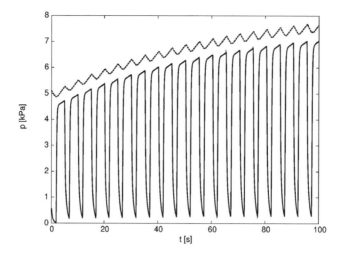

Figure 8.18. p_{CO_2} *in a lung with low compliance* $(C_1 = C_{1,std}/2)$. *Solid line: central compartment; dashed line: alveolar compartment.*

the alveoli is much less, which is not surprising since the alveoli make up most of the lung volume and since the alveolar volume is about six times the tidal volume. The graphs show how the excess carbon dioxide is eventually removed and the oxygen level is increased to the normal levels.

When the same test is simulated in a lung with half compliance, the result is very different. The ventilation of this lung is not sufficient to remove the carbon dioxide or supply the oxygen. During anesthesia, this can be counteracted using a respirator with an increased pressure, but the graphs in Figures 8.18 and 8.19 illustrate the effect of poor ventilation. The level of CO_2 rises and the O_2 level drops.

A final view of the dynamics of our gas model can be found in Figure 8.20, where the respirator pressure is shown together with the carbon dioxide partial pressure in the central compartment. When expiration starts, at the peak of the respirator pressure curve, the analyzer will start to receive air from the central compartment. This air will show the characteristic increase in carbon dioxide level when the air from the alveoli has filled the central compartment to an extent where most of the air expired originates in the alveoli.

8.4.2 Dissociation Curves

In this section we show output from the models based on the dissociation curves of Siggaard-Andersen et al. (1988) under different conditions. The output is compared with data from Nunn (1987). Other literature (Widdicombe and Davies, 1991; Grodins and Yamashiro, 1978; West, 1979) was consulted as well, but was found to be in agreement with Nunn (1987).

The dissociation curves are curves of the gas concentration in blood as a function of the tension of the gas, but since the respiratory gases dissociate into blood in a nontrivial way, due to chemical reactions, the curves depend on several other parameters. This section has two aims. One aim is to run scenarios with the models, which produce dissociation curves

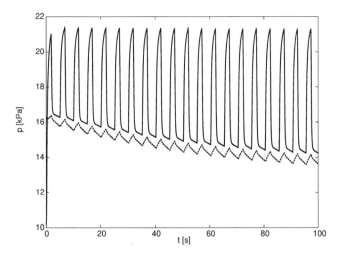

Figure 8.19. p_{O_2} *in a lung with low compliance. Solid line: central compartment; dashed line: alveolar compartment.*

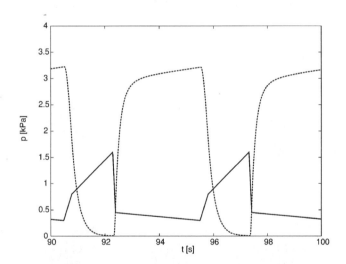

Figure 8.20. *The respirator pressure U_m (solid line) and the CO_2 pressure (dashed line) in the central compartment.*

under different circumstances, and discuss whether the models of Siggaard-Andersen et al. (1988) are in accordance with the literature. The other aim is to evaluate the importance of the different effects included in the model.

Dissociation of Carbon Dioxide

The models (Siggaard-Andersen et al., 1988) for dissociation of carbon dioxide (see p. 224ff.) were developed to be used for analyzing measured data. The pH of blood is a parameter of

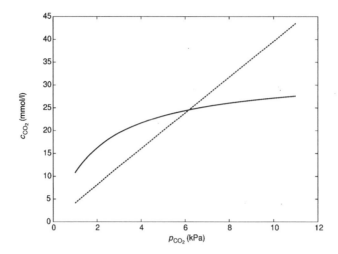

Figure 8.21. *The carbon dioxide model of Siggaard-Andersen et al.* (1988), *with (solid line) and without (dashed line) the pH model.*

the models that is usually measured. Thus adapting the models of the dissociation curves from Siggaard-Andersen et al. (1988), we have added the model of pH in blood developed by Chiari et al. (1994). The graph in Figure 8.21 shows the dissociation curve of carbon dioxide with constant pH 7.4, compared to the curve with a dynamic pH varying with the carbon dioxide concentration. The intersection of the curves shows the concentration of carbon dioxide at which blood pH is 7.4. Both curves are plotted with constant oxygen tension.

The interaction of carbon dioxide with hydrogen ions is the main reason for the nonlinearity of the CO_2 dissociation curve. The higher the tension of p_{CO_2}, the lower the pH, which limits the solubility of the carbon dioxide. In conclusion to Figure 8.21 we can say that the pH dependence of the CO_2 dissociation curve is very important and that the model with constant pH value is not adequate.

In Figure 8.22 we have varied the tension of oxygen from 13.3 kPa (normal arterial tension) to 2 kPa, so that the saturation of hemoglobin with oxygen is lowered from approximately 100% to a saturation that at pH 7.8 is about 40% according to the model of O_2 dissociation. A small leftward shift of the curve at low oxygen saturation is found. According to the values in Table 8.5 from Nunn (1987), the difference in CO_2 transport of venous and arterial blood is about 1.8 mmol/l. According to Nunn (1987), about one third of the reported difference is due to the Haldane effect, which is the difference in the quantity of CO_2 carried, at constant p_{CO_2}, in oxygenated and reduced blood. The rest of the difference is caused by the change in p_{CO_2} in arterial and venous blood.

Thus the difference in carbon dioxide concentration in arterial blood at $(p_{CO_2}, p_{O_2}) = $ (5.3 kPa, 13.3 kPa) and in mixed venous blood at $(p_{CO_2}, p_{O_2}) = $ (6.1 kPa, 5.3 kPa) ought to be about 1.8 mmol/l. The model found it to be about 1.2 mmol/l. We do not consider this discrepancy to be crucial for the models, as the Haldane effect is rather small and thus not significant for the behavior of the model.

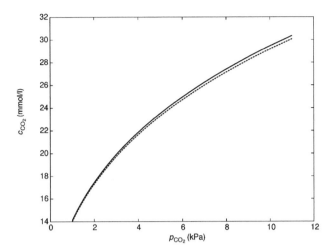

Figure 8.22. *The Haldane effect yields different dissociation of carbon dioxide for changed oxygen tensions. Solid line:* $p_{O_2} = 2$ *kPa; dashed line:* $p_{O_2} = 13.3$ *kPa.*

Table 8.5. *Normal values for carbon dioxide in blood.*

Whole blood	Arterial blood Hb 95% sat.	Mixed venous blood Hb 70% sat.	Arterial/venous difference
pH	7.4	7.367	−0.033
p_{CO_2} [kPa]	5.3	6.1	+0.8
c_{CO_2} [mmol/l]	21.5	23.3	+1.8

Dissociation of Oxygen

The curves of oxygen dissociation are, under normal circumstances, dominated by hemoglobin transport, since only a negligible part of the oxygen is dissolved in blood. Therefore, there are no visible differences between the forms of the oxygen dissociation curve and the oxygen saturation curve of hemoglobin when a normal content of hemoglobin is found in the blood. In this section we will show the saturation curves for scenarios with normal hemoglobin content in blood.

When we plot the oxygen saturation curve under different circumstances, it is not always possible to change one parameter at a time, since some parameters of the model influence each other. Normally, the literature does not address this problem when graphs illustrating changes in one parameter are shown. These curves have no straightforward physiological interpretation and thus we do not have the exact information to reproduce them. We keep this fact in mind when comparing our model with the literature.

We have investigated the influence of a change in pH on the oxygen saturation curve. The pH value changes with the carbon dioxide concentration, which can be calculated as a function of the different tensions of carbon dioxide and oxygen. Hence a change in pH may be caused by either a variation in the CO_2 tension or some other change that will in turn

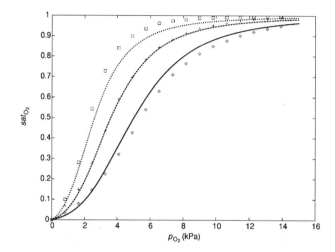

Figure 8.23. *The oxygen dissociation with different pH values compared with data from Nunn (1987). Solid line model and $\Diamond$ data: pH = 7.0; dashed line model and $\square$ data: pH = 7.4; dotted line model and + data: pH = 7.8.*

change the balance between oxygen and carbon dioxide. Hence without information about the cause of the change in pH, it is not possible to deduce in a unique way the CO_2 tension along the curves.

Figure 8.23 shows the saturation of hemoglobin as a function of oxygen tension for different values of pH. The carbon dioxide tension is constantly 5.3 kPa. The curve is shifted to the left with increased pH values. Compared with the data from Nunn (1987, p. 265), the curves lie a little too close together, but the graphs in Figure 8.23 are qualitatively in agreement with the data. It is not stated which values of p_{CO_2} the curves from Nunn (1987) represent.

The saturation curves of our model are shown for other values of p_{CO_2} in Figure 8.24. We have plotted the dissociation curves for oxygen at various tensions of carbon dioxide, which correspond to the pH values indicated. This situation is not physiologically realistic because the carbon dioxide tension exceeds the physiological range. The drastic changes in the pH values shown in the graphs are only found in situations with other disturbances of the pH value. The curves coincide better with the points from Nunn (1987).

Other parameters relevant to changes in the model of oxygen dissociation are the concentration of hemoglobin in blood and the fraction of hemoglobin bound with carbon monoxide. The remaining parameters of the model are not subject to changes in the simulator. The concentration of hemoglobin is normally 7.0 mmol/l, corresponding to a content of 14.4 g/dl. At lower concentrations the blood is called anemic and the amount of oxygen transported is reduced. Another way to obtain a reduction in oxygen carriage is by an increased fraction of hemoglobin in combination with carbon monoxide. The presence of carbon monoxide will displace the hemoglobin from the combination with oxygen.

In Figure 8.25 the dissociation curves of the two cases of anemia are shown (together with the normal condition). We have plotted the concentration of oxygen in blood as

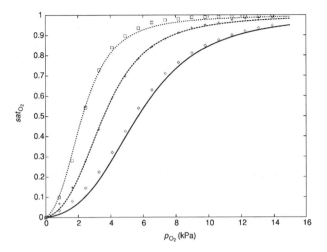

Figure 8.24. *The oxygen saturation curves with adjusted* p_{CO_2}. *Our model is compared with data from Nunn (1987). Solid line model and* $\diamond$ *data: pH = 7.0; dashed line model and* $\square$ *data: pH = 7.4; dotted line model and + data: pH = 7.8.*

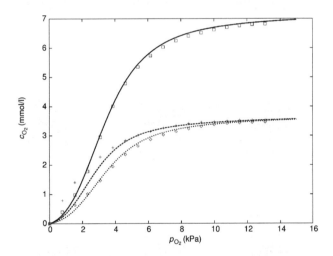

Figure 8.25. *Oxygen concentration in normal, anemic, and CO-poisoned blood. Our model compared with data from Nunn (1987). Solid line model* $\square$ *data: pH = 7.0; dashed line model* $\diamond$ *data: pH = 7.4; dotted line model + data: pH = 7.8.*

a function of the oxygen tension, with hemoglobin concentration set to 50% below the normal hemoglobin concentration and with a carboxyhemoglobin level set at 50% of the total hemoglobin concentration. The output of the model for anemic blood is in good agreement with the data from Nunn (1987). The output shows that the concentration of oxygen in blood is halved as compared to the situation with a normal content of hemoglobin in the blood.

The modeled reduction of oxygen transport by means of a high level of carboxy-hemoglobin does not correspond to data obtained by Nunn (1987). In Nunn (1987) it is stated that the leftward shift is due to a change in the level of c_{DPG} (concentration of 2,3-diphosphoglycerate), which is found when carbon monoxide is present. We have not modeled carbon monoxide's influence on c_{DPG}; hence the model does not produce a realistic output of oxygen tensions below 4 kPa, under CO poisoning.

8.4.3   Blood Transport Model

The transport model keeps track of both anesthetic agents and respiratory gases. Here we present only the respiratory dynamic of the respiratory gases.

Blood Transport of Respiratory Gases

In this section we present the output of the transport model concerning transport of the respiratory gases. The dynamic of the model is presented step by step, taking one effect into account at a time. First, we will show the simplest possible version of the transport model. This simulation does not show a physiological situation. The initial simulations are made with the metabolism set to zero, with constant partial pressures in the alveoli of 13.3 kPa in oxygen tension and 5.3 kPa in carbon dioxide tension (these are standard arterial values), and without a pulmonary shunt. The oxygen and carbon dioxide tensions are initiated at 10 kPa in all other compartments to show how the oxygen is distributed into the compartments, and carbon dioxide is removed.

Figures 8.26 and 8.27 show the curves of oxygen tension and carbon dioxide tension, respectively, for venous and arterial blood pools and for several selected organs. The figures also show the constant curves of the pulmonary capillaries. Since the metabolism is zero, all tensions will tend to the same value as the pressure in the alveoli, and hence the graphs show how the compartments in different stages reach the tension of the capillaries.

The arterial compartment is the only compartment connected directly to the pulmonary compartment, and hence the increase in oxygen and decrease in carbon dioxide both begin in the arterial pool. The venous pool does not change until the tensions in the flow out of the body compartments change.

The states of the body compartment depend on the perfusion compared to the size of the compartment. Thus at first the kidney compartment has the greatest increase in O_2 and decrease in CO_2 tension. Next is the brain compartment and at last the low perfused muscle compartment. The perfusion of the muscle compartment is much lower than the viscera compartments, and therefore the change in the tensions in the venous blood is higher than that in the muscle compartment, even though part of the blood flow to the venous pool arrives from the muscle compartment.

When the oxygen tension of a compartment increases toward the tension of the pulmonary compartment, it will increase to a level above the oxygen tension of the compartment from which the blood supplying this compartment came. This might seem strange, but it is explained by the Bohr effect, which causes the shift of the dissociation curves of oxygen at different tensions of carbon dioxide.

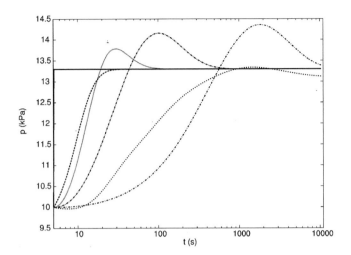

Figure 8.26. *Oxygen tension in various compartments without metabolism. Solid line: pulmonary capillary; long dashed line: arterial compartment; short dashed line: venous compartment; dotted line: kidney; long dash-dotted line: brain; short dash-dotted line: muscle.*

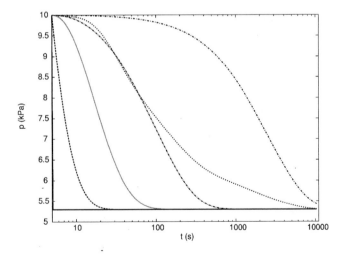

Figure 8.27. *Carbon dioxide tension in various compartments without metabolism. Solid line: pulmonary capillary; long dashed line: arterial compartment; short dashed line: venous compartment; dotted line: kidney; long dash-dotted line: brain; short dash-dotted line: muscle.*

The Bohr Effect

Because body compartments have a higher tension of carbon dioxide than arterial blood, the ability of the blood to carry oxygen is lower in these compartments. Thus a lower concentration is found in blood flowing out of a body compartment than in blood flowing in, even though there is a higher oxygen tension in the compartment than in the inflowing blood.

The Bohr effect is seen in the overshoot of tensions in all the body compartments and in the venous pool. In this section we will focus on the kidney. We will take a close look at what happens in the kidney compartment during the first 2 min of the simulation. In Figure 8.28 we have plotted the oxygen and carbon dioxide tensions in the arterial and kidney compartments from Figures 8.26 and 8.27 on a linear time scale.

The two oxygen curves intersect at about 12 s at an oxygen tension of about 13 kPa. The concentrations of oxygen in the two compartments at the moment the two curves intersect are found in Figure 8.29. Even though the oxygen tensions are equal in the two compartments, the shift of the oxygen dissociation curve by the difference in $p_{\mathrm{CO_2}}$ results in different oxygen concentrations in the blood in the two compartments. Thus the concentration of oxygen in the kidneys is lower than in the arterial blood. The total amount of oxygen in the kidneys continues to increase as more oxygen is brought with the blood into the compartment than is removed by the outflowing blood.

The oxygen curve for the kidneys peaks at the moment when the oxygen concentration equals the concentration of arterial blood. This happens because the concentration of carbon dioxide is reduced significantly.

The Bohr effect causes the oxygen tension of all the body compartments and the venous blood pool to overshoot in Figure 8.26. When the tension in the venous blood decreases to below the tension of oxygen in the body compartment, the overshoot happens again but with opposite sign.

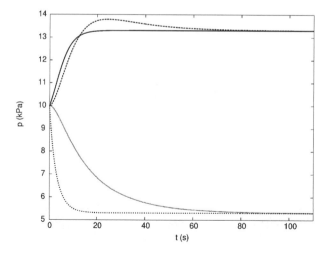

Figure 8.28. *Oxygen and carbon dioxide tension in the arterial pool and kidney. Solid lines: arterial pool; dashed lines: kidney. The upper curves refer to oxygen and the lower curves refer to carbon dioxide.*

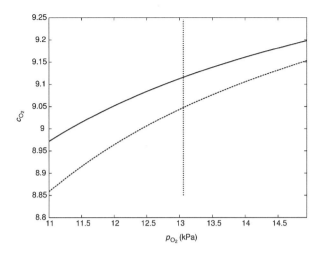

Figure 8.29. *Oxygen concentrations in the two compartments are marked at the moment of intersection of the tension curves. Solid line:* $p_{CO_2} = 5.4 \ kPa$; *dashed line:* $p_{CO_2} = 7.82 \ kPa$; *dotted line:* $p_{O_2} = 13 \ kPa$.

Steady State of the Respiratory Model

In this final section we introduce the metabolic functions and thus observe oscillations toward a more realistic equilibrium than in the previously shown results. Section 8.2.3 shows the metabolic rates we found for each compartment in the transport model. These metabolic rates were found to agree with published values and further calculated in such a way that a physiological steady state of the model was obtained.

In this section we have used the metabolic rates given in Table 8.6, and in Figures 8.30 and 8.31 it is shown how the system reaches a steady state. The Bohr effect is still causing oscillations in the curves of the oxygen tension (Figure 8.31), especially for the kidney compartment. The equilibriums of the gas tensions of the various compartments are no longer identical because the metabolic rates of the compartments differ. The high equilibrium

Table 8.6. *Metabolic rates. These metabolic rates are obtained by separating the body into compartments.*

	$M \ CO_2$ [ml/min]	$M \ O_2$ [ml/min]
Liver	143	88.9
Kidney	11.3	6.99
Heart	12.2	7.56
Remaining viscera	8.15	5.05
Brain	50	31
Connective tissue	8.52	5.28
Muscles	26.5	16.4
Adiapose	0	0

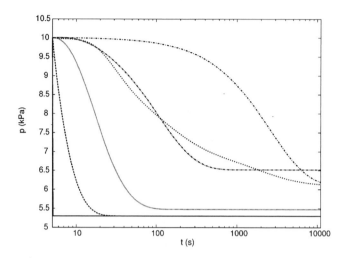

Figure 8.30. *Carbon dioxide concentrations in various compartments with metabolism. Solid line: pulmonary capillary; long dashed line: arterial compartment; short dashed line: venous compartment; dotted line: kidney; long dash-dotted line: brain; short dash-dotted line: muscle.*

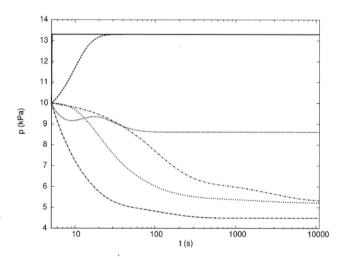

Figure 8.31. *Oxygen concentrations in various compartments with metabolism. Solid line: pulmonary capillary; long dashed line: arterial compartment; short dashed line: venous compartment; dotted line: kidney; long dash-dotted line: brain; short dash-dotted line: muscle.*

tension of oxygen in the kidney compartment reflects the fact that the metabolic rates have been assigned in accordance with the size of the compartment and not with respect to the blood perfusion. A small compartment with a high perfusion content will therefore reach a high tension of oxygen at equilibrium. For the same reason the carbon dioxide tension of the kidneys is low.

Acknowledgments

The authors were supported by High Performance Parellel Computing, Software Engineering Applications, Eureka Project 1063 (SIMA—SIMulation in Anesthesia).

Appendix A

The SIMA Simulator

J.T. Ottesen, M.S. Olufsen, and J.K. Larsen

A.1 Introduction

This appendix gives a brief description of anesthesia in general and then a general description of the models in the SIMA (SIMulation in Anesthesia) simulator, showing the relationship between the models actually implemented in the simulator and the reference model reported in this book. A detailed description of the actual models is beyond the scope of this book.

A.2 Anesthesia Simulation

Anesthesia is a highly complex procedure, where drugs are given to keep the patient unconscious, relaxed, and free from pain. Drugs are typically injected directly into the bloodstream, from which they spread to the brain and the muscles where they act (and where in time they are decomposed). The concentration of drugs can, under normal circumstances, be measured in the bloodstream only. However, during surgery, the concentration of some drugs, for example, pancuronium, should be kept within a given range in muscle tissue: high enough to function adequately and low enough not to cause damage. These observations should be guided by questions such as what dosing strategy, if any, can achieve this balance, and should be chosen based on thinking in terms of mathematical models. Models for uptake, distribution, and decay of pharmaceuticals during anesthesia involve both individual- and drug-specific parameter values. Therefore, to maintain a correct level of the drug one must first estimate the parameters, i.e., solve an inverse problem.

Another issue related to anesthesia is the complex interaction between the anesthesiologists administering the drugs and the other people (surgeons, nurses, and technical staff) in the operating theater. To enhance their communication the surgery is controlled by a large number of monitors and measuring equipment. As a result the operators—the anesthesiologists, nurses, and technical staff—must be highly skilled. Operators should be able to monitor and control a vast number of dynamic variables simultaneously and in real time. It is required of the personnel that they be able to diagnose and remedy critical incidents dynamically by picking up and processing huge amounts of information from monitors and measuring equipment and communicating the results to other people working with them. Often the amount of information that has to be evaluated is so vast that it approaches the cognitive limits of well-trained operators. Therefore, it is necessary to find technical solutions to the cognitive problem of processing such comprehensive and dynamically changing systems of information.

Using mathematical terminology, the modern operating room can be viewed as a complex dynamical system where human agents interact with advanced equipment and processes. The anesthesiologist, like a pilot or a process control engineer, plays an important role in this dynamical system. In this setting the anesthesiologist should be capable of diagnosing critical incidents, predicting the outcome of various possible interventions, devising plans, and making decisions. His or her performance depends to a large extent on how well he or she comprehends the available information. He or she must synthesize relevant information into a coherent picture of the state of the patient. He or she must separate false alarms from real ones. Today, it is known that a considerable fraction of all accidents during anesthesia are due to human errors, not malfunctioning of technical equipment (Cook, Woods, and McDonald, 1991). Clearly, operating room safety and performance may be improved by improving the skill of the anesthesiologist via training or presenting better operating room information.

Complex cognitive judgments can be enhanced using simulators based on mathematical models. Simulators have proved successful in aviation and navigation, for example. Simulators can be based on fundamental physical laws, on curve fitting, or on expert systems that learn as the system gains knowledge. If simulators are based on fundamental laws, they can present data and predict future events in a fast and reliable way. Outcomes of calculations can be animated and presented in visual form. In this way complex information can be quickly accessed and easily comprehended. Therefore, it may be better used in the operator's planning and decision making. Moreover, full-scale simulators, such as flight simulators and the SIMA anesthesia simulator, can be used for cognitive training, for testing new equipment (e.g., monitors) and levels of monitoring, and for studies of an operator's cognitive behavior.

In order to improve the performance of humans using technically complex simulators, the simulation designer needs to address several issues. First, the designer of the simulator needs a good scientific understanding of the system being manipulated. Hence in anesthesia simulators mathematical models of physiological processes based on first principles should be used. Such models are necessary if we want to predict the outcome of various interventions. Second, the operators of the system should have a thorough knowledge of the system (the combined system of the patient and the monitoring and control systems hooked to the patient), its various modes of behavior, and its possible malfunctions. Third, operators should be skilled enough to make appropriate interventions in critical situations. Therefore, to develop a simulator the task is to develop several kinds of mathematical models: science-based models of physiological processes, ad hoc prediction models, and models for information processing and presentation.

A.3 The Models of SIMA

The SIMA simulator is a near replica of the anesthetic work environment. An overview of the different components is shown in Figure A.1.

A complete simulation is a play with several actors. In order to make this play as realistic as possible, we created a scenario closely resembling what happens in an operating theater. The main players in a simulation are the anesthesiologist (the trainee), possibly a nurse, one or two surgeons, an instructor who is controlling the simulation, and an operator who is controlling the computer. The latter two people will not be present during a real anesthetic procedure. To make the simulation more real all the players dress as medical personnel. And it is not unusual to have a number of people observing a simulation or additional hands available in case something goes wrong. Of course the training scenario is carried out on a mannequin, not a real patient. The mannequin, a modified resuscitation dummy, cannot provide the electrocardiogram (ECG), blood pressures, and other signals to the monitors. It is here that the mathematical models come into play: All signals that need to be generated for the monitors are generated by mathematical models running on

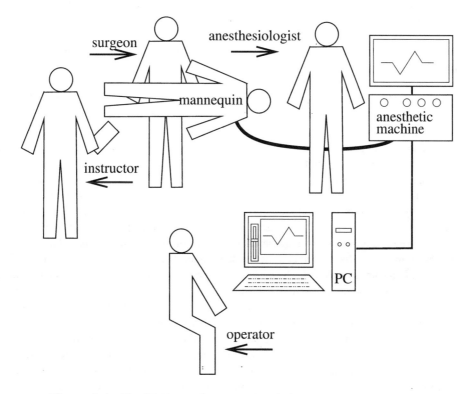

Figure A.1. *The SIMA simulator. (A detailed description of the various components can be found in the text.) From J. Larsen and S.A. Pedersen (2000), Mathematical models behind advanced simulators in medicine, in* Mathematical Modelling in Medicine, *J.T. Ottesen and M. Danielsen (eds.), SHTI 71, IOS Press, Amsterdam, pp. 203–216. Used by permission of the publisher.*

a high-end PC. Then an interface box converts these signals into a form compatible with various anesthetic monitors. The interface box receives digital signals from the computer and converts these to analog, the form the monitors would receive from a real patient. Most often the computer and its operator sit in a room adjacent to the operating theater, and the operator communicates with the instructor via an intercom.

The configuration described above is diagrammed in Figure A.1, which also shows how the various machines are connected to each other. During a simulation, a predefined sequence of events, designated by a script, occurs. The trainee and the computer operator must respond dynamically to the scripted events. The computer operator is able to monitor the simulation through a graphical user interface that shows all the variable states. He or she controls all parameters in the models either from the user interface or via scripts (predefined sequences of events). For example, when drugs need to be administered to the "patient," the amount and the type of the drug are communicated by the instructor to the computer operator, who then modifies the model parameters accordingly, or the computer can be pre-programmed to change certain parameters either at a given point in time or over a given period of time. The instructor monitors the trainee and communicates relevant information between the computer operator and the trainee. For example, the instructor might tell the trainee that the patient is blue due to a lack of oxygen, the patient's pupils are dilated, the patient's temperature is elevated, etc. This aspect of the simulation is somewhat unrealistic, but in our experience the trainee adapts to the nonphysiological state of the mannequin and engages in the scenario in question (Gaba, 1994).

The major models of SIMA and the connections among the models are shown in Figure A.2. The models can be divided into the following groups: an ECG model; two circulatory models, a cardiovascular model and a baroreceptor model; two transport models, a respiratory model and a pharmacokinetic model; two models related to metabolism, a metabolistic model per se and a related temperature model; a fluid and electrolyte model; and finally the pharmacodynamic models.

Each of these models is shown as a box in Figure A.2. The arrows in the figure correspond to data transport between the models. Data transport to one of the three bigger boxes (the transport box, the circulation box, or the metabolism box) indicates that the same data go to all models within that bigger box. Besides the transport of data between the models there exist a number of parameters common for all the submodels. These shared parameters are illustrated by the common outer frame. The pharmacodynamics should be interpreted in a somewhat broader frame since the process reacts both to the concentrations of anesthetic gases and injections and to the natural gases: oxygen and carbon dioxide. Since the nitrogen concentrations are so low, they are not taken into account by the respiratory model. Additionally, the respiration model accounts for blood pH.

Some of these concepts could be simulated using the zero-dimensional models described in Chapters 4, 6, 7, and 8. By zero-dimensional models we mean models based on ordinary differential equations that include time but have no spatial information. Models including one, two, or three spatial dimensions are denoted one-, two-, or three-dimensional models. The latter models are based on partial differential equations. The models presented in the remaining chapters may be used to provide parameters and validate some of the zero-dimensional models. For example, the 1D model in Chapter 5 of the arterial system could be used to specify the profile of the flow and pressure wave or to determine some of the parameters of the zero-dimensional cardiovascular model presented in Chapter 6. The

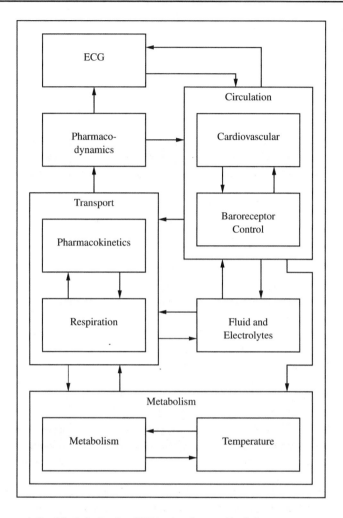

Figure A.2. *Models in the SIMA simulator. Each box represents a model, and the three larger boxes representing the circulation, transport models, and metabolism share common parameters in such a way that changes in certain circulatory parameters affect, e.g., all transport models. The larger box enclosing all models means that they all use some common basic parameters. The arrows between the boxes represent flow of data. From J. Larsen and S.A. Pedersen (2000),* Mathematical models behind advanced simulators in medicine, in Mathematical Modelling in Medicine, *J.T. Ottesen and M. Danielsen (eds.), SHTI 71, IOS Press, Amsterdam, pp. 203–216. Used by permission of the publisher.*

outflow into the aorta from the 2D heart model (Chapter 3) could be used as input to the 1D model (Chapter 5).

Using a reference model for the validation and calibration of a zero-dimensional model requires that the model be a true subset of the more detailed higher order model. In other words, mechanisms represented in the zero-dimensional model should also be present in the higher order model. For example, the mechanism responsible for the ejection effect of the heart (see Chapter 4) is believed to be a constitutive relationship providing detailed information on pressure and muscle fibers. This mechanism is not described in the more detailed 2D heart model that has been created as a tool to study flow patterns in a plane section through the left side of the heart (see Chapter 3). Consequently, the 2D heart model cannot be used for calibration and validation of the zero-dimensional model describing the ejection effect. In contrast, it would be possible to construct and validate the arterial part of the zero-dimensional cardiovascular model described in Chapter 6 using the 1D model described in Chapter 5. To sum up, there are many ways to develop models of varying complexity, but the key is that a given model should be as simple as possible for its intended application.

The models presented in this book show a sample of the interests and expertise of members of the BioMath group at Roskilde University. There are many other models that could have been studied and would have added valuable information to the simulator. The pharmacodynamic model is an example of an empirical model. This model consists of a fitted correlation between concentrations of drugs and the states of the drugs' target organs, such as the heart and the brain. A shortcoming of empirical models is that one cannot obtain reliable predictions from such models as soon as one moves outside the range in which they were fitted.

In addition, the SIMA simulator developed by Math-Tech and the BioMath group also included profile models, which are epistemologically even weaker than empirical models. The ECG models and the fluid and electrolyte models are of this type. They consist of a database of predesigned profiles together with rules for selecting the profile that at each instant corresponds best to the condition of the heart. Of course, it is impossible to base any predictions on such models. They are only able to supply predefined profiles. They do not give any insight into the electrochemical processes that control the heartbeat.

In the SIMA simulator a variety of models are used, from detailed reference models, based on fundamental physiological laws and principles, to simple script-based models. Common to all the models is that they model in a satisfactory way biological processes occurring in an anesthetized patient. The reference models are more detailed, since they could contain more spatial information, making them too complex to run in real time. Therefore, they serve mainly to validate and scale the real-time models. The more empirically fitted mathematical relations (models of pharmacodynamic and metabolic processes) as well as the pure script-based models (e.g., the ECG model) are all included in the simulator.

Acknowledgments

The authors were supported by High Performance Parallel Computing, Software Engineering Applications, Eureka Project 1063 (SIMA—SIMulation in Anesthesia).

Appendix B

Momentum Equation for a Small Artery

M.S. Olufsen

B.1 Introduction

Equations for blood flow in the smaller arteries can be derived from studying

- the motion of the fluid,

- the motion of the vessel wall, and

- the interaction between the fluid and the wall.

The motion of the fluid can be described using the axisymmetric Navier–Stokes equations; see section B.2. The motion of the vessel wall can be described by studying the forces acting on the vessel wall, i.e., by balancing internal and external forces; see section B.5. Internal forces are those found in the vessel wall itself, and external forces are due to inertia, tethering, and surface forces. Finally, the interaction between the fluid and the wall can be described by the no-slip condition. This condition ensures that the velocity of the fluid at the wall, in both the radial and the longitudinal direction, is balanced by the corresponding movement of the wall. These equations are linearized in section B.6; simplified; and solved in section B.7.

B.2 Motion of the Fluid

The fluid dynamic equations are derived using cylindrical coordinates (r, x, θ). We assume that the flow is axisymmetric (no dependence on θ) and without swirl (no θ component).

Hence the Navier–Stokes equations take the form

$$\frac{\partial u}{\partial t} + u\frac{\partial u}{\partial r} + w\frac{\partial u}{\partial x} = -\frac{1}{\rho}\frac{\partial p}{\partial r} + \nu\left(\frac{\partial^2 u}{\partial r^2} + \frac{1}{r}\frac{\partial u}{\partial r} + \frac{\partial^2 u}{\partial x^2} - \frac{u}{r^2}\right), \tag{B.1}$$

$$\frac{\partial w}{\partial t} + u\frac{\partial w}{\partial r} + w\frac{\partial w}{\partial x} = -\frac{1}{\rho}\frac{\partial p}{\partial x} + \nu\left(\frac{\partial^2 w}{\partial r^2} + \frac{1}{r}\frac{\partial w}{\partial r} + \frac{\partial^2 w}{\partial x^2}\right), \tag{B.2}$$

$$\frac{1}{r}\frac{\partial}{\partial r}(ru) + \frac{\partial w}{\partial x} = 0, \tag{B.3}$$

where u is the radial velocity, w is the longitudinal velocity, p is pressure, ρ is the density of blood (constant), and $\nu = \mu/\rho$ is the kinematic viscosity (also constant).

B.3 Motion of the Vessel Wall

The movement of the vessel wall can be described by balancing internal and external forces on a surface element of the vessel wall in its deformed state. It is convenient to change the variables to a coordinate system connected to the surface of the vessel. This is shown in the top part of Figure B.1. Let H be any vector pointing to the middle surface, as shown in Figure B.1:

$$H = x\hat{x} + R\hat{r},$$

where $\hat{x}$ and $\hat{r}$ are unit vectors in the cylindrical coordinate system in the longitudinal and radial directions, respectively, and $R(x, t)$ is the radius of the vessel. The new coordinates (n, t, θ) can be determined from H. By assuming axial symmetry, all equations can be expressed in terms of $\hat{t}$ and $\hat{n}$ given by

$$\hat{t} = \frac{\frac{\partial H}{\partial x}}{\left|\frac{\partial H}{\partial x}\right|} = \frac{\hat{x} + \frac{\partial R}{\partial x}\hat{r}}{\sqrt{1 + \left(\frac{\partial R}{\partial x}\right)^2}} \quad \text{and} \quad \hat{n} = \frac{\hat{r} - \frac{\partial R}{\partial x}\hat{x}}{\sqrt{1 + \left(\frac{\partial R}{\partial x}\right)^2}} \tag{B.4}$$

because $\hat{t}$ and $\hat{n}$ are orthogonal. Solving for $\hat{x}$ and $\hat{r}$ gives

$$\hat{x} = \frac{\hat{t} - \hat{n}\frac{\partial R}{\partial x}}{\sqrt{1 + \left(\frac{\partial R}{\partial x}\right)^2}} \quad \text{and} \quad \hat{r} = \frac{\hat{n} + \hat{t}\frac{\partial R}{\partial x}}{\sqrt{1 + \left(\frac{\partial R}{\partial x}\right)^2}}. \tag{B.5}$$

B.3.1 Internal Forces

The internal forces on the infinitesimal surface element $(dx \times r d\theta)$ have three components: a force N across the vessel wall, a shearing force S on the sides of the element, and a force T normal to each of the edges; see the bottom part of Figure B.1. Most of these components are zero. The vessel wall is thin, and so any variation in the force across the wall can be neglected; i.e., $N_t = N_\theta = 0$. The flow is axisymmetric and without swirl. Hence no shearing force will act on the side of the element; i.e., $S_t = S_\theta = 0$. Thus the only forces left are T_t and T_θ, the normal forces to each of the edges.

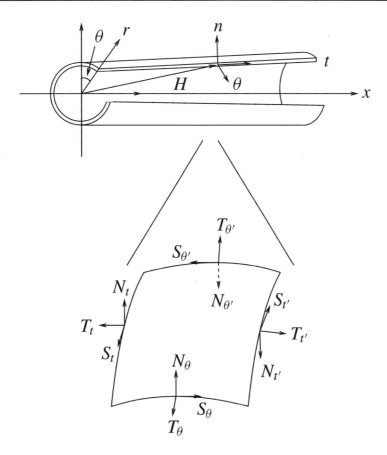

Figure B.1. *The top part shows the original (r, x, θ) and the new (n, t, θ) coordinates. The bottom part shows the forces on an infinitesimal surface element cut out of the vessel. N is the force acting across the vessel wall, S is the shearing force acting on the side of the element, and T is the force acting normal to each of the edges. The subscripts t and θ indicate the direction according to the coordinate system following the surface of the vessel, and the superscript $'$ indicates that the force is acting in the negative direction (e.g., $N_{t'} = -N_t$).*

B.3.2 External Forces

The internal forces must be balanced by external forces acting on the element. Let total external force be denoted by

$$P = P_t \hat{t} + P_n \hat{n}, \tag{B.6}$$

where P_t and P_n are the tangential and normal components, respectively. P can be split into inertial forces, tethering forces, and surface forces. In the following sections, these will be analyzed separately.

Inertial Force

Let $\xi(r, x, t)$ and $\eta(r, x, t)$ be the longitudinal and radial displacements of the wall. The inertial force per unit area is given by (see Atabek and Lew (1966))

$$T_{F_I} = -\rho_0 h \left(\frac{\partial^2 \xi}{\partial t^2} \hat{x} + \frac{\partial^2 \eta}{\partial t^2} \hat{r} \right), \tag{B.7}$$

where ρ_0 is the density and h is the thickness of the wall. Because of the thin wall assumption, h must be small compared to the vessel radius. We assume that both ρ_0 and h are constant along any vessel of a given radius. The inertial force is the force ensuring that the internal and external forces are balanced. The inertial force must be included because the system is not steady, so it is necessary to take acceleration into account. In physics, this is known as d'Alambert's principle.

Tethering Force

The tethering force T_{F_T} can be modeled using a simple mechanical model consisting of a spring, a dash pot, and some lumped additional mass (Atabek, 1968). The tethering force (per unit area) acting in the radial and longitudinal directions is given by

$$T_{F_T} = -\left(M_a \frac{\partial^2 \xi}{\partial t^2} + L_x \frac{\partial \xi}{\partial t} + K_x \xi \right) \hat{x} - \left(M_a \frac{\partial^2 \eta}{\partial t^2} + L_r \frac{\partial \eta}{\partial t} + K_r \eta \right) \hat{r}, \tag{B.8}$$

where K_i and L_i, $i = x, r$, are the spring and frictional coefficients of the dash pot in the ith direction and M_a is the additional mass of the system. These are assumed to be the same in both directions.

Since both inertial and tethering forces act in the same direction, it is convenient to add them before projecting the forces in the normal and tangential directions. Let

$$M_0 = M_a + \rho_0 h.$$

The resultant inertial and tethering force in the tangential and normal directions, respectively, then yield

$$T_{F_{T_{res}}} \cdot \hat{t} \tag{B.9}$$

$$= -\left[\left(M_0 \frac{\partial^2 \xi}{\partial t^2} + L_x \frac{\partial \xi}{\partial t} + K_x \xi \right) + \left(M_0 \frac{\partial^2 \eta}{\partial t^2} + L_r \frac{\partial \eta}{\partial t} + K_r \eta \right) \frac{\partial R}{\partial x} \right] \Big/ \sqrt{1 + \left(\frac{\partial R}{\partial x} \right)^2},$$

$$T_{F_{T_{res}}} \cdot \hat{n} \tag{B.10}$$

$$= \left[\left(M_0 \frac{\partial^2 \xi}{\partial t^2} + L_x \frac{\partial \xi}{\partial t} + K_x \xi \right) \frac{\partial R}{\partial x} - \left(M_0 \frac{\partial^2 \eta}{\partial t^2} + L_r \frac{\partial \eta}{\partial t} + K_r \eta \right) \right] \Big/ \sqrt{1 + \left(\frac{\partial R}{\partial x} \right)^2}.$$

Surface Force

The surface force is a result of fluid interaction with the vessel wall. If the stress tensor of the fluid is given by $\mathbf{T_{F_S}}$, then interaction with the inner vessel wall (at $r = R - h/2 = a$) is given by $-\mathbf{T}_{\mathbf{F_S}} \cdot \hat{n}$. Assume that the stress tensor can be separated into radial and longitudinal directions

$$(-\mathbf{T}_{\mathbf{F_S}} \cdot \hat{n}) \cdot \hat{t} \quad \text{and} \quad (-\mathbf{T}_{\mathbf{F_S}} \cdot \hat{n}) \cdot \hat{n}. \tag{B.11}$$

The stress tensor for incompressible flow is given by Ockendon and Ockendon (1995):

$$\sigma_{ij} = -p\delta_{ij} + \mu \left(\frac{\partial u_i}{\partial x_j} + \frac{\partial u_j}{\partial x_i} \right).$$

In cylindrical coordinates the stress tensor becomes

$$\mathbf{T}_{\mathbf{F_S}} = \begin{bmatrix} T_{rr} & T_{rx} \\ T_{rx} & T_{xx} \end{bmatrix}_a = \begin{bmatrix} -p + 2\mu\dfrac{\partial u}{\partial r} & \mu \left(\dfrac{\partial w}{\partial r} + \dfrac{\partial u}{\partial x} \right) \\ \mu \left(\dfrac{\partial w}{\partial r} + \dfrac{\partial u}{\partial x} \right) & -p + 2\mu\dfrac{\partial w}{\partial x} \end{bmatrix}_a. \tag{B.12}$$

The fluid stress in the $\hat{t}$ and $\hat{n}$ directions can be found as

$$(-\mathbf{T}_{\mathbf{F_S}} \cdot \hat{n}) \cdot \hat{t} = \left[(T_{xx} - T_{rr})\frac{\partial R}{\partial x} + T_{rx}\left(\left(\frac{\partial R}{\partial x}\right)^2 - 1 \right) \right]_a \bigg/ \left(1 + \left(\frac{\partial R}{\partial x}\right)^2 \right), \tag{B.13}$$

$$(-\mathbf{T}_{\mathbf{F_S}} \cdot \hat{n}) \cdot \hat{n} = \left[2T_{rx}\frac{\partial R}{\partial x} - T_{rr} - T_{xx}\left(\frac{\partial R}{\partial x}\right)^2 \right]_a \bigg/ \left(1 + \left(\frac{\partial R}{\partial x}\right)^2 \right). \tag{B.14}$$

Total External Force

The total external force can be found by adding the inertial and tethering forces (B.9) and (B.10) as well as the surface forces (B.13) and (B.14). Generally, these forces are not estimated at the same point, but because of the thin wall assumption the resulting error in the total external force is negligible. Equation (B.6) gives

$$P = P_t\hat{t} + P_n\hat{n} = \left(-\mathbf{T}_{\mathbf{F_S}} \cdot \hat{n} + T_{F_{Tres}}\right) \cdot \hat{t} + \left(-\mathbf{T}_{\mathbf{F_S}} \cdot \hat{n} + T_{F_{Tres}}\right) \cdot \hat{n}.$$

The tangential component is

$$P_t = \left[(T_{xx} - T_{rr})\frac{\partial R}{\partial x} + T_{rx}\left(\left(\frac{\partial R}{\partial x}\right)^2 - 1 \right) \right]_a \bigg/ \left(1 + \left(\frac{\partial R}{\partial x}\right)^2 \right) \tag{B.15}$$

$$- \left(M_0\frac{\partial^2 \xi}{\partial t^2} + L_x\frac{\partial \xi}{\partial t} + K_x\xi + \left(M_0\frac{\partial^2 \eta}{\partial t^2} + L_r\frac{\partial \eta}{\partial t} + K_r\eta \right)\frac{\partial R}{\partial x} \right) \bigg/ \sqrt{1 + \left(\frac{\partial R}{\partial x}\right)^2}$$

and the normal component is

$$P_n = \left[2T_{rx}\frac{\partial R}{\partial x} - T_{rr} - T_{xx}\left(\frac{\partial R}{\partial x}\right)^2 \right]_a \bigg/ \left(1 + \left(\frac{\partial R}{\partial x}\right)^2 \right) \tag{B.16}$$

$$+ \left(\left(M_0\frac{\partial^2 \xi}{\partial t^2} + L_x\frac{\partial \xi}{\partial t} + K_x\xi \right)\frac{\partial R}{\partial x} - \left(M_0\frac{\partial^2 \eta}{\partial t^2} + L_r\frac{\partial \eta}{\partial t} + K_r\eta \right) \right) \bigg/ \sqrt{1 + \left(\frac{\partial R}{\partial x}\right)^2}.$$

B.3.3 Balancing Internal and External Forces

When a wave is propagated along a vessel, the vessel will dilate. Hence the surface will appear as shown in Figure B.2. Considering this surface, we can derive the equilibrium equations. Balancing of internal and external forces will also be carried out in two parts: one for tangential contributions and one for normal contributions.

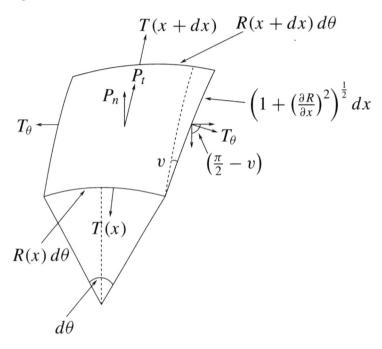

Figure B.2. *A volume element and its internal T_i and external P_i forces.*

Balancing Tangential Components of Internal and External Forces

The area of the surface in Figure B.2 is given by $Rd\theta\sqrt{1 + (\partial R/\partial x)^2}\,dx$, and the tangential part P_{tan} of the external strain P_t is given by

$$P_{tan} = P_t Rd\theta\sqrt{1 + \left(\frac{\partial R}{\partial x}\right)^2}\,dx.$$

The pressure load on any given volume element is $-P_{ext}$. This should be balanced by the internal stress over the surface element projected in the tangential direction. Thus the stress over the surface in the tangential direction is given by

$$T_{tan_1} = -T_t(x)R(x)d\theta + T_t(x + dx)R(x + dx)d\theta \approx \frac{\partial}{\partial x}(T_t R)dxd\theta,$$

where the last equality is approximated using the first order Taylor expansion for $T_t(x + dx)R(x + dx)$.

Furthermore, the stress from the radial tension also contributes. As seen on the right-hand side of the surface element in Figure B.2, the radial tension T_θ gives contributions in both the tangential and the radial directions. Since we have axial symmetry, the net tension around the vessel at any location is zero. The part of T_θ pointing backward in the tangential direction is given by

$$T_{tan_2} = -T_\theta \cos\left(\frac{\pi}{2} - v\right)\sqrt{1 + \left(\frac{\partial R}{\partial x}\right)^2}\, dx = -T_\theta \frac{\partial R}{\partial x} d\theta dx,$$

where v is defined as shown in Figure B.2. Balancing T_{tan_1} and T_{tan_2} with P_{tan} and dividing by $d\theta dx$ gives

$$-T_\theta \frac{\partial R}{\partial x} + \frac{\partial}{\partial x}(RT_t) + P_t R\sqrt{1 + \left(\frac{\partial R}{\partial x}\right)^2} = 0. \qquad (B.17)$$

Balancing Normal Components of Internal and External Forces

Balancing normal internal stresses with the normal external strain gives

$$P_n = \kappa_\theta T_\theta + \kappa_t T_t,$$

where κ_i, $i = \theta, t$, is the curvature in the i direction. As seen in Figure B.3, the curvatures in the longitudinal and angular directions are given by

$$\kappa_\theta = \frac{1}{R}\Bigg/ \sqrt{1 + \left(\frac{\partial R}{\partial x}\right)^2} \quad \text{and} \quad \kappa_t = -\frac{\partial^2 R}{\partial x^2}\Bigg/ \sqrt{1 + \left(\frac{\partial R}{\partial x}\right)^2}^{\,3}.$$

Hence the balancing equation becomes

$$\kappa_\theta T_\theta + \kappa_t T_t - P_n = 0$$

$$\Leftrightarrow \frac{T_\theta}{R} - T_t \frac{\partial^2 R}{\partial x^2}\Bigg/ \left(1 + \left(\frac{\partial R}{\partial x}\right)^2\right) - P_n \sqrt{1 + \left(\frac{\partial R}{\partial x}\right)^2} = 0. \qquad (B.18)$$

Inserting (B.15) and (B.16) into (B.17) and (B.18) gives

$$-T_\theta \frac{\partial R}{\partial x} + \frac{\partial}{\partial x}(RT_t)$$

$$-R\left(M_0 \frac{\partial^2 \xi}{\partial t^2} + L_x \frac{\partial \xi}{\partial t} + K_x \xi + \left(M_0 \frac{\partial^2 \eta}{\partial t^2} + L_r \frac{\partial \eta}{\partial t} + K_r \eta\right)\frac{\partial R}{\partial x}\right)$$

$$+R\left[(T_{xx} - T_{rr})\frac{\partial R}{\partial x} + T_{rx}\left(\left(\frac{\partial R}{\partial x}\right)^2 - 1\right)\right]_a \Bigg/ \sqrt{1 + \left(\frac{\partial R}{\partial x}\right)^2} = 0, \qquad (B.19)$$

A

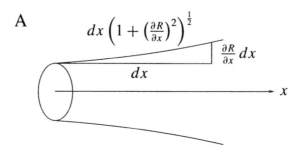

B

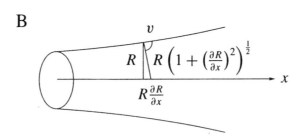

Figure B.3. *Curvature of the vessel. The longitudinal curvature (in A) is given by* κ_l, *and the tangential curvature normal to the surface (in B) is given by* κ_θ.

$$\frac{T_\theta}{R} - T_l \frac{\partial^2 R}{\partial x^2} \Bigg/ \left(1 + \left(\frac{\partial R}{\partial x}\right)^2\right)$$

$$- \left(M_0 \frac{\partial^2 \xi}{\partial t^2} + L_x \frac{\partial \xi}{\partial t} + K_x \xi\right) \frac{\partial R}{\partial x} + M_0 \frac{\partial^2 \eta}{\partial t^2} + L_r \frac{\partial \eta}{\partial t} + K_r \eta$$

$$- \left[2T_{rx} \frac{\partial R}{\partial x} - T_{rr} - T_{xx} \left(\frac{\partial R}{\partial x}\right)^2\right]_a \Bigg/ \sqrt{1 + \left(\frac{\partial R}{\partial x}\right)^2} = 0. \qquad (\text{B.20})$$

B.4   Elasticity Relations

The purpose of this section is to set up stress-strain relations such that the stress components T_i can be related to the displacements of the wall (ξ, η). These are measured from some reference state where vessels are stretched to their in vivo length. The reason is that a loose piece of artery (unstressed) requires very large deformations to be brought to its original stressed state. However, the general theory of elasticity applies only for small deformations; see, e.g., Landau and Lifshitz (1986). This problem can be avoided by making the derivations orginate from some initial stressed state. Hence it is assumed that, when a wave moves along an artery, it undergoes small deformations from its reference state. The initial state is chosen to be the state where the transmural pressure of the artery is zero. Furthermore, it is assumed that it is adequate to apply a linear relation between stress and strain.

Let the reference state of stresses in the longitudinal and circumferential directions be denoted by T_{t_0} and T_{θ_0}. Then the following relations can be obtained:

$$T_\theta - T_{\theta_0} = \frac{E_\theta h}{1 - \sigma_\theta \sigma_x}(\epsilon_r + \sigma_x \epsilon_x) \quad \text{and} \quad T_t - T_{t_0} = \frac{E_x h}{1 - \sigma_\theta \sigma_x}(\epsilon_x + \sigma_\theta \epsilon_r), \quad \text{(B.21)}$$

where E_i, $i = \theta, t$, is Young's modulus in the ith direction; h is the wall thickness; σ_i, $i = \theta, x$, is the Poisson ratio in the ith direction; and ϵ_i, $i = \theta, x$, is the displacement relative to the reference state; see, e.g., Landau and Lifshitz (1986). The relative circumferential and longitudinal displacements are given by

$$\epsilon_r = \frac{\eta}{R} \quad \text{and} \quad \epsilon_x = \frac{\partial \xi}{dx}.$$

B.5 Balancing Fluid and Wall Motions

Boundary conditions linking the velocity of the wall to the velocity of the fluid remain to be specified. Assume that the fluid particles are at rest at the wall. Hence

$$[u]_{r=a} = \frac{\partial \eta}{\partial t} \quad \text{and} \quad [w]_{r=a} = \frac{\partial \xi}{\partial t}. \quad \text{(B.22)}$$

Furthermore, assume that the component of the fluid velocity normal to the wall is equal to the normal velocity of the inner surface of the vessel wall. Hence the normal velocity of the wall, at $a = R(x + \xi, t) - h/2$, is given by

$$\frac{d}{dt}\left(r - R + \frac{h}{2}\right) = 0 \quad \Leftrightarrow \quad [u]_{r=a} - [w]_{r=a}\frac{\partial R}{\partial x} - \frac{\partial R}{\partial t} = 0.$$

B.6 Linearization

In principle the correct number of equations and boundary conditions are present. However, in their present form these equations are too complicated to solve analytically. As discussed earlier, the purpose was to set up a simple system of equations for the smaller arteries. Therefore, following Atabek and Lew (1966), we have chosen to linearize them.

The linearization is based on expansion of the dependent variables in power series of a small parameter ϵ around a known solution. This is defined by a situation where the fluid is at rest and the vessel is inflated and stretched. Furthermore, if $\epsilon = 0$, then all dependent variables give the known solution. The expansion is given by

$$s = s_1 \epsilon + s_2 \epsilon^2 + \cdots \qquad \text{for } s = u, w, \xi, \eta, T_{rx}, \qquad \text{(B.23)}$$

$$\tilde{s} = \tilde{s}_0 + \tilde{s}_1 \epsilon + \tilde{s}_2 \epsilon^2 + \cdots \qquad \text{for } \tilde{s} = p, R, T_\theta, T_t, T_{rr}, T_{xx}, \qquad \text{(B.24)}$$

where s_0 is a constant defining the reference state (at zero transmural pressure). Let $f(r, x, t)$ be either of the functions in (B.23) or (B.24). In order to accomplish the linearization,

$f(r, x, t)$ must be evaluated at $r = a = R - h/2$. The power series expansion together with the Taylor series expansion to first order yields

$$
\begin{aligned}
f(r, x, t) &\approx f(a, x, t) + f'(a, x, t)(r - a) \\
&= f_0(a, x, t) + f_1(a, x, t)\epsilon \\
&\quad + (f_0'(a, x, t) + f_1'(a, x, t)\epsilon)\,(r - (R_0 + R_1\epsilon - h/2)) \\
&= f_0(a, x, t) + k f_0'(a, x, t) \\
&\quad + \epsilon \left(f_1(a, x, t) - R_1 f_0'(a, x, t) + k f_1'(a, x, t) \right),
\end{aligned}
\tag{B.25}
$$

where $k = r - R_0 + h/2$. Using (B.23) to (B.25), the zeroth and first order equations can be obtained by assembling terms to the respective powers of ϵ from the nonlinear equations (B.1) to (B.3), (B.19), and (B.20).

B.6.1   Terms of Zeroth Order Approximations

From the fluid equations (B.1) to (B.3) only the pressure terms contribute since the expansions for u and w have no zero order terms:

$$
\frac{\partial p_0}{\partial r} = 0 \quad \text{and} \quad \frac{\partial p_0}{\partial x} = 0.
$$

From the shell equations only (B.20) contributes:

$$
\frac{T_{\theta_0}}{R_0} + T_{rr_0} = \frac{T_{\theta_0}}{R_0} - p_0 = 0,
\tag{B.26}
$$

where we have used the stress tensor (B.12) for the zeroth order approximation of T_{rr_0}.

B.6.2   Terms of First Order Approximations

In this case all equations contribute. Equations (B.1) to (B.3) give

$$
\frac{\partial u_1}{\partial t} = -\frac{1}{\rho}\frac{\partial p_1}{\partial r} + \nu\left(\frac{\partial^2 u_1}{\partial r^2} + \frac{1}{r}\frac{\partial u_1}{\partial r} + \frac{\partial^2 u_1}{\partial x^2} - \frac{u_1}{r^2} \right),
\tag{B.27}
$$

$$
\frac{\partial w_1}{\partial t} = -\frac{1}{\rho}\frac{\partial p_1}{\partial x} + \nu\left(\frac{\partial^2 w_1}{\partial r^2} + \frac{1}{r}\frac{\partial w_1}{\partial r} + \frac{\partial^2 w_1}{\partial x^2} \right),
\tag{B.28}
$$

$$
\frac{1}{r}\frac{\partial}{\partial r}(r u_1) + \frac{\partial w_1}{\partial x} = 0.
\tag{B.29}
$$

The first order terms of the shell equation (B.19) give

$$
-T_{\theta_0}\frac{\partial R_1}{\partial x} + \frac{\partial}{\partial x}(R_0 T_{t_1} + R_1 T_{t_0})
\tag{B.30}
$$

$$
-R_0\left(M_0\frac{\partial^2 \xi_1}{\partial t^2} + L_x\frac{\partial \xi_1}{\partial t} + K_x\xi_1 - \left[(T_{xx_0} - T_{rr_0})\frac{\partial R_1}{\partial x} - T_{rx_1} \right]_a \right) = 0
$$

$$
\Leftrightarrow M_0\frac{\partial^2 \xi_1}{\partial t^2} + L_x\frac{\partial \xi_1}{\partial t} + K_x\xi_1 = \frac{\partial T_{t_1}}{\partial x} + \frac{T_{t_0} - T_{\theta_0}}{R_0}\frac{\partial R_1}{\partial x} - \mu\left[\frac{\partial w_1}{\partial r} + \frac{\partial u_1}{\partial x} \right]_a.
$$

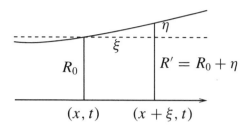

Figure B.4. *Estimation of $R(x + \xi, t)$ using the definitions of ξ and η.*

The last equation is obtained using the stress tensor (B.12) for the first order approximation of T_{rx_1} and the zeroth order approximation of $T_{xx_0} - T_{rr_0}$, which cancel. The first order terms of the shell equation (B.20) give

$$\frac{T_{\theta_1}}{R_0} - T_{\theta_0}\frac{R_1}{R_0^2} - T_{t_0}\frac{\partial^2 R_1}{\partial x^2} + M_0\frac{\partial^2 \eta_1}{\partial t^2} + L_r\frac{\partial \eta_1}{\partial t} + K_r\eta_1 + T_{rr_1} = 0$$

$$\Leftrightarrow M_0\frac{\partial^2 \eta_1}{\partial t^2} + L_r\frac{\partial \eta_1}{\partial t} + K_r\eta_1 = -\frac{T_{\theta_1}}{R_0} + T_{\theta_0}\frac{R_1}{R_0^2} + T_{t_0}\frac{\partial^2 R_1}{\partial x^2} + \left[p_1 - 2\mu\frac{\partial u_1}{\partial r}\right]_a,$$

(B.31)

where we have again used (B.12) for the first order approximation T_{rr_1}.

Assuming that the second order approximations can be neglected, ϵ can be incorporated into the dependent variables and we can set $\epsilon = 1$. For any (x, t) the first order Taylor expansion of $R(x + \xi, t)$ gives

$$R(x + \xi, t) = R(x, t) + \frac{\partial R}{\partial x}\xi = R_0 + \eta,$$

as seen in Figure B.4. The first order expansion of R from (B.24) is given by

$$R(x, t) = R_0 + R_1\epsilon + \mathcal{O}(\epsilon^2) = R_0 + \eta_1\epsilon + \mathcal{O}(\epsilon^2) \quad \Leftrightarrow \quad \eta_1 = R_1,$$

since η has no zeroth order term. Furthermore, we approximate R_0 by the inner radius $a = R_0 - h/2$. Since the walls are assumed to be thin compared with the vessel radius, i.e., $h \ll a$, the error is negligible. Finally, the indices 1 are dropped and the definitions in (B.21) are used for T_{θ_1} and T_{t_1}. The linearized equations can be obtained from their first order approximations; i.e., (B.30) and (B.31) become

$$M_0\frac{\partial^2 \xi}{\partial t^2} + L_x\frac{\partial \xi}{\partial t} + K_x\xi$$

$$= \frac{E_x h}{1 - \sigma_\theta \sigma_x}\left(\frac{\partial^2 \xi}{\partial x^2} + \frac{\sigma_x}{a}\frac{\partial \eta}{\partial x}\right) + \frac{\partial \eta}{\partial x}\frac{T_{t_0} - T_{\theta_0}}{a} - \mu\left[\frac{\partial w}{\partial r} + \frac{\partial u}{\partial x}\right]_a,$$

(B.32)

$$M_0\frac{\partial^2 \eta}{\partial t^2} + L_r\frac{\partial \eta}{\partial t} + K_r\eta$$

$$= -\frac{E_\theta h}{1 - \sigma_\theta \sigma_x}\left(\frac{\eta}{a^2} + \frac{\sigma_\theta}{a}\frac{\partial \xi}{\partial x}\right) + T_{\theta_0}\frac{\eta}{a^2} + T_{t_0}\frac{\partial^2 \eta}{\partial x^2} + \left[p - 2\mu\frac{\partial u}{\partial r}\right]_a.$$

(B.33)

B.7   Solution of the Linearized Equations

Equations (B.27) to (B.29) for the fluid motion must be solved first because the equations for the wall motion (B.32) and (B.33) couple to the fluid equations. Propagation of blood flow and pressure in human arteries is periodic, and hence the solutions can be constructed from simple harmonic functions; i.e., u, w, p, ξ, and η can be written as propagating waves harmonic in both x and t. Therefore, we seek solutions of the form

$$\tilde{s}(r, x, t) = s(r)\, e^{i\omega(t - x/c)}, \tag{B.34}$$

where $\omega = 2\pi/T$ is the angular frequency, T is the period of one heartbeat, and c is the wave propagation speed. Furthermore, we assume that s_r is dependent on r for u, w, p but constant for ξ, η. More general solutions can be found by superposition of solutions of this form (Whitham, 1974). Inserting these into (B.27) to (B.29) gives

$$i\omega u = -\frac{1}{\rho}\frac{dp}{dr} + \nu\left(\frac{d^2 u}{dr^2} + \frac{1}{r}\frac{du}{dr} + \frac{i^2\omega^2 u}{c^2} - \frac{u}{r^2}\right),$$

$$i\omega w = \frac{i\omega\, p}{c\rho} + \nu\left(\frac{d^2 w}{dr^2} + \frac{1}{r}\frac{dw}{dr} + \frac{i^2\omega^2 w}{c^2}\right),$$

$$\frac{1}{r}\frac{d}{dr}(ur) = \frac{i\omega w}{c}.$$

Let $\beta_0 = ia\omega/c$ and $\mathrm{w}_0^2 = i^3 a^2\omega/\nu = i^3\mathrm{w}$ be dimensionless parameters, where w is the Womersley number. Multiplying the equations above by a^2/ν and using $y = r/a$ then gives the following inhomogeneous Bessel equations:

$$\frac{a}{\mu}\frac{dp}{dy} = \frac{d^2 u}{dy^2} + \frac{1}{y}\frac{du}{dy} + (\mathrm{w}_0^2 + \beta_0^2)u - \frac{u}{y^2}, \tag{B.35}$$

$$-\frac{a\beta_0}{\mu}p = \frac{d^2 w}{dy^2} + \frac{1}{y}\frac{dw}{dy} + (\mathrm{w}_0^2 + \beta_0^2)w, \tag{B.36}$$

$$\frac{1}{y}\frac{d}{dy}(uy) - \beta_0 w = 0. \tag{B.37}$$

The solutions to the corresponding homogeneous equations are

$$u = u_c\frac{J_1\left(\frac{r}{a}\sqrt{\mathrm{w}_0^2 + \beta_0^2}\right)}{J_0(\mathrm{w}_0)} \quad \text{and} \quad w = w_c\frac{J_0\left(\frac{r}{a}\sqrt{\mathrm{w}_0^2 + \beta_0^2}\right)}{J_0(\mathrm{w}_0)}, \tag{B.38}$$

where $J_0(x)$ and $J_1(x)$ are the zeroth and first order Bessel functions. The divergence of (B.28) gives

$$\nabla \cdot \frac{\partial w}{\partial t} = -\frac{1}{\rho}\nabla \cdot \nabla p + \nu\nabla \cdot (\nabla^2 w) \quad \Leftrightarrow \quad \nabla^2 p = 0$$

because of the no-slip condition, which states that $\nabla \cdot w = 0$. Expanding the Laplacian of p yields

$$\nabla^2 p = \frac{1}{r}\frac{d}{dr}\left(r\frac{dp}{dr}\right) + \frac{d^2 p}{dx^2} = 0$$

$$\Leftrightarrow \frac{d^2 p}{dr^2} + \frac{1}{r}\frac{dp}{dr} + \frac{\beta_0^2}{a^2}p = 0,$$

which has the solution

$$p = p_c J_0\left(\frac{\beta_0 r}{a}\right). \tag{B.39}$$

A particular solution of (B.35) and (B.36) has the form

$$u = u_{c_p} J_1\left(\frac{\beta_0 r}{a}\right) \quad \text{and} \quad w = w_{c_p} J_0\left(\frac{\beta_0 r}{a}\right). \tag{B.40}$$

Inserting (B.39) together with (B.40) into (B.35) and (B.36) gives

$$u = -\frac{a p_c \beta_0}{\mu w_0^2} J_1\left(\frac{\beta_0 r}{a}\right) + u_c \frac{J_1\left(\frac{r}{a}\sqrt{w_0^2 + \beta_0^2}\right)}{J_0(w_0)}, \tag{B.41}$$

$$w = -\frac{a p_c \beta_0}{\mu w_0^2} J_0\left(\frac{\beta_0 r}{a}\right) + w_c \frac{J_0\left(\frac{r}{a}\sqrt{w_0^2 + \beta_0^2}\right)}{J_0(w_0)}. \tag{B.42}$$

Equation (B.37) gives

$$\beta_0 w_c = u_c \sqrt{w_0^2 + \beta_0^2}.$$

According to Pedley (1980), we assume that

$$w_c = \frac{p_c A}{c_0 \rho}\frac{\sqrt{w_0^2 + \beta_0^2}}{w_0},$$

where c_0 is a scale for c (the wave propagation speed). For convenience we let $c_0 = E_\theta h/(2a\rho)$, the Moens–Korteweg wave propagation factor. Finally, A and p_c are constants that still have to be determined. Inserting these conditions into (B.41) and (B.42) gives

$$u = -\frac{a p_c \beta_0}{\mu w_0^2} J_1\left(\frac{\beta_0 r}{a}\right) + \frac{p_c A \beta_0}{c_0 \rho w_0}\frac{J_1\left(\frac{r}{a}\sqrt{w_0^2 + \beta_0^2}\right)}{J_0(w_0)}, \tag{B.43}$$

$$w = -\frac{a p_c \beta_0}{\mu w_0^2} J_0\left(\frac{\beta_0 r}{a}\right) + \frac{p_c A \sqrt{w_0^2 + \beta_0^2}}{c_0 \rho w_0}\frac{J_0\left(\frac{r}{a}\sqrt{w_0^2 + \beta_0^2}\right)}{J_0(w_0)}. \tag{B.44}$$

Finally, we insert the expressions above (for u, w) into the boundary conditions (B.22), which gives

$$-\frac{ap_c\beta_0}{\mu w_0^2} J_1(\beta_0) + \frac{p_c A\beta_0}{c_0\rho w_0} \frac{J_1\left(\sqrt{w_0^2 + \beta_0^2}\right)}{J_0(w_0)} - i\omega\eta_1 = 0, \qquad (B.45)$$

$$-\frac{ap_c\beta_0}{\mu w_0^2} J_0(\beta_0) + \frac{p_c A\sqrt{w_0^2 + \beta_0^2}}{c_0\rho w_0} \frac{J_0\left(\sqrt{w_0^2 + \beta_0^2}\right)}{J_0(w_0)} - i\omega\xi_1 = 0. \qquad (B.46)$$

The system has four unknowns: p_c, A, ξ_1, and η_1. Therefore, two more equations are needed. They stem from the equations for motion of the wall (B.32) and (B.33). Expanding ξ and η using (B.23) and (B.34) and inserting u and w from (B.43) and (B.44) gives

$$-\frac{2p_c\beta_0^2}{w_0^2} J_1(\beta_0) + \frac{p_c A\mu}{ac_0\rho} \frac{w_0^2 + 2\beta_0^2}{w_0} \frac{J_1\left(\sqrt{w_0^2 + \beta_0^2}\right)}{J_0(w_0)} - \frac{(B_{21} + T_{t_0} - T_{\theta_0})\beta_0\eta_1}{a^2}$$

$$+ \xi_1\left(\frac{B_{22}\beta_0^2}{a^2} + M_0\omega^2 - iL_x\omega - K_x\right) = 0, \qquad (B.47)$$

$$p_c\left(J_0(\beta_0) + \frac{\beta_0^2}{w_0^2}(J_0(\beta_0) - J_2(\beta_0))\right)$$

$$-\frac{p_c A\mu\beta_0}{ac_0\rho}\left(\frac{\sqrt{w_0^2 + \beta_0^2}}{w_0}\right) \frac{J_0\left(\sqrt{w_0^2 + \beta_0^2}\right) - J_2\left(\sqrt{w_0^2 + \beta_0^2}\right)}{J_0(w_0)}$$

$$+ \eta_1\left(\frac{T_{t_0}\beta_0^2 + T_{\theta_0} - B_{11}}{a^2} + M_0\omega^2 - iL_r\omega - K_r\right) + \frac{B_{12}\beta_0\xi_1}{a^2} = 0, \qquad (B.48)$$

where

$$B_{11} = \frac{E_\theta h}{1 - \sigma_x\sigma_\theta}, \qquad B_{22} = \frac{E_x h}{1 - \sigma_x\sigma_\theta}, \qquad B_{12} = \frac{E_\theta h\sigma_x}{1 - \sigma_x\sigma_\theta}, \qquad B_{21} = \frac{E_x h\sigma_\theta}{1 - \sigma_x\sigma_\theta}.$$

These provide the last two equations for the four unknowns ξ_1, η_1, p_c, and A. They are, nevertheless, still very complicated. However, they can be simplified further using a long wave approximation; i.e.,

$$|\beta_0| = \left|\frac{a\omega}{c}\right| \ll 1.$$

Hence

$$J_0(\beta_0) \approx 1, \qquad J_1(\beta_0) \approx \frac{\beta_0}{2}, \qquad J_2(\beta_0) \approx \frac{\beta_0^2}{8}. \qquad (B.49)$$

Furthermore, we notice that

$$\left|\frac{\beta_0^2}{w_0^2}\right| = \left|\frac{2\pi v}{Lc}\right| = \mathcal{O}\left(\frac{1}{\mathcal{R}}\frac{a}{L}\frac{v_c}{c}\right) \ll 1 \quad \Leftrightarrow \quad w_0^2 + \beta_0^2 \approx w_0^2, \tag{B.50}$$

where $c = L/T$ is the wave propagation velocity, L is the wavelength, and T is the period. The above approximation applies because the reciprocal Reynolds number $1/\mathcal{R}$, the ratio of characteristic velocity to wave propagation velocity v_c/c, and the ratio of radius to wavelength a/L are all small (Caro et al., 1978). Inserting the approximations (B.49) and (B.50) into (B.45) to (B.47) gives

$$\frac{p_c\beta_0}{2}\left(-\frac{a\beta_0}{\mu w_0^2} + \frac{AF_J}{c_0\rho}\right) - i\omega\eta_1 = 0, \tag{B.51}$$

$$p_c\left(-\frac{a\beta_0}{\mu w_0^2} + \frac{A}{c_0\rho}\right) - i\omega\xi_1 = 0, \tag{B.52}$$

$$p_c - \frac{p_c A\mu\beta_0}{ac_0\rho}(2 - F_J) + \frac{B_{12}\beta_0\xi_1}{a^2}$$

$$+ \eta_1\left(\frac{T_{t_0}\beta_0^2 + T_{\theta_0} - B_{11}}{a^2} + M_0\omega^2 - iL_r\omega - K_r\right) = 0, \tag{B.53}$$

$$-\frac{p_c\beta_0^3}{w_0^2} + \frac{p_c A\mu w_0^2 F_J}{2ac_0\rho} - \frac{(B_{21} + T_{t_0} - T_{\theta_0})\beta_0\eta_1}{a^2}$$

$$+ \xi_1\left(\frac{B_{22}\beta_0^2}{a^2} + M_0\omega^2 - iL_x\omega - K_x\right) = 0, \tag{B.54}$$

where the recursion formulas

$$J_{n+1}(x) = \frac{2n}{x}J_n(x) - J_{n-1}(x)$$

and

$$F_J = \frac{2J_1(w_0)}{w_0 J_0(w_0)}$$

have been used. Equations (B.51) to (B.53) make up a complete system for the variables p_c, $Ap_c/(c_0\rho)$, ξ_1, and η_1. They have nontrivial solutions if the determinant is zero. When computing the determinant, it is possible to obtain an equation that has the complex propagation velocity as the only unknown. The determinant is given by

$$\begin{vmatrix} -\frac{a\beta_0}{\mu w_0^2} & 1 & 0 & -i\omega \\ -\frac{a\beta_0^2}{2\mu w_0^2} & \frac{\beta_0 F_J}{2} & -i\omega & 0 \\ 1 & -\frac{\mu\beta_0}{a}(2 - F_J) & \frac{T_{t_0}\beta_0^2 + T_{\theta_0} - B_{11}}{a^2} + \omega^2 K_r' & \frac{B_{12}\beta_0}{a^2} \\ -\frac{\beta_0^3}{w_0^2} & \frac{\mu w_0^2 F_J}{2a} & -\frac{(B_{21} + T_{t_0} - T_{\theta_0})\beta_0}{a^2} & \frac{B_{22}\beta_0^2}{a^2} + \omega^2 K_x' \end{vmatrix}, \tag{B.55}$$

where $K'_j = M_0 - iL_r/\omega - K_r/\omega^2$ for $j = x, r$. For convenience we have rearranged the rows. In order to simplify (B.55), we neglect small terms and perform a number of row/column operations on the determinant.

First, $T_{t_0}\beta_0^2/a^2$ can be neglected since it is much smaller than $(T_{\theta_0} - B_{11})/a^2$, which is of order one. Second, $w_0^2 = i^3 w^2$ and $\beta_0 = i\beta$ are inserted, and the following operations are performed:

1. Multiply the first column by $\mu w^2/(\beta a)$, the third column by β/w, and the fourth column by i/ω.

2. Multiply the second row by $1/(i\beta)$, the third row by $\beta a/(\mu w^2)$, and the fourth row by $ia/(\mu w^2)$.

3. Replace the second row with the first row minus twice the second row, and the third row with the first row minus the third row.

The determinant becomes

$$
\begin{vmatrix}
0 & 1 - F_J & 2 & 1 \\[2mm]
1 & 1 & 0 & 1 \\[2mm]
0 & 1 + \dfrac{i\beta^2(2-F_J)}{w^2} & -\dfrac{(T_{\theta_0}-B_{11})\beta^2}{a\omega\mu w^2} - \dfrac{a\omega\beta^2 K'_r}{\mu w^2} & 1 + \dfrac{B_{12}\beta^2}{a\omega\mu w^2} \\[3mm]
-\dfrac{\beta^2 i}{w^2} & \dfrac{F_J}{2} & \dfrac{(B_{21}+T_{t_0}-T_{\theta_0})\beta^2}{a\omega\mu w^2} & \dfrac{B_{22}\beta^2}{a\omega\mu w^2} - \dfrac{\omega a K'_x}{\mu w^2}
\end{vmatrix}. \tag{B.56}
$$

A further analysis of the magnitude of the terms reveals that $1 + i\beta^2(2 - F_J)/w^2 \approx 1$. Expanding the determinant after the first column gives two subdeterminants with the same order of magnitude. Hence the one arising from $-i\beta^2$ can be neglected.

Let

$$
B'_{11} = \frac{B_{11} - T_{\theta_0}}{ac_0^2\rho}, \qquad\qquad B'_{12} = \frac{B_{12}}{ac_0^2\rho},
$$

$$
B'_{21} = \frac{B_{21} + T_{t_0} - T_{\theta_0}}{ac_0^2\rho}, \qquad\qquad B'_{22} = \frac{B_{22}}{ac_0^2\rho},
$$

$$
K'_x = \frac{a\omega^2}{c_0^2\rho}\left(-M_0 + \frac{iL_x}{\omega} + \frac{K_x}{\omega^2}\right), \qquad K'_r = \frac{a\omega^2}{c_0^2\rho}\left(-M_0 + \frac{iL_r}{\omega} + \frac{K_r}{\omega^2}\right),
$$

where K'_r and K'_x have been redefined. Inserting these and expanding $w^2 = a^2\omega/\nu$, $\beta = a\omega/c$ gives

$$
\begin{vmatrix}
1 - F_J & 2 & 1 \\[2mm]
1 & (k')^2(B'_{11} + K'_r) & 1 + (k')^2 B'_{12} \\[2mm]
\dfrac{F_J}{2} & (k')^2 B'_{21} & (k')^2 B'_{22} + \dfrac{c_0^2 K'_x}{a^2\omega^2}
\end{vmatrix} = 0, \tag{B.57}
$$

where $k' = c_0/c$ is the wave number and $\rho = \mu/\nu$. The resulting determinant is

$$(k')^4(1 - F_J)\left(B'_{22}(B'_{11} + K'_r) - B'_{12}B'_{21}\right)$$

$$+ (k')^2 \left(F_J\left(B'_{12} + B'_{21} - \frac{B'_{11} + K'_r}{2}\right) - 2B'_{22} + \frac{c_0^2}{a^2\omega^2}K'_x(B'_{11} + K'_r)(1 - F_J)\right)$$

$$+ F_J - \frac{2c_0^2}{a^2\omega^2}K'_x = 0. \tag{B.58}$$

This is the dispersion relation for the waves and is equivalent to the result obtained by Pedley (1980).

The original equations can be simplified by neglecting the terms corresponding to the simplifications that led to (B.58) and using B'_{ij} and K'_j defined on page 270. This gives the following system of equations, which can also be found in Pedley (1980):

$$\frac{ap_c}{c_0^2\rho}\left((k')^2 + AF_Jk'\right) - 2\eta_1 = 0, \tag{B.59}$$

$$\frac{p_c}{c_0\rho}\left(k' + A\right) - i\omega\xi_1 = 0, \tag{B.60}$$

$$\frac{ap_c}{c_0^2\rho} + \eta_1\left(B'_{11} + K'_r\right) - 2\eta_1 = 0, \tag{B.61}$$

$$\frac{iap_cAF_J}{2c_0^2\rho} + ik'B'_{21}\eta_1 + \left(\frac{a\omega\xi_1}{c_0}\right)\left(B'_{22}(k')^2 + \frac{c_0^2}{\omega^2a^2}K'_x\right) = 0. \tag{B.62}$$

Generally, $k' = k'_{re} + ik'_{im}$ is a complex number, so we define the wave number k from

$$\frac{c_0}{k'_{re}} = \frac{\omega}{k}.$$

Hence

$$\exp\left(i\omega(t - x/c)\right) = \exp\left(i\omega(t - xk'_{re}/c_0)\right)\exp\left(-\omega xk'_{im}/c_0\right) \tag{B.63}$$
$$= \exp\left(i\omega(t - xk'_{re}/c_0)\right)\exp\left(-2\pi xk'_{im}/(\lambda k'_{re})\right),$$

where $\lambda = 2\pi/k$ is the wavelength. The last exponential represents the transmission per wavelength. The fourth order equation (B.58) gives two solutions for $(k')^2$, but only one of them represents the pressure wave. In any case the solutions depend on w, the Womersley number, through F_J. The ratio of vessel radius to thickness of the oscillating Stokes boundary layer is proportional to w. If w is large, the boundary layer is thin and the velocity profile is almost flat across the core vessel; if w is small, the boundary layer is thick (the vessel is fully occupied by the innermost portion of the boundary layer) and the flow becomes quasi-steady. For $w \to 0$ a Poiseuille flow is obtained. The following asymptotic expansion can be derived for F_J for large and small values of w, respectively:

$$F_J(w) = \begin{cases} 2/(wi^{1/2})\left(1 + (2w)^{-1} + \mathcal{O}(w^{-2})\right) & \text{for } w \to \infty, \\ 1 - i(w^2/8) - (w^4/48) + \mathcal{O}(w^6) & \text{for } w \to 0. \end{cases} \tag{B.64}$$

The equations in (B.59) to (B.62) are still too complicated to solve for any real applications. Therefore, they must be simplified further. This can be done by assuming longitudinal tethering, i.e., $T_{t_0} = 0$, $T_{\theta_0} = 0$, and inserting the Moens–Korteweg wave propagation factor $c_0 = E_\theta h/(2a\rho)$. Thus

$$B'_{11} = \frac{2}{1 - \sigma_x \sigma_\theta}, \quad B'_{12} = \frac{2\sigma_x}{1 - \sigma_x \sigma_\theta},$$

$$B'_{21} = \frac{2E_x/E_\theta \sigma_\theta}{1 - \sigma_x \sigma_\theta}, \quad B'_{22} = \frac{2E_x/E_\theta}{1 - \sigma_x \sigma_\theta}.$$

These quantities are of order one. According to experimental results by Bergel (1972), good estimates are $\sigma_x = \sigma_\theta = 0.29$ and $E_x/E_\theta = 1.2$. This yields

$$B'_{11} = 2.18, \quad B'_{22} = 2.62, \quad B'_{12} = 0.63, \quad B'_{21} = 0.76.$$

Bergel also gives values for the tethering constants:

$$K_{r,x} \approx 33 \times 10^3 \text{ kg/(s}^2\text{m}^2), \quad L_{r,x} \approx 17 \times 10^3 \text{ kg/(s·m}^2), \quad M_0 \approx 4 \text{ kg/m}^2.$$

Consequently, M_0 is negligible compared to $K_{r,x}$ and $L_{r,x}$. Hence

$$K'_{r,x} \approx \frac{(33 + 17i\omega)\, a}{c_0^2 \rho} \times 10^3 \text{ kg/(s}^2\text{m}^2).$$

In the smaller arteries $a \approx 10^{-3}$ m, $\rho \approx 10^3$ kg/m^3, $c_0 \approx 5$ m/s, and $\omega \approx 4\pi \text{s}^{-1}$; hence $|K'_{r,x}| \approx 0.00987$. These data are as estimated by Pedley (1980). From these estimates we see that $K'_r \ll B'_{ij}$, and hence K'_r can be neglected in (B.58). The term involving K'_x appears as $c_0^2/(\omega^2 a^2)K_x \approx 1560$, which is large compared to the other terms. Therefore, the dispersion relation (B.58) reduces to

$$a_0(k')^4 + \left(a_1 K - \frac{a_2}{1 - F_J} \right)(k')^2 - \frac{2K + F_J}{1 - F_J} = 0, \tag{B.65}$$

where

$$a_0 = B'_{22}B'_{11} - B'_{12}B'_{21},$$

$$a_1 = B'_{11},$$

$$a_2 = F_J \left(B'_{12} + B'_{21} - \frac{B'_{11}}{2} \right) - 2B'_{22}, \quad \text{and} \quad K = \frac{c_0^2}{a^2\omega^2}K'_x.$$

Since K is large, $a_2/(1 - F_J)$ and $F_J/(1 - F_J)$ can be neglected as well, and we are left with

$$a_0(k')^4 + a_1 K(k')^2 - \frac{2K}{1 - F_J} = 0,$$

which has the approximate solutions

$$(k')^2 \approx \frac{2}{(1 - F_J)a_1}, \quad -\frac{a_1 K}{a_0}. \tag{B.66}$$

The first of these solutions is the pressure wave. This can be compared to the result found by Lighthill (1975), where the wall is assumed to be an isotropic elastic solid; i.e., $a_1 = B'_{11} = 2/(1 - \sigma^2)$ and $w \to \infty$ such that $F_J \approx 0$. Then (B.66) gives $(k')^2 = 1 - \sigma^2$ or equivalently $c^2 = c_0^2/(1 - \sigma^2)$.

The solution can also be found by letting $K \to \infty$ directly in (B.59) to (B.62). However, this is possible only if $\xi = 0$. Hence from (B.61) one finds (still neglecting K'_r) that

$$B'_{11}\eta_1 = \frac{ap_c}{c_0^2\rho}.$$

From (B.59) and (B.60)

$$A = -k' = -\sqrt{\frac{2}{(1 - F_J)B'_{11}}},$$

which is consistent with (B.66). Finally, from (B.61)

$$B'_{21}\eta_1 = -\frac{F_J a p_c}{2c_0^2\rho}.$$

We are now able to state the final equations for the longitudinal velocity. From (B.44) we had

$$w = -\frac{ap_c\beta_0}{\mu w_0^2} J_0\left(\frac{\beta_0 r}{a}\right) + \frac{p_c A}{c_0\rho}\left(\frac{\sqrt{w_0^2 + \beta_0^2}}{w_0}\right)\frac{J_0\left(\frac{r}{a}\sqrt{w_0^2 + \beta_0^2}\right)}{J_0(w_0)}$$

$$\approx \frac{p_c k'}{c_0\rho}\left(1 - \frac{J_0\left(\frac{w_0 r}{a}\right)}{J_0(w_0)}\right). \tag{B.67}$$

In the last equation we inserted the definitions for β_0 and w_0 and used $A = -k'$, $\rho = \mu/\nu$, and the long wave approximation (page 268) for $J_0(\beta_0 r/a) \approx 1$.

Finally, the pressure gradient can be found from the long wave approximation, the assumption of harmonic solutions (B.34), and the pressure (B.39):

$$p_r = p_c J_0\left(\frac{\beta_0 r}{a}\right) \approx p_c \quad \Leftrightarrow \quad -\frac{i\omega p_c k'}{c_0} = \frac{\partial P}{\partial x}. \tag{B.68}$$

Acknowledgments

The author was supported by High Performance Parallel Computing, Software Engineering Applications Eureka Project 1063 (SIMA—SIMulation in Anesthesia), and by the Danish Academy for Technical Sciences.

Bibliography

Acierno, L. (1994). *The History of Cardiology*, The Parthenon Publishing Group, New York.

Adeler, P.T. (2001). *Hemodynamic Simulation of the Heart using a 2D model and MR Data*, Ph.D. thesis, Department of Informatics and Mathematical Modelling, Danish Technical University, Lyngby, Denmark.

Alonso, M. and Finn, E. (1979). *Mechanics and Thermodynamics*, Vol. 1 of *Fundamental University Physics*, Addison-Wesley, Reading, Massachusetts.

Andreasen, T., Christensen, B., Green, C., Hansen, A., and Helmgaard, L. (1994). *Model 10 – En matematisk model af intravenøse anæstetikas farmakokinetik*, Technical report, IMFUFA, Roskilde University, Denmark. Text No. 274.

Anliker, M. (1977). Current and future aspects of biomedical engineering, *Triangle* **16**: 129–140.

Anliker, M., Rockwell, R., and Ogden, E. (1971). Nonlinear analysis of flow pulses and shock waves in arteries, *ZAMP* **22**: 217–246.

Atabek, H. (1968). Wave propagation through a viscous fluid contained in a tethered, initially stressed, orthotropic elastic tube, *Biophys. J.* **8**: 626–649.

Atabek, H. and Lew, H. (1966). Wave propagation through a viscous incompressible fluid contained in an initially stressed elastic tube, *Biophys. J.* **6**: 481–503.

Atkins, P. (1990). *Physical Chemistry*, Oxford University Press, Oxford, U.K.

Avolio, A. (1980). Multi-branched model of the human arterial system, *Med. Biol. Eng. Comput.* **18**: 709–718.

Avolio, A. (1992). Ageing and wave reflection, *J. Hypertension* **10 (suppl 6)**: S83–S86.

Baan, J. (1992). Ventricular pressure-volume relation in vivo, *Euro. Heart J.* **13**: 2–6.

Bardoczky, G., de Vries, J., Meriläinen, P., Schofield, J., and Tuomaala, L. (1996). *Side stream spirometry*, Datex Division Instrumentarium Corp., Helsinki, Finland.

Bardos, J., Golse, F., and Levermore, C. (1991). Fluid dynamic limits of kinetic equations, *J. Stat. Phys.* **63**: 323–344.

Barnard, A., Hunt, W., Timlake, W., and Varley, E. (1966). A theory of fluid flow in compliant tubes, *Biophys. J.* **6**: 717–724.

Bassingthwaighte, J., Liebovitch, L., and West, B. (1994). *Fractal Physiology*, The American Physiological Society Methods in Physiology Series, Oxford University Press, New York.

Batchelor, G. (1992). *Fluid Dynamics*, Cambridge University Press, Cambridge, U.K.

Beneken, J. (1965). *A Mathematical Approach to Cardio-Vascular Function*, Ph.D. thesis, University of Utrecht, The Netherlands.

Bergel, D. (1972). *Cardiovascular Fluid Dynamics*, Vol. 2, Academic Press, London, U.K.

Berger, D., Li, J., Lasky, W., and Noordergraaf, A. (1993). Repeated reflection of waves in the systemic arterial system, *Am. J. Physiol.* **264**: H269–H281.

Beyer, R. (1992). A computational model of the cochlea using the immersed boundary method, *J. Comput. Phys.* **98**: 145–162.

Beyer, R. and LeVeque, R. (1992). Analysis of a one-dimensional model for the immersed boundary method, *SIAM J. Numer. Anal.* **29**: 332–364.

Bischoff, K. and Dedrick, R. (1968). Thiopental pharmacokinetics, *J. Pharm. Sci.* **57**: 1346–1351.

Bolter, C. and Ledsome, J. (1976). The effect of cervical sympathetic nerve stimulation on canine carotid sinus reflex, *Am. J. Physiol.* **230**: 1026–1030.

Börgers, C. and Peskin, C. (1987). A Lagrangian fractional step method for the incompressible Navier-Stokes equations on a periodic domain, *J. Comput. Phys.* **70**: 397–438.

Bronk, D. and Stella, G. (1932). Afferent impulses in the carotid sinus nerve, *Am. J. Physiol.* **1**: 113–130.

Bronk, D. and Stella, G. (1935). The response to steady pressure of single end organs in the isolated carotid sinus, *Am. J. Physiol.* **110**: 708–714.

Brown, A. (1980). Receptors under pressure, an update on baroreceptors, *Circ. Res.* **46**: 1–10.

Burkhoff, D., De Tombe, P., and Hunter, W. (1993). Impact of ejection on magnitude and time course of ventricular pressure-generating capacity, *Am. J. Physiol.* **265**: H899–H909.

Caflisch, R., Majda, G., Peskin, C., and Strumolo, G. (1980). Distortion of the arterial pulse, *Math. Biosci.* **51**: 229–260.

Campbell, K., Ringo, J., Knowlen, G., Kirkpatrick, R., and Schmidt, S. (1986). Validation of optional elastance-resistance left ventricle pump models, *Am. J. Physiol.* **251**: H382–H397.

Canic, S. and Mikelic, A. (2002). Effective equations describing the flow of a viscous incompressible fluid through a long elastic tube. *Comptes Rendus Mechanique Acad. Sci.* **330**: 661–666.

Canic, S. and Mikelic, A. (2003). Effective equations modeling the flow of a viscous incompressible fluid through a long elastic tube arising in the study of blood flow through small arteries. *SIAM J. Appl. Dyn. Syst.* **2**(3): 431–463.

Canic, S. and Kim, E.H. (2003). Mathematical analysis of the quasilinear effects in a hyperbolic model of blood flow through compliant axisymmetric vessels. *Meth. Appl. Sci.* **26**(13): 1–26.

Cappello, A., Gnudi, G., and Lamberti, C. (1995). Identification of the three-element Windkessel model incorporating a pressure-dependent compliance, *Ann. Biomed. Eng.* **23**: 164–177.

Caro, C., Pedley, T., Schroter, R., and Seed, W. (1978). *The Mechanics of the Circulation*, Oxford University Press, Oxford, U.K.

Cecchini, A., Melbin, J., and Noordergraaf, A. (1981). Set-point: Is it a distinct structural entity in biological control?, *J. Theor. Biol.* **93**: 387–394.

Cecchini, A., Tiplitz, K., Melbin, J., and Noordergraaf, A. (1982). Baroreceptor activity related to cell properties, *Proceeding of the 35th Annual Conference on Engineering in Medicine and Biology*, Vol. 24, p. 20.

Chapleau, M. and Abboud, F. (1987). Contrasting effects of static and pulsatile pressure on carotid baroreceptor activity in dogs, *Circ. Res.* **61**: 648–658.

Chiari, L., Avanzolini, G., Gnudi, G., and Grandi, F. (1995). A non-linear simulator of the human respiratory chemostat., in H. Power and R.T. Hart (eds.), *Computer Simulations in Biomedicine*, pp. 87–98. Computer Mechanics Publications, Boston, Massachussetts.

Chiari, L., Avanzolini, G., Grandi, F., and Gnudi, G. (1994). A simple model of the chemical regulation of acid-base balance in blood, in N. Sheppard, M. Eden, and G. Kantor (eds.), *Engineering Advances/94. Proceedings of the 16th International Conference of the IEEE-EMBS*, pp. 1025–1026. Baltimore, Maryland.

Chorin, A. (1968). Numerical solution of the Navier-Stokes equations, *Math. Comp.* **22**: 745–762.

Chorin, A. (1969). On the convergence of discrete approximations to the Navier-Stokes equations, *Math. Comp.* **23**: 341–353.

Chorin, A.J. and Marsden, J.E. (1998). *A Mathematical Introduction to Fluid Mechanics, 3rd ed.*, Springer-Verlag, New York.

Christiansen, T. and Dræby, C. (1996). *Modelling the Respiratory System*, Technical report, IMFUFA, Roskilde University, Denmark. Text No. 318.

Cook, R., Woods, D., and McDonald, J. (1991). *Human Performance in Anesthesia. A Corpus of Cases*, CSEL Report SCEL 09.003.

Cox, R. and Bagshaw, R. (1975). Baroreceptor reflex control of arterial hemodynamics in the dog, *Circ. Res.* **37**(6): 772–786.

Danielsen, M. (1998). *Modeling of Feedback Mechanisms Which Control the Heart Function in a View to an Implementation in Cardiovascular Models*, Technical report, IMFUFA, Roskilde University, Denmark. Ph.D. thesis—Text No. 358.

Danielsen, M. and Ottesen, J. (1997). A dynamical approach to the baroreceptor regulation of the cardiovascular system, *Proceedings of the 5th International Symposium, Symbiosis '97*, pp. 25–29.

Danielsen, M. and Ottesen, J. (2001). Describing the pumping heart as a pressure source, *J. Theor. Biol.* **212**: 71–81.

Danielsen, M., Palladino, J., and Noordergraaf, A. (2000a). The left ventricular ejection effect, in J. Ottesen and M. Danielsen (eds.), *Mathematical Modeling in Medicine*, IOS Press, Philadelphia, Pennsylvania, pp. 13–28.

Danielsen, M., Palladino, J., and Noordergraaf, A. (2000b). Positive and negative effects of ventricular ejection, in J. Enderle and L.L. Marctarlance (eds.), *IEEE 26th Annual Northeast Bioengineering Conference*, University of Connecticut, Storrs, Connecticut, pp. 33–34.

Davis, P. (1995). The art of the heart valve, *SIAM News* **28**(5): 1 and 13.

Dawber, T., Thomas, H., and McNamara, P. (1973). Characteristics of the dicrotic notch of the arterial pulse wave in coronary heart disease, *Angiology* **24**(4): 244–255.

De Tombe, P. and Little, W. (1994). Inotropic effects of ejection are myocardial properties, *Am. J. Physiol.* **266**: H1202–H1213.

Despopoulos, A. and Silbernagl, S. (1991). *Color Atlas of Physiology*, 4th ed., Thieme, Stuttgart, Germany.

Ding, T. and Schoephoerster, R. (1997). Evaluation of global left ventricular function based on simulated flow dynamics computed from regional wall motion, *ASME BED* **35**: 193–194.

Donald, D. and Edis, A. (1970). Comparison of aortic and carotid baroreflexes in the dog, *J. Physiol.* **215**: 521–538.

Ducas, J., Schick, U., Girling, L., and Prewitt, R. (1985). Effects of reduced resistive afterload on left ventricular pressure volume relationship, *Am. J. Physiol.* **248**: H163–H169.

Elliott, G., Rome, E., and Spencer, M. (1970). A type of contraction hypothesis applicable to all muscles, *Nature* **226**: 417–420.

Evans, J., Wagner, P., and West, J. (1974). Conditions for reduction of pulmonary gas transfer by ventilation-perfusion inequality, *J. Appl. Physiol.* **36**: 533–537.

Feinberg, A. and Lax, H. (1958). Studies of the arterial pulse wave, *Circulation* **18**: 1125–1130.

Fincham, W. and Tehrani, F. (1983). A mathematical model of the human respiratory system, *J. Biomed. Eng.* **5**: 125–133.

Forbes, L. (1981). On the evolution of shock-waves in mathematical models of the aorta, *Aust. Math. Soc. (B)* **22**: 257–269.

Franz, G. (1969). Non-linear rate sensitivity of carotid sinus reflex as a consequence of static and dynamic nonlinearities in baroreceptor behavior, *Ann. NY Acad. Sci.* **156**: 811–824.

Fung, Y. (1993). *Biomechanics. Mechanical Properties of Living Tissues*, 2nd ed., Springer-Verlag, New York.

Gaba, D. (1994). Human work environment and simulators, in R. Miller (ed.), *Anesthesia*, Churchill Livingstone, Edinburgh, Scotland, Chap. 85, pp. 2635–2679.

Galster, G. (1995). *Source Code for a Model of Pulmonary Mechanics*, unpublished manuscript, Bispebjerg Hospital, København, Denmark.

Ganong, W. (1975). *Review of Medical Physiology*, 17th ed., LANGE Medical Publications, Los Altos, California.

Georgiadis, J., Wang, M., and Pasipoularides, A. (1992). Computational fluid dynamics of left ventricular ejection, *Ann. Biomed. Eng.* **20**: 81–97.

Gibaldi, M. and Perrier, D. (1982). *Pharmacokinetics. Drugs and the Pharmaceutical Sciences*, Vol. 15, 2nd ed., Marcel Dekker, New York.

Golden, J., Clark, J., and Stevens, P. (1973). Mathematical modeling of pulmonary airway dynamics, *IEEE-TBME* **20**: 397–404.

Gonzalez, E. and Schoephoerster, R. (1996). A simulation of three-dimensional systolic flow dynamics in a spherical ventricle: Effects of abnormal wall motion, *Ann. Biomed. Eng.* **24**: 48–57.

Granger, W.M., Miller, D.A., Erhart, I.C., and Hofman, W.F. (1987). The effect of blood flow and diffusion impairment on pulmonary gas exchange: A computer model, *Comput. Biomed. Res.* **20**: 497–506.

Greenberg, S., McQueen, D., and Peskin, C. (1987). Three-dimensional fluid dynamics in a two-dimensional amount of central memory, in A.J. Chorin and A.J. Majda (eds.), *Wave Motion: Theory, Modelling and Computation. Proceedings of a Conference in Honor of the 60th Birthday of Peter D. Lax*, Springer-Verlag, New York, pp. 85–146.

Greene, H.D. (1986). Changes in canine cardiac function and venous return curves by the carotid sinus baroreflex, *Am. J. Physiol.* **251**: H288–H296.

Gregg, D. (1966). Dynamics of blood and lymph flow, in C. Best and N. Taylor (eds.), *The Physiological Basis of Medical Practice*, 8th ed., Williams and Wilkins, New York.

Grodins, F. (1959). Integrative cardiovascular physiology: A mathematical synthesis of cardiac and blood vessel hemodynamics, *Q. Rev. Biol.* **34**: 93–116.

Grodins, F. and Yamashiro, S. (1978). *Respiratory Function of the Lung and Its Control*, Macmillan, New York.

Groen, J. and van Dijk, P. (1987). Design of flow adjustable gradient waveforms, in J.T.E. der Kleef (ed.), *Society of Magnetic Resonance in Medicine, 6th Annual Meeting*, p. 868.

Guasp, F. (1980). La estructuración macroscópica del miocardio ventricular, *Revista Española de Cardiologio* **33**(3): 265–287.

Guyton, A. (1981). The relationship of cardiac output and arterial pressure control, *Circulation* **64**(6): 1079–1088.

Guyton, A. (1991). *Textbook of Medical Physiology*, 8th ed., W.B. Saunders, Philadelphia, Pennsylvania.

Hill, A. (1938). The heat of shortening and the dynamic constants of muscle, *Proc. R. Soc. London Ser. B* **126**: 136–195.

Hoppenstaedt, F. and Peskin, C. (1992). *Mathematics in Medicine and the Life Sciences*, Springer-Verlag, New York.

Horsfield, K. and Woldenberg, M. (1989). Diameters and cross-sectional areas of branches in the human pulmonary arterial tree, *Anat. Rec.* **223**: 245–251.

Hosomi, H. and Sagawa, K. (1979). Effect of pentobarbital anesthesia on hypotension after 10% hemorrhage in the dog, *Am. J. Physiol.* **236**: H607–H612.

Houlind, K., Pedersen, E., Oyre, S., Kim, W., Walker, P., Egeblad, H., and Yoganathan, A. (1994). Left ventricular blood flow patterns assessed by magnetic resonance velocity mapping in patients with ischemic heart disease, *Am. J. Noninvas Cardiol.* **8**: 317–325.

Hull, C. (1979). Pharmacokinetics and pharmacodynamics, *Br. J. Anaesth.* **51**: 579–594.

Hunter, W. (1989). End-systolic pressure as a balance between opposing effects of ejection, *Circ. Res.* **64**: 265–275.

Hunter, W., Janicki, J., Weber, K., and Noordergraaf, A. (1983). Systolic mechanical properties of the left ventricle, *Circ. Res.* **52**: 319–327.

Iberall, A. (1967). Anatomy and steady flow characteristics of the arterial system with an introduction to its pulsatile characteristics, *Math. Biosci.* **1**: 375–385.

Jackson, A. and Milhorn, H. (1973). Digital computer simulation of respiratory mechanics, *Comp. Biomed. Res.* **6**: 27–56.

Jacobsen, J., Adeler, P. T., Kim, W., Houlind, K., Pedersen, E., and Larsen, J. (2001). Evaluation of a 2D heart model against magnetic resonance velocity mapping, *Cardiovasc. Eng.* **1**: 59–76.

Jacquez, J. (1985). *Compartmental Analysis in Biology and Medicine*, 2nd ed., University of Michigan Press, Ann Arbor, Michigan.

Jakobsen, B. and Niss, K. (2000). *Behandling af Impuls ved Kilder og dræn i C.S. Peskins 2D-Hjertemodel*, Technical report (in Danish), IMFUFA Roskilde University, Denmark. Tekst Nr. 388.

Jensen, P. (1994–1998). *Personal communication*, Department of Anaesthesiology, Copenhagen University Hospital, Rigshospitalet, Denmark.

Jin, Z. and Qin, J. (1993). An electric model with time varying resistance for a pneumatic membrane blood pump, *ASAIO J.* **39**: 56–61.

Jones, T. and Metaxas, D. (1998). Patient-specific analysis of left ventricular blood flow, in W. Wells, A. Colchester, and S. Delp (eds.), *Medical Image Computing and Computer-Assisted Intervention - MICCAI '98. Proceedings*, Vol. 1498 of *Lecture Notes in Computer Science*, Springer-Verlag, New York, pp. 156–166.

Kamiya, A., Ando, J., Shibata, M., and Masuda, H. (1988). Roles of fluid shear stress in physiological regulation of vascular structure and function, *Biorheol.* **25**: 271–278.

Kannel, W., Wolf, P., McGee, D., Dawber, T., McNamara, P., and Castelli, W. (1981). Systolic blood pressure, arterial rigidity, and risk of stroke, *JAMA* **245**: 1225–1229.

Kassab, G. and Fung, Y. (1995). The pattern of coronary arteriolar bifurcations and the uniform shear hypothesis, *Ann. Biomech. Eng.* **23**: 13–20.

Keener, J. and Sneyd, J. (1998). *Mathematical Physiology*, Springer-Verlag, New York.

Kelman, R. (1966). Digital computer subroutine for the conversion of oxygen tension into saturation, *J. Appl. Physiol.* **21**: 1375.

Kim, W. (1996–1998). *Personal communication* (magnetic resonance measurement of blood flow in the ascending aorta), Department of Cardiothoracic and Vascular Surgery and Institute for Experimental Clinical Research, Aarhus University Hospital, Skejby, Denmark.

Kim, W., Walker, P., Pedersen, E., Paulsen, J., Oyre, S., Houlind, K., and Yoganathan, A. (1995). Left ventricular blood flow patterns in normal subjects: A quantitative analysis by three-dimensional magnetic resonance velocity mapping, *Am. Coll. Cardiol.* **26**: 224–238.

Korner, P. (1971). Integrative neural cardiovascular control, *Physiol. Rev.* **51**: 50–51.

Korner, P. (1974). "Steady state" properties of the baroreceptor-heart rate reflex in essential hypertension in man, *Clin. Exp. Pharm. Phys.* **1**: 65–75.

Kumada, M., Azuma, T., and Matsuda, K. (1967). The cardiac output-heart rate relationship under different conditions, *Jpn. J. Physiol.* **17**: 538–555.

Kumada, M., Schmidt, R., Sagawa, K., and Tan, K. (1970). Carotid sinus reflex in response to hemorrhage, *Am. J. Physiol.* **219**: 1373–1379.

Landau, L. and Lifshitz, E. (1986). *Theory of Elasticity*, Vol. 7 of *Course of Theoretical Physics*, 3rd ed., Pergamon Press, Oxford, U.K.

Landgren, S. (1952). On the excitation mechanism of the carotid sinus baroreceptors, *Acta Physiol. Scan.* **26**: 1–34.

Langewouters, G., Wesseling, K., and Goedhard, W. (1984). The static elastic properties of 45 human thoracic and 20 abdominal aortas in vitro and the parameters of a new model, *J. Biomech.* **17**(6): 425–435.

Lax, H. and Feinberg, A. (1959). Abnormalities of the arterial pulse wave in young diabetic subjects, *Circulation* **20**: 1106–1110.

Lax, H., Feinberg, A., and Cohen, B. (1956). The normal pulse wave and its modification in the presence of human atherosclerosis, *J. Chronic. Dis.* **3**: 618–631.

Leaning, M., Pullen, H., Carson, E., Al-Dahan, M., Rajkumart, N., and Finkelstein, L. (1983a). Modelling a complex biological system: The human cardiovascular system. 2. Model validation, reduction and development, *Trans. Inst. M. C.* **5**: 71–86.

Leaning, M., Pullen, H., Carson, E., and Finkelstein, L. (1983b). Modelling a complex biological system: The human cardiovascular system. 1. Methodology and model description, *Trans. Inst. M. C.* **5**: 71–86.

Lerou, J., Dirksen, R., Kolmer, H., and Booij, L. (1991). A system model for closed-circuit inhalation anesthesia, *Anesthesiology* **75**: 345–355.

Li, J.-J. (1987). *Arterial System Dynamics*, Biomedical Engineering Series, New York University Press, New York.

Lighthill, J. (1975). *Mathematical Biofluiddynamics*, CBMS-NSF Regional Conf. Ser. in Appl. Math. 17, SIAM, Philadelphia, Pennsylvania.

Longobardo, G., Cherniack, N., and Fishman, A. (1966). Cheyne-Stokes breathing produced by a model of the human respiratory system, *J. Appl. Physiol.* **21**: 1839–1846.

Martin, J., Schneider, A., Mandel, J., Prutow, R., and Smith, N. (1986). A new cardiovascular model for real-time applications, *Transactions* **3**: 31–65.

Mayo, A.A. and Peskin, C.S. (1993). An implicit numerical method for fluid dynamics problems with immersed elastic boundaries, in A.Y. Cheer and C.P. van Dam (eds.), *Fluid Dynamics in Biology*, Vol. 141, American Mathematical Society, Providence, Rhode Island, pp. 261–277.

McCacken, M. and Peskin, C. (1980). A vortex method for blood flow through heart valves, *J. Comput. Phys.* **35**: 183–205.

McDonald, D. (1974). *Blood Flow in Arteries*, 2nd ed., Edward Arnold, London, U.K.

McQueen, D. and Peskin, C. (1983). Computer-assisted design of pivoting disc prosthetic mitral valves, *J. Thor. Cardiovasc. Surg.* **86**: 126–135.

McQueen, D. and Peskin, C. (1985). Computer-assisted design of butterfly bileaflet valves for the mitral position, *Scand. J. Thor. Cardiovasc. Surg.* **19**: 139–148.

McQueen, D. and Peskin, C. (1989). A three-dimensional computational method for blood flow in the heart. II. Contractile fibers, *J. Comput. Phys.* **89**: 289–297.

McQueen, D. and Peskin, C. (1997). Shared-memory parallel vector implementation of the immersed boundary method for the computation of blood flow in the beating mammalian heart, *J. Supercomput.* **11**: 213–236.

McQueen, D. and Peskin, C. (2000). Heart simulation by an immersed boundary method with formal second-order accuracy and reduced numerical viscosity, in H. Aref and I.W. Phillips (eds.), Mechanics for a New Millenium, *Proceedings of ICTAM* 2000, Kluwer Academic Publishers, Norwell, Massachusetts.

McQueen, D., Peskin, C., and Yellin, E. (1982). Fluid dynamics of the mitral valve: Physiological aspects of a mathematical model, *Am. J. Physiol.* **242**: H1095–H1110.

Meisner, J., McQueen, D., Ishida, Y., Vetter, H., Bortolotti, U., Strom, J., Frater, R., Peskin, C., and Yellin, E. (1985). Effects of timing of atrial systole on LV filling and mitral valve closure: Computer and dog studies, *Am. J. Physiol.* **249**: H604–H619.

Melbin, J., Detweiler, D., Riffle, R., and Noordergraaf, A. (1982). Coherence of cardiac output with rate changes, *Am. J. Physiol.* **243**: H499–H504.

More, W. (1972). *Physical Chemistry*, 5th ed., Longman, London, U.K.

Mulier, J.P. (1994). *Ventricular Pressure as a Function of Volume and Flow*, Ph.D. thesis, University of Leuven, Belgium.

Murray, C. (1926a). The physiological principle of minimum work. I. The vascular system and the cost of blood volume, *Proc. Nat. Acad. Sci.* **12**: 207–214.

Murray, C. (1926b). The physiological principle of minimum work. II. Oxygen exchange in capillaries, *Proc. Nat. Acad. Sci.* **12**: 299–304.

Netter, F. (1991). *Heart*, Vol. 5 of *The Ciba Collection of Medical Illustrations*, CIBA, Basel, Germany.

Neumann, S. (1996). *Modeling Acute Hemorrhage in the Human Cardiovascular System*, Ph.D. thesis, University of Pennsylvania, Philadelphia, Pennsylvania.

Nichols, W. and O'Rourke, M. (1998). *McDonald's Blood Flow in Arteries*, 4th ed., Edward Arnold, London, U.K.

Noordergraaf, A. (1978). *Circulatory System Dynamics*, Academic Press, San Diego, California.

Nunn, J. (1987). *Nunn's Applied Respiratory Physiology*, Butterworths, London, U.K.

Ockendon, H. and Ockendon, J. (1995). *Viscous Flow*, Cambridge Texts in Applied Mathematics, Cambridge University Press, Cambridge, U.K.

Olofsen, E. (1994). *Modelling Arterial Blood Desaturation During Apnoea*, Technical report, Department of Anaesthesiology, University Hospital, Leiden, The Netherlands.

Olsen, J. and Shapiro, A. (1967). Large-amplitude unsteady flow in liquid-filled elastic tubes, *J. Fluid. Mech.* **29**: 513–538.

Olufsen, M. (1998). *Modeling the Arterial System with Reference to an Anesthesia Simulator*, Technical report, IMFUFA, Roskilde University, Denmark. Ph.D. thesis—Text No. 345.

Olufsen, M. (1999). Structured tree outflow condition for blood flow in larger systemic arteries, *Am. J. Physiol.* **276**: H257–H268.

Olufsen, M., Nadim, A., and Lipsitz, L. (2002). Cerebral blood flow regulation explained using a lumped parameter model, *Am. J. Physiol.* **282**: R611–R622.

Olufsen, M. and Ottesen, J. (1995a). *A Fluid-Dynamical Model of the Aorta with Bifurcations*, Technical report, IMFUFA, Roskilde University, Denmark, Text No. 297.

Olufsen, M. and Ottesen, J. (1995b). Outflow conditions in human arterial flow, in H. Power and R. Hart (eds.), *Computer Simulations in Biomedicine*, Computational Mechanics Publications, Southampton, U.K., pp. 249–256.

Olufsen, M., Peskin, C., Larsen, J., and Nadim, A. (2000). Numerical simulation and experimental validation of blood flow in arteries with structured-tree outflow conditions, *Ann. Biomed. Eng.* **28**: 1281–1299.

Olufsen, M.S., Nadim, A., and Lipsitz, L.A. (2001). Dynamics of cerebral blood flow regulation explained using a lumped parameter model. *Am. J. Physiol.* **282**: R611–R622.

Olufsen, M.S., Tran, H.T., and Ottesen, J.T. (2003). Modeling cerebral blood flow control during postural change from sitting to standing. *J. Cardiovasc. Eng.*, to appear.

O'Rourke, M., Kelly, R., and Avolio, A. (1992). *The Arterial Pulse*, Lea & Febiger, Philadelphia.

Ottesen, J. (1997a). Modelling the baroreflex-feedback mechanism with time-delay, *J. Math. Biol.* **36**: 41–63.

Ottesen, J. (1997b). Nonlinearity of baroreceptor nerves, *Surv. Math. Ind.* **7**: 187–201.

Ottesen, J. (2000). General compartment models of the cardiovascular system, in J.T. Ottesen and M. Danielsen (eds.), *Mathematical Modelling in Medicine*, IOS Press, Philadelphia, Pennsylvania, pp. 121–138.

Ottesen, J. (2003). Valveless pumping in a fluid-filled closed elastic tube-system; One-dimensional theory with experimental validation. *J. Math. Biol.* **46**: 309–332.

Ottesen, J., Danielsen, M., Palladino, J., and Noordergraaf, A. (1999). The impact of ejection on ventricular performance, *Proc. First Joint BMES/EMBS Conf.*, Omnipress, Atlanta, Georgia, p. 245.

Ottesen, J.T. and Danielsen, M. (2003). Modeling ventricular contraction with heart rate changes. *J. Theoret. Biol.* **222**: 337–346.

Palladino, J. (1990). *Models of the Cardiac Muscle Contraction and Relaxation*, Ph.D. thesis, University of Pennsylvania, Philadelphia, Pennsylvania.

Palladino, J., Mulier, J., and Noordergraaf, A. (1997). Closed-loop circulation model based on the Frank mechanism, *Surv. Math. Ind.* **7**: 177–186.

Palladino, J., Mulier, J., and Noordergraaf, A. (2000). The changing views of the heart through the centuries, in J. Ottesen and M. Danielsen (eds.), *Mathematical Modeling in Medicine*, IOS Press, Philadelphia, Pennsylvania, pp. 3–11.

Palladino, J., Rabbany, S., Mulier, J., and Noordergraaf, A. (1997). A perspective on myocardial contractility, *Surv. Math. Ind.* **5**: 135–144.

Palladino, J., Ribeiro, L., and Noordergraaf, A. (2000). Human circulatory system model based on Frank's mechanism, in J. Ottesen and M. Danielsen (eds.), *Mathematical Modeling in Medicine*, IOS Press, Philadelphia, Pennsylvania, pp. 29–40.

Papageorgiou, G., Jones, B., Redding, V., and Hudson, N. (1990). The area ratio of normal arterial junctions and its implications in pulse wave reflections, *Cardiovasc. Res.* **24**: 478–484.

Park, J., Metaxas, D., and Axel, L. (1996). Analysis of left ventricular wall motion based on volumetric deformable models and MRI-SPAMM, *Med. Imag. Anal.* **1**: 53–71.

Pasipoularides, A., Murgo, J., Miller, J., and Craig, E. (1987). Nonobstructive left ventricular ejection pressure gradients in man, *Circ. Res.* **61**: 220–227.

Patel, D., DeFreitas, F., and Greenfield, J. (1963). Relationship of radius to pressure along the aorta in living dogs. *J. Appl. Physiol.* **18**: 1111–1117.

Pedersen, E. (1993). *In Vitro and In Vivo Studies of Blood Flow in the Normal Abdominal Aorta and Aorta Bifurcation (in Danish)*, Ph.D. thesis, Department of Cardiothoracic and Vascular Surgery and Institute for Experimental Clinical Research, Århus University Hospital, Skejby, Denmark.

Pedersen, E., Sung, H.-W., Burlson, A., and Yoganathan, A. (1993). Two-dimensional velocity measurements in a pulsatile flow model of the abdominal aorta simulating different hemodynamic conditions, *J. Biomech.* **26**(10): 1237–1247.

Pedley, T. (1980). *The Fluid Mechanics of Large Blood Vessels*, Cambridge University Press, Cambridge, U.K.

Peskin, C. (1972a). *Flow Patterns around Heart Valves*, Ph.D. thesis, Albert Einstein College of Medicine, New York.

Peskin, C. (1972b). Flow patterns around heart valves: A numerical method, *J. Comput. Phys.* **10**: 252–271.

Peskin, C. (1975). *Mathematical Aspects of Heart Physiology*, Courant Institute of Mathematical Sciences, New York University, New York.

Peskin, C. (1976). *Partial Differential Equations in Biology*, Courant Institute of Mathematical Sciences, New York University, New York.

Peskin, C. (1977). Numerical analysis of blood flow in the heart, *J. Comput. Phys.* **25**: 220–252.

Peskin, C. (1981). Mathematical aspects of physiology, in F.C. Hoppensteadt (ed.), *Mathematical Aspects of Physiology. Lectures in Applied Mathematics*, Vol. 19, American Mathematical Society, Providence, Rhode Island, pp. 69–93.

Peskin, C. and McQueen, D. (1980). Modeling prosthetic heart valves for numerical analysis of blood flow in the heart, *J. Comput. Phys.* **37**: 113–132.

Peskin, C. and McQueen, D. (1989). A three-dimensional computational method for blood flow in the heart, *J. Comput. Phys.* **81**: 372–405.

Peskin, C. and McQueen, D. (1992). Cardiac fluid dynamics, *CRC Crit. Rev. Biomed. Eng.* **20**: 451–459.

Peskin, C. and McQueen, D. (1993a). Cardiac fluid dynamics, in T.C. Pilkington, B. Loftis, T. Palmer, and T.F. Budinger (eds.), *High-Performance Computing in Biomedical Research*, CRC Press, Boca Raton, Florida.

Peskin, C. and McQueen, D. (1993b). Computational biofluid dynamics, in A.Y. Cheer and C.P. van Dam (eds.), *Fluid Dynamics in Biology*, Vol. 141, American Mathematical Society, Providence, Rhode Island, pp. 161–186.

Peskin, C. and McQueen, D. (1994). Mechanical equilibrium determines the fractal fiber architecture of aortic heart valve leaflets, *Am. J. Physiol.* **266**: H319–H328.

Peskin, C. and McQueen, D. (1996). Fluid dynamics of the heart and its valves, in H. Othmer, F.R. Adler, M.A. Lewis, and J.C. Dallon (eds.), *Case Studies in Mathematical Modeling—Ecology, Physiology, and Cell Biology*, Prentice–Hall, Englewood Cliffs, New Jersey, Chap. 14, pp. 309–337.

Peskin, C. and Printz, B. (1993). Improved volume conservation in the computation of flows with immersed boundaries, *J. Comput. Phys.* **105**: 33–46.

Piiper, J. and Scheid, P. (1981). Model for capillary-alveolar equilibrium with special reference to O_2 uptake in hypoxia, *Respir. Physiol.* **46**: 193.

Pollanen, M. (1992). Dimensional optimization at different levels at the arterial hierarchy, *J. Theor. Biol.* **159**: 267–270.

Poon, C.-S. and Wiberg, D. (1981). Dynamics of gaseous uptake in the lungs: The concentrations and second gas effects, *IEEE Trans. Biomed. Eng.* **28**: 823–831.

Raines, J. (1972). *Diagnosis and Analysis of Atherosclerosis in the Lower Limb*, Ph.D. thesis, Massachusetts Institute of Technology, Boston, Massachusetts.

Raines, J., Jaffrin, M., and Shapiro, A. (1974). A computer simulation of arterial dynamics in the human leg, *J. Biomech.* **7**: 77–91.

Remington, J. and Wood, E. (1956). Formation of peripheral pulse contour in man, *J. Appl. Physiol.* **9**: 433–442.

Rideout, V. (1991). *Mathematical and Computer Modeling of Physiological Systems*, Biophysics and Bioengineering Series, Prentice–Hall, Englewood Cliffs, New Jersey.

Riley, R. and Cournand, A. (1949). "Ideal" alveolar air and the analysis of ventilation-perfusion relationships in the lungs, *J. Appl. Physiol.* **1**: 825.

Robinson, J. and Sleight, P. (1980). Single carotid sinus baroreceptor adaptation in normatensive and hypertensive dogs, in P. Sleight (ed.), *Arterial Baroreceptors and Hypertension*, Oxford University Press, New York, pp. 45–52.

Rockwell, R., Anliker, M., and Elsner, J. (1974). Model studies of the pressure and flow pulses in a viscoelastic arterial conduit, *J. Franklin Inst.* **297**: 405–427.

Rossitti, S. and Löfgren, J. (1993). Vascular dimensions of the cerebral arteries follow the principle of minimum work, *Stroke* **24**: 371–377.

Rothe, C. (1983). Reflex control of veins and vascular capacitance, *Physiol. Rev.* **63**: 1281–1342.

Saber, N., Gosman, A., Wood, N., Kilner, P., Charrier, C., and Firmin, D. (2001). Computational flow modeling of the left ventricle based on in vivo MRI data: Initial experiment, *Ann. Biomed. Eng.* **29**: 275–283.

Saunders, K., Bali, H., and Carson, E. (1980). A breathing model of the respiratory system: The controlled system, *J. Theor. Biol.* **84**: 135–161.

Schaaf, B. and Abbrecht, P. (1972). Digital computer simulation of human systemic arterial pulse wave transmission: A nonlinear model, *J. Biomech.* **5**: 345–364.

Secher, A. and Young, A. (1973). Servoanalysis of carotid sinus reflex effects on peripheral resistance, *Circ. Res.* **12**: 152–162.

Schmidt, R., Kumada, M., and Sagawa, K. (1972). Cardiovascular responses to various pulsatile pressures in the carotid sinus, *Am. J. Physiol.* **223**: 1–7.

Schmidt, R. and Thews, G. (1976). *Einfurung in de Physiologie des Menschen* (in German), Springer-Verlag, Berlin.

Schoephoerster, R., Silva, C., and Ray, G. (1993). Finite analytic model for left ventricular systolic flow dynamics, *J. Eng. Mech.* **119**: 733–747.

Schoephoerster, R., Silva, C., and Ray, G. (1994). Evaluation of left ventricular function based on simulated systolic flow dynamics computed from regional wall motion, *J. Biomech.* **27**: 125–136.

Schreiner, W. and Buxbaum, P. (1993). Computer-optimization of vascular trees, *J. Biomed. Eng.* **40**: 482–490.

Segers, P., Dubois, F., DeWachter, D., and Verdonck, P. (1998). Role and relevancy of a cardiovascular simulator, *J. Cardiovasc. Eng.* **3**: 48–56.

Shoukas, A. and Brunner, M. (1980). Epinephrine and the carotid sinus baroreceptor reflex, *Circ. Res.* **47**(2): 249–257.

Shoukas, A. and Sagawa, K. (1973). Control of total systemic vascular capacity by the carotid sinus baroreceptor reflex, *Circ. Res.* **33**: 22–33.

Siggaard-Andersen, M. and Siggaard-Andersen, O. (1995). Oxygen status algorithm, version 3, with some applications, *Acta Anaesthesiol. Scan.* **39**: 13–20.

Siggaard-Andersen, O. (1974). *The Acid-Base Status of the Blood*, 4th ed., Munksgaard, Copenhagen, Denmark.

Siggaard-Andersen, O. and Gøthgen, I. (1989). *The Oxygen Status of the Arterial Blood*, Radiometer, Copenhagen, Denmark.

Siggaard-Andersen, O., Wimberley, P., Fogh-Andersen, N., and Gøthgen, I. (1988). Measured and derived quantities with modern pH and blood gas equipment: Calculation algorithms with 54 equations, *Scan. J. Clin. Lab. Invest.* **48**: 7–15.

Siggaard-Andersen, O., Wimberley, P., Gøthgen, I., and Siggard-Andersen, M. (1984). A mathematical model of the hemoglobin-oxygen dissociation curve of the human blood and of oxygen partial pressure as a function of temperature, *Clin. Chem.* **30**: 1646–1651.

Singer, R. and Hastings (1948). An improved clinical method for the estimation of disturbances of the acid-base balance of human blood, *Medicine* **27**: 223–242.

Solomon, E., Smidt, R., and Adragna, P. (1990). *Human Anatomy & Physiology*, 2nd international ed., Saunders College Publishing, Philadelphia, Pennsylvania.

Spencer, M. and Worthington, C. (1960). A hypothesis of contraction in straited muscle, *Nature* **187**: 388–391.

Spickler, J. and Kezdi, P. (1967). Dynamic response characteristics of carotid sinus barore-ceptors, *Am. J. Physiol.* **212**: 472–476.

Srinivasan, R. and Nudelman, H. (1972). Modeling the carotid sinus baroreceptor, *Biophys. J.* **12**: 1171–1182.

Stergiopulos, N., Meister, J., and Westerhof, N. (1996). Determinants of stroke volume and systolic and diastolic aortic pressure, *Am. J. Physiol.* **270**: H2050–H2059.

Stergiopulos, N., Young, D., and Rogge, T. (1992). Computer simulation of arterial flow with applications to arterial and aortic stenosis, *J. Biomech.* **25**: 1477–1488.

Stettler, J., Niederer, P., and Anliker, M. (1981). Theoretical analysis of arterial hemodynam-ics including the influence of bifurcations, Part I: Mathematical model and prediction of normal pulse patterns, *Ann. Biomech. Eng.* **9**: 145–164.

Stockie, J.M. and Wetton, B.T.R. (1995). Stability analysis for the immersed fiber problem, *SIAM J. Appl. Math.* **55**: 1577–1591.

Streeter, V., Keitzer, W., and Bohr, D. (1963). Pulsatile pressure and flow through distensible vessels, *Circ. Res.* **13**(3): 3–20.

Suga, H., Sagawa, K., and Kostiuk, D. (1976). Controls of ventricular contractility assessed by pressure-volume ratio, e_{max}, *Cardiovasc. Res.* **10**: 582–592.

Suga, H., Sagawa, K., and Shoukas, A. (1973). Load independence of the instantaneous pressure-volume ratio of the canine left ventricle and effects of epinephrine and heart rate on the ratio, *Circ. Res.* **32**: 314–322.

Sun, Y., Beshara, M., Lucariello, R., and Chiaramida, S. (1997). A comprehensive model for right-left heart interaction under the influence of pericardium and baroreflex, *Am. J. Physiol.* **272**: H1499–1515.

Suwa, N., Niwa, T., Fukasawa, H., and Sasaki, Y. (1963). Estimation of intravascular blood pressure gradients by mathematical analysis of arterial casts. *Tohoku J. Exp. Med.* **79**: 168–198.

Taher, M., Cecchini, A., Allen, M., Gobran, S., Gorman, R., Guthrie, B., Lingenfelter, K., Rabbany, S., Rolchigo, P., Melbin, J., and Noordergraaf, A. (1988). Baroreceptor responses derived from a fundamental concept, *Ann. Biomed. Eng.* **16**: 429–443.

Tardy, Y., Meiseter, J., Perret, F., Brunner, H., and Arditi, M. (1991). Non-invasive estimate of the mechanical properties of peripheral arteries from ultrasonic and photoplethys-mographic measurements, *Clin. Phys. Physiol. Meas.* **12**: 39–54.

Taylor, T., Goto, Y., Hata, K., Takasago, T., Saeki, A., Nishioka, T., and Suga, H. (1993). Comparison of the cardiac force-time integral with energetics using a cardiac muscle model, *J. Biomech.* **26**: 1217–1225.

Taylor, T. and Yamaguchi, T. (1995a). Flow patterns in three-dimensional left ventricular systolic and diastolic flows determined from computational fluid dynamics, *Biorheol.* **32**: 61–71.

Taylor, T. and Yamaguchi, T. (1995b). Realistic three-dimensional left ventricular ejection determined from computational fluid dynamics, *Med. Eng. Phys.* **17**: 602–608.

Tham, R.-Y. (1988). *A Study of Effects of Halothane on Canine Cardiovascular System and Baroreceptor Control*, Ph.D. thesis, University of Wisconsin, Madison, Wisconsin.

Tortora, G. (1999). *Principles of Human Anatomy*, 8th ed., Addison Wesley Longman, Menlo Park, California.

Tortora, G. and Anagnostakos, N. (1990). *Principles of Anatomy and Physiology*, 6th ed., Harper and Row Publishers, New York.

Toy, S., Melbin, J., and Noordergraaf, A. (1985). Reduced models of arterial system, *IEEE Trans. Biomed. Eng.* **32**: 174–177.

Tu, C. and Peskin, C.S. (1992). Stability and instability in the computation of flows with moving immersed boundaries: A comparison of three methods, *SIAM J. Sci. Stat. Comput.* **13**: 1361–1376.

Ursino, M. (1997). A mathematical model of the carotid-baroreflex control in pulsatile conditions, *Sur. Math. Ind.* **7**: 203–220.

Ursino, M. (2000). Modelling the interaction among several mechanisms in the short-term arterial pressure control, in M. Danielsen and J. Ottesen (eds.), *Mathematical Modeling in Medicine*, IOS Press, Philadelphia, Pennsylvania, pp. 139–162.

Ursino, M., Antonucci, M., and Belardinelli, E. (1994). Role of active changes in venous capacity by the carotid baroreflex: Analysis with a mathematical model, *Am. J. Physiol.* **36**(6): H2531–H2546.

Ursino, M., Artioli, E., and Gallerani, M. (1994). An experimental comparison of different methods of measuring wave propagation in viscoelastic tubes, *J. Biomech.* **27**(7): 979–990.

Uylings, H. (1977). Optimization of diameters and bifurcation angles in lung and vascular tree structures, *Bull. Math. Biol.* **39**: 509–520.

Vaartjes, S. and Herman, B. (1987). Left ventricular internal resistance and unloaded ejection flow assessed from pressure-flow relations: A flow-clamp study on isolated rabbit hearts, *Circ. Res.* **60**: 727–737.

Vander, A., Sherman, J., and Luciano, D. (1990). *Human Physiology*, 5th ed., McGraw-Hill Publishing Company, New York.

Vesier, C. and Yoganathan, A. (1992). A computer method for simulation of cardiovascular flow fields: validation of approach, *J. Comput. Phys.* **99**: 271–287.

Walker, P., Cranney, G., Grimes, R., Delatore, J., Rectenwald, J., Pohost, G., and Yoganathan, A. (1996). Three-dimensional reconstruction of the flow in a human left heart by using magnetic resonance phase velocity encoding, *Ann. Biomed. Eng.* **24**: 139–147.

Walker, P., Cranney, G., Scheidegger, M., Waseleski, G., Pohost, G., and Yoganathan, A. (1993). Semi-automated method for noise reduction and background phase error reduction of MR phase velocity data, *J. Magn. Reson. Imag.* **3**: 521–530.

Wan, J., Steele, B., Spicer, S.A., Stohband, S., Feijoo, G.R., Hughes, T.J., and Taylor, C.A. (2002). A one-dimensional finite element method for simulation-based medical planning for cardiovascular disease. *Comput. Methods Biomech. Biomed. Engin.* **5**(3): 195–206.

Wang, A., Brandle, M., and Zucker, I. (1993). Influence by vagotomy on the baroreflex sensitivity in anesthetized dogs with experimental heart failure, *Am. J. Physiol.* **265**: H1310–H1317.

Warberg, J. (1995). *Human Fysiologi*, Polyteknisk Forlag, Lyngby, Denmark.

Warner, H. (1958). The frequency dependent nature of blood pressure regulation by carotid sinus studied with an electric analog, *Circ. Res.* **6**: 35–40.

Warner, H. (1959). The use of an analog computer for analysis of control mechanisms in the circulation, *Proc. IRE.* **47**: 1913–1916.

Wemple, R. and Mockros, L. (1972). Pressure and flow in the systemic arterial system, *J. Biomech.* **5**: 629–641.

Werff, T.V. (1974). Significant parameters in arterial pressure and velocity development, *J. Biomech.* **7**: 437–447.

West, B., Bhargava, V., and Goldberger, A. (1986). Beyond the principle of similitude: Renormalization in the bronchial tree, *J. Appl. Physiol.* **60**: 1089–1097.

West, J. (1974). Blood flow to the lung and gas exchange, *Anesthesiology* **41**: 124.

West, J. (1979). *Respiratory Physiology*, Williams & Wilkins, Baltimore, Maryland.

Westerhof, N., Bosman, F., DeVries, C., and Noordergraaf, A. (1969). Analog studies of the human systemic arterial tree, *J. Biomech.* **2**: 121–143.

Westerhof, N. and Stergiopulos, N. (1998). Models of the arterial tree, in J. Ottesen and M. Danielsen (eds.), *Mathematical Modeling in Medicine*, IOS Press, Philadelphia, Pennsylvania, pp. 65–78.

Wheater, P., Burkitt, H., and Daniels, V. (1987). *Functional Histology, A Text and Colour Atlas*, 2nd ed., Churchill Livingstone, Edinburgh, Scotland.

Whitham, G. (1974). *Linear and Nonlinear Waves*, John Wiley & Sons, New York.

Widdicombe, J. and Davies, A. (1991). *Respiratory Physiology*, 2nd ed., Physiological Principles of Medicine Series, Edward Arnold, London, U.K.

Womersley, J. (1957). Oscillatory flow in arteries: The constrained elastic tube as a model of arterial flow and pulse transmission, *Phys. Med. Biol.* **2**: 178–187.

Yellin, E., Peskin, C., Yoran, C., Koenigsberg, M., Matsumoto, M., Laniado, S., McQueen, D., Shore, D., and Frater, R. (1981). Mechanisms of mitral valve motion during diastole, *Am. J. Physiol.* **241**: H389–H400.

Yoganathan, A., Lemmon, J., Kim, Y., Levine, R., and Vesier, C. (1995). A three-dimensional computer investigation of intraventricular fluid dynamics: Examination into the initiation of systolic anterior motion of the mitral valve leaflets, *J. Biomech. Eng.* **117**: 94–102.

Yoganathan, A., Lemmon, J., Kim, Y., Walker, P., Levine, R., and Vesier, C. (1994). A computational study of a thin-walled three-dimensional left ventricle during early systole, *J. Biomech. Eng.* **116**: 307–314.

Zamir, M. (1978). Nonsymmetrical bifurcations in arterial branching, *J. Gen. Physiol.* **72**: 837–845.

Zamir, M., Langille, B., and Wonnacott, T. (1984). Branching characteristics of coronary arteries in rats, *Can. J. Physiol. Pharmacol.* **62**: 1453–1459.

Index